Seabirds: feeding ecology and role in marine ecosystems

SEABIRDS

feeding ecology and role in marine ecosystems

EDITED BY J. P. CROXALL

British Antarctic Survey, Cambridge

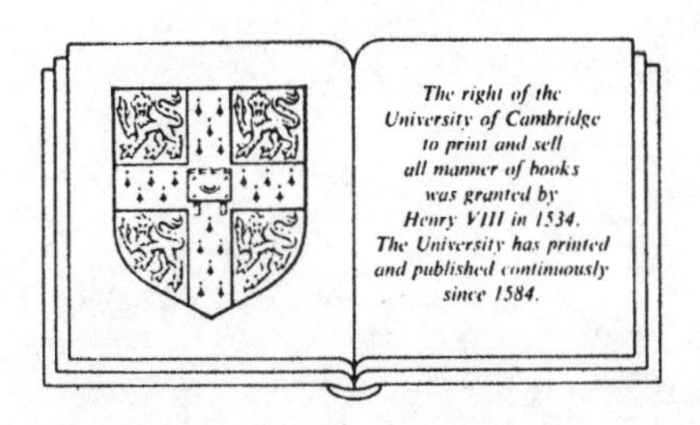
The right of the University of Cambridge to print and sell all manner of books was granted by Henry VIII in 1534. The University has printed and published continuously since 1584.

CAMBRIDGE UNIVERSITY PRESS

Cambridge

New York Port Chester

Melbourne Sydney

CAMBRIDGE UNIVERSITY PRESS
Cambridge, New York, Melbourne, Madrid, Cape Town, Singapore, São Paulo, Delhi

Cambridge University Press
The Edinburgh Building, Cambridge CB2 8RU, UK

Published in the United States of America by Cambridge University Press, New York

www.cambridge.org
Information on this title: www.cambridge.org/9780521105101

First published 1987
Reprinted 1990
This digitally printed version 2009

A catalogue record for this publication is available from the British Library

Library of Congress Cataloguing in Publication data

Seabirds: feeding, ecology and role in marine ecosystems

Includes index.
1. Sea birds-Food. 2. Seabirds-Ecology.
3. Birds–Foods. 4. Birds-Ecology. I. Croxall, J. P.
QL698.3.S4 1987 598.4 86–8236

ISBN 978-0-521-30178-7 hardback
ISBN 978-0-521-10510-1 paperback

CONTENTS

CONTRIBUTORS

Kenneth T. Briggs Center for Marine Studies, University of California, Santa Cruz, California 95064, USA.

Ellen W. Chu Editor, *Bioscience*, American Institute of Biological Sciences, 730 11th Street N.W., Washington, D.C., 20001–4584, USA.

Roger B. Clapp Museum Section, US Fish and Wildlife Service, National Museum of Natural History, Washington, D.C. 20560, USA.

J. P. Croxall British Antarctic Survey, Natural Environment Research Council, High Cross, Madingley Road, Cambridge CB3 0ET, UK.

R. W. Davis Physiological Research Laboratory, Scripps Institution of Oceanography, University of California, San Diego, La Jolla, California 92093, USA.

David Cameron Duffy Percy FitzPatrick Institute of African Ornithology, University of Cape Town, Rondebosch 7700, South Africa.

R. W. Furness Zoology Department, Glasgow University, Glasgow G12 8QQ, U.K.

Craig S. Harrison Environment and Policy Institute, East–West Center, 177 East West Road, Honolulu, Hawaii 96848, USA.

G. L. Hunt, Jr Department of Ecology and Evolutionary Biology, University of California, Irvine, California 92717, USA.

G. L. Kooyman Physiological Research Laboratory, Scripps Institution of Oceanography, University of California, San Diego, La Jolla, California 92093, USA.

G. S. Lishman British Antarctic Survey, Natural Environment Research Council, High Cross, Madingley Road, Cambridge, CB3 0ET, UK.

R. A. Morgan British Antarctic Survey, Natural Environment Research Council, High Cross, Madingley Road, Cambridge CB3 0ET, UK.

C. J. Pennycuick Department of Biology, P.O. Box 249118, University of Miami, Coral Gables, Florida 33124, USA.

K. D. Powers Manomet Bird Observatory, Manomet, Massachusetts 02345, USA.

P. A. Prince British Antarctic Survey, Natural Environment Research Council, High Cross, Madingley Road, Cambridge CB3 0ET, UK.

Gerald A. Sanger U.S. Fish and Wildlife Service, 1101 East Tudor Road, Anchorage, Alaska 99503, USA.

D. C. Schneider Newfoundland Institute for Cold Ocean Science, Memorial University, St John's, Newfoundland, A1B 3X7, Canada.

Ralph W. Schreiber Los Angeles County Museum of Natural History, 900 Exposition Boulevard, Los Angeles, California 90007, USA.

Spencer G. Sealy Department of Zoology, University of Manitoba, Winnipeg, Manitoba, R3T 2N2, Canada.

Michael P. Seki Southwest Fisheries Center Honolulu Laboratory, National Marine Fisheries Service, National Oceanic and Atmospheric Administration, P.O. Box 3830, Honolulu, Hawaii 96812, USA.

W. Roy Siegfried Percy FitzPatrick Institute of African Ornithology, University of Cape Town, Rondebosch 7700, South Africa.

Kees Vermeer Canadian Wildlife Service, P.O. Box 340, Delta, British Columbia, V4K 3Y3, Canada.

1

Introduction

J. P. CROXALL

British Antarctic Survey, Natural Environment Research Council, High Cross, Madingley Road, Cambridge CB3 OET, UK

The last decade or so has been a period of unrivalled progress in our quantitative understanding of the processes of energy flux in marine systems. Not surprisingly, much of this research has been directed to questions and topics relevant to the management of commercial fisheries. Nevertheless, this has also served to focus attention on those natural components of the system, such as seabirds and seals, which might compete with, or prey upon, commercially important species (e.g Furness, 1982; Nettleship, Sanger & Springer, 1984).

Understanding predator–prey relationships requires much information about the diet, distribution and bioenergetics of both predators and prey and on the detailed nature of their interactions. Such data are not easy to acquire, even for terrestrial systems where direct, simultaneous observation and experimentation is possible. In pelagic marine systems, wide-ranging predators operate in a vast, ever-changing three-dimensional environment. Study and sampling under these conditions are severely constrained by the limitations of operating from research vessels and shore stations. However, a number of major studies have recently been undertaken to assess the role of top predators, especially seabirds, as consumers in productive marine systems.

This book, which originated from a symposium on 'Seabirds and nutrient cycles' at the XVIII International Ornithological Congress in Moscow (Croxall, 1986), presents results from a selection of these studies, covering a broad geographical range from Arctic (Chapters 10, 11), north temperate (Chapter 11), tropical–subtropical (Chapters 12–14), south temperate (Chapter 14) and sub-Antarctic (Chapter 15) regions. These illustrate both the general features, and different views, of the role of seabirds in a variety of marine ecosystems.

To complement these case-studies of particular seabird communities and to provide a broader perspective concerning many of the species involved, the first part of the book treats in detail the feeding ecology of the main seabird groups and reviews a number of topics of fundamental importance to understanding the interactions of seabirds with their marine environments.

Most of this book – and especially the reviews of feeding ecology – is concerned with four seabird groups: penguins (Sphenisciformes); albatrosses, petrels, storm petrels and diving petrels (Procellariiformes); pelicans, gannets and boobies, frigatebirds, tropicbirds and cormorants (Pelecaniformes); and auks, or alcids (Alcidae). This assemblage totals nearly 200 species; about another 100 species (mainly terns, gulls and sea ducks), which derive nearly all their food from the sea, are not considered in any detail here. This was partly an arbitrary decision to keep the volume within bounds but also reflects the fact that few of these species both venture far from land and are also sufficiently abundant to have any major impact on marine resources.

There are a number of major questions relating to predators in the context of their interactions with prey. 1. How many are there? 2. What do they eat? 3. How much energy (i.e. food) do they need? 4. When, where (and how) do they take their prey?

This volume seeks to deal with the nature of seabird diets, when and how they catch prey and, in respect of the case-studies, where these activities are concentrated, at least at certain times of the year. Detailed information on the current status of all the world's seabirds (Croxall, Evans & Schreiber, 1984) and on seabird energetics (Whittow & Rahn, 1984) has recently been published. These topics are not covered here.

Because the early chapters of this book review the characteristics of the marine environment (Chapter 2), seabirds' adaptations and abilities for flight (Chapter 3) and diving (Chapter 4) and their adaptations for feeding at sea (Chapters 6–9), this introduction need only briefly summarise various aspects of the life style of seabirds.

Although oceans cover 60% of the earth's surface, only 2–3% of the world's birds have been able to exploit them effectively. This suggests that it is no simple matter for birds to live at sea especially perhaps because they must spend considerable time ashore when they breed. It is not surprising, therefore, that amongst birds, seabirds show a number of quite extreme biological and ecological adaptations and features.

In general, seabirds are larger than land birds and lay eggs that are also larger, but not disproportionately so. Seabirds also lay small clutches and in three groups (Procellariiformes, Fregatidae, Phaethontidae) all species produce only a single egg, proportionately larger than the eggs of species with larger clutches. Large eggs tend to take a long time to hatch and seabirds, in particular many tropical species and all Procellariiformes, show very extended incubation periods compared with other birds. This feature and the relations between egg size, yolk content, embryonic growth rate and other aspects of incubation are treated in Whittow & Rahn (1984). Egg formation is also a lengthy process in seabirds (Grau, 1984) with yolk deposition taking up to 30 days and the lag period (between yolk completion and laying) lasting up to 10 days (cf one day in chickens).

Similarly, rearing chicks to independence lasts a long time in seabirds, often 2 months or more and up to about one year in the King Penguin *Aptenodytes patagonicus*. Only for the auks are short periods typical; in this family a few species have highly precocial young, which depart to sea with their parents when only 2–3 days old. However, such chicks receive extensive parental care at sea. Post-fledging parental care is also typical, even for well-grown offspring that may be capable of some independent feeding, in the case of pelicans, boobies, cormorants, frigatebirds and at least one species of tropicbird. It is very rare in penguins, Procellariiformes (only occurring in tropical albatrosses) and auks (except for the precocial murrelets and possibly the Little Auk (Dovekie) *Alle alle*). The topic is reviewed by Burger (1980).

It will be clear from this that seabirds tend to have long breeding seasons. In extreme cases, this period, from arrival at the colony in order to commence pre-breeding courtship activities until the time when chicks become independent, may last more than a year (e.g. in King Penguins, several albatrosses and frigatebirds). If successful in rearing a chick such species breed less frequently than annually. Most other species, however, breed once each year.

Seabirds are long-lived, with an adult survivorship that is rarely less than 80% per year, typically around 90% and attains 95% in many Procellariiformes, some gannets and some auks. They also show considerably deferred sexual and/or social maturity and the mean ages at which reproduction starts in albatrosses (and possibly frigatebirds) are the oldest recorded in birds. Sexual maturity in albatrosses is attained at earlier ages but the acquisition of a partner (which involves spending substantial periods ashore – and hence not feeding) takes several further years. All this suggests that the acquisition of experience, presumably initially learning to find and capture food and later forming pair-bonds with a partner, is a key feature of the early life of seabirds. While the majority

Table 1.1. *Some mean demographic and biological characteristics of the main families of seabirds*

Group	No. of species	Adult weight (kg)	Age (years) at first breeding	Adult annual survival rate (%)	Clutch size	Incubation period (days)	Chick-rearing period (days)
Sphenisciformes							
Spheniscidae (penguins)	16–18	4–12(–40)[a]	4–8	75–85(–95)[a]	1–2	33–62	50–80(–350)[b]
Procellariiformes							
Diomedeidae (albatrosses)	14	2–4(–12)[c]	7–13	92–97	1	60–79	116–150(–280)[c]
Procellariidae (petrels)	*c.* 70	0·15–2·3(–7)[d]	4–10	90–96	1	43–62	42–120
Hydrobatidae (storm petrels)	20	0·03–0·05	*c.* 4–5	*c.* 90+	1	40–50	55–70
Pelecanoididae (diving petrels)	4	0·1	2–3	75–80	1	46–56	45–55
Pelecaniformes							
Pelecanidae (pelicans)	2[e]	3·5	3–4	*c.* 85	2–3	30	55–60
Sulidae (gannets, boobies)	9	1–3·5	3–5	90–95	1–2(–3)[f]	41–56	90–120(–170)[g]
Phaethontidae (tropicbirds)	3	0·5–0·8	?	?	1	41–44	60–90
Fregatidae (frigatebirds)	5	0·7–1·5	?*c.* 9–10	?	1	44–55	140–170+
Phalacrocoracidae (cormorants)	29	1–5	4–5	85–90	2–3	25–30	60–90
Charadriiformes							
Alcidae (auks)	22	0·1–1·0	2–5	80–93	1–2	29–42	(2[h]–)15–50

[a] Emperor Penguin *Aptenodytes forsteri*. [b] Emperor Penguin 170 days; King Penguin *A. patagonicus* 350 days. [c] Great albatrosses. [d] Giant petrels *Macronectes* spp. [e] There are 6 other, essentially non-marine, species. [f] Boobies. [g] Red-footed Booby *Sula sula* to 140 days; Abbott's Booby *S. abbotti* to 170+ days; [h] *Synthliboramphus* murrelets.

of seabirds are monogamous and pair for life (although some divorces occur even in such species), there is evidence that most species of booby and frigatebird change partners more regularly.

Some of these features are summarised for each of the families treated in this book in Table 1.1. Within each family there is a considerable range of adaptions, perhaps particularly in the auks, but there is not space to treat this in detail here. Many aspects of the general biology and ecology of seabirds have been reviewed by Nelson (1980); the emphasis in the rest of this volume must remain with feeding ecology.

Acknowledgements I am particularly grateful to all the authors who contributed to this volume, not least for their patience with delays in its production mainly imposed by my other commitments. I am indebted to Christine Thulbourn and most especially to Susan Norris for typing (and retyping) the book and to Christine Phillips for obtaining and checking numerous references and for much help during the final stages of production. Steve Gardiner, Helen Mills and Martin Walters, of Cambridge University Press, provided help throughout. I should also like to thank Dr Fenella Wrigley for compiling the index with meticulous care.

References

Burger, J. (1980). The transition to independence and post-fledging parental care in seabirds. In *Behavior of marine animals: Vol. 4. Marine birds*, ed. J. Burger, B.L. Olla & H.E. Winn, pp. 367–447. New York: Plenum Press.

Croxall, J.P. (1986) Seabirds and nutrient cycles. *Proc. XVIII Int. Ornithol. Congr.*, **2**, 1063–6.

Croxall, J.P., Evans, P.G.H. & Schreiber, R.W. (eds) (1984). *Status and conservation of the world's seabirds*. Cambridge: ICBP.

Furness, R.W. (1982). Competition between fisheries and seabird communities. *Adv. Mar. Biol.*, **20**, 225–307.

Grau, C.R. (1984). Egg formation. In *Seabird energetics*, ed. G.C. Whittow & H. Rahn, pp. 33–57. New York: Plenum Press.

Nelson, J.B. (1980). *Seabirds, their biology and ecology*. London: Hamlyn.

Nettleship, D.N., Sanger, G.A. & Springer, P.F. (1984). *Marine birds: their feeding ecology and commercial fisheries relationships*. Ottawa: Can. Wildl. Serv., Spec. Publ.

Whittow, G.C. & Rahn, H. (1984). *Seabird energetics*. New York: Plenum Press.

2

Scale-dependent processes in the physical and biological environment of marine birds

G. L. HUNT, Jr and D. C. SCHNEIDER
Department of Ecology and Evolutionary Biology, University of California, Irvine, California 92717, USA

Introduction

A major goal of pelagic bird studies is to understand the determinants of bird distribution and abundance. Seabirds inhabit a highly heterogeneous environment, in which physical processes are a function of scale (Stommel, 1963). Large-scale processes include major circulation patterns in ocean basins; intermediate-scale processes include differentiation of water masses and movements of large gyres; and small-scale processes include eddies that may be only a few meters across. Because most chemical and biological processes in the ocean are driven by physical processes, they are also·a function of scale. In this chapter we discuss seabirds in the context of scale-dependent variability in their physical and biological environment.

Prey distribution is one likely determinant of bird distribution, and investigators have attempted to measure and correlate bird and prey numbers. Jesperson (1930) made daily counts of birds while zooplankton were sampled over 1000 km × 1000 km segments of the North Atlantic. These bird and plankton abundances were highly correlated (Ricklefs, 1973), with a remarkable explained variance ($r^2 = 0.76$) for a pelagic bird study. On a smaller scale, Woodby (1984) used hydroacoustic surveys of euphausiids (*Thysanoessa* sp.) in the southeastern Bering Sea and found a correlation between regions with acoustic echoes and the presence of murres *Uria* sp. in one of two years. However, he did not demonstrate that the murres were foraging on the source of the echoes. In the Antarctic, Obst (1985) used a depth sounder for a hydroacoustic survey and correlated the presence of birds and swarms of *Euphausia superba*, their principal food. Identification of the echo source was made by net hauls and SCUBA investigations. In the Bay of Fundy phalaropes concentrate near an upwelling off Brier Island (Brown, Barker & Gaskin, 1979). The phalaropes feed on copepods which are brought to the surface by tidal currents sweeping over a shallow sill and are subsequently concentrated by a convergence front.

In contrast to these investigations, most pelagic bird studies have not included direct correlations between bird and prey abundance. There are several reasons. Bird studies are usually made from ships on other business and there is no chance to stop and sample prey or to assess bird diets. Many prey taken by birds are hard to sample using traditional plankton nets and trawls (Ashmole & Ashmole, 1968). Fish, squid and even euphausiids move too rapidly to be sampled quantitatively by this gear. Hydroacoustic survey methods promise to alleviate this problem. Additionally, because most seabirds are opportunistic predators that partition marine resources on the basis of habitat (water masses) rather than by prey type (Ainley, 1980), it is difficult to know which prey to sample. Once we have a better knowledge of prey distribution, it will then be crucial to understand how physical and chemical processes control prey productivity, patchiness and availability to birds. Conversely, because bird distribution and abundance reflect oceanic conditions, and because concentrations of birds are relatively easily observed, information on birds may be used by individuals in other oceanographic disciplines for locating features of interest (Gudkov, 1962; Ashmole & Ashmole, 1968; Zelickman & Golovkin, 1972).

While the pelagic distribution of birds may be closely tied to prey distribution, other behavioral, morphological or energetic constraints can disrupt this relation. Flight performance, related to wing loading and aspect ratios, may set morphological limits on distribution related to the predictability of winds of some minimum (or perhaps maximum) velocity (Bailey, 1968; Pennycuick, 1982; Ainley & Boekelheide, 1983). Cormorants are limited in their seaward dispersal by the wetability of their feathers (Rijke, 1968), while the distribution of birds with different styles of foraging (plunge diving *versus* pursuit diving) depends not only on food availability, but also on water clarity (Ainley, 1977). Finally, birds commuting between foraging areas or to and from a colony pass over water masses not used for foraging, thereby diminishing our ability to determine avian use of specific water masses (Ainley, 1980; Ainley, O'Connor & Boekelheide, 1984). This problem is exacerbated by birds that forage nocturnally but are censused during the day. Thus, in considering the role of physical events in patterning bird distribution, we cannot ignore patterns, unrelated to water-mass properties, generated by the other aspects of seabird biology (Pielou, 1960).

Spatial and temporal scales of variability

It is difficult to interpret observations of an isolated part of a system without some idea of the scale and structural organization of the whole (Hutchinson, 1953). Without prior experience, we cannot tell whether a particular observation is typical, or is a rare ocurrence. Likewise, unexplained variance, noise at one scale, may be due to pattern at a different scale.

Stommel (1963) recognized the importance of scale and illustrated the relation between the temporal and spatial scales of several different physical processes in the ocean. Subsequently, Haury, McGowan & Wiebe (1978) modified Stommel's type of diagram to show the time/space scales of variability of zooplankton biomass (Fig. 2.1). The patterns of greatest concern to marine ornithologists fall between B (swarms) and H (biogeographic provinces).

If we are to document quantitative relations between seabirds and oceanographic processes our sampling effort must reflect the scales of these processes (Crowley, 1977). When there is little spatial or temporal variation in a phenomenon, relatively few samples may describe it adequately. If there is considerable variation occurring at different scales then sampling must be increased (Stommel, 1963). Interactions between events at different scales confuse our ability to identify simple patterns, and these interactions may have biological importance.

Fig. 2.1. Temporal and spatial scales of physical and biological events in the ocean (after Haury *et al.*, 1978, with permission of the authors and Plenum Press).

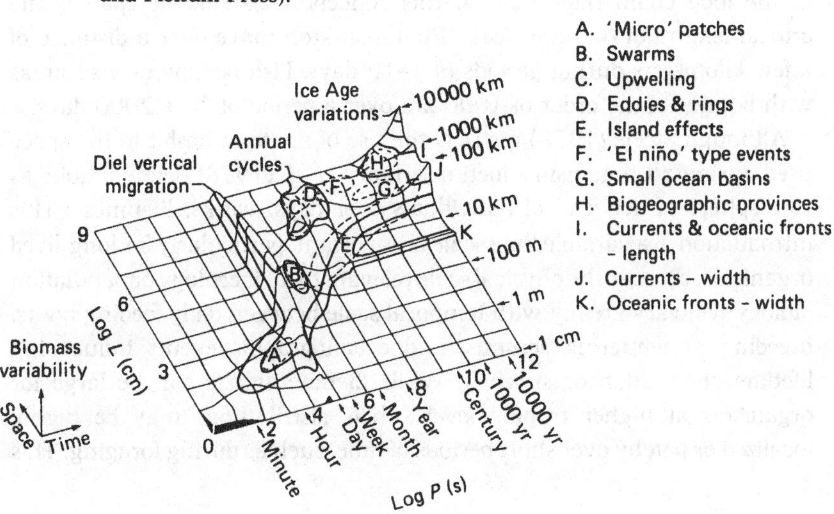

Sampling program design has recently gained prominence as oceanographers found that sampling effort affects their conclusions concerning the scale and variance of phenomena (Kelley, 1976; Dayton & Tegnor, 1984). Continuous sampling or at least frequent sampling at a scale much less than the smallest periodicity of interest (space or time) is critical (Platt & Denman, 1975). When the sampling interval is too large, the shortest fluctuations in the data cannot be resolved, and an apparent periodicity larger than the true periodicity will be obtained. Failure to perceive patches due to sampling at too large a scale can reduce apparent variability because variance generated at smaller scales is overlooked (Cassie, 1959; Crowley, 1977). In a computer simulation study and subsequent field study, Wiebe (1971, 1972) found that increasing the sampling effort (in his case, net size and tow length), increased the precision of population estimates because the variance of the estimates was an inverse function of the number of plankters 'caught'.

Patchiness in plankton is likely to be important in understanding patchiness in bird distribution. Birds can prey directly on plankton; in other cases prey organisms, such as anchovies, can respond to plankton patchiness (Visser *et al.*, 1973). It is important to know the scales of prey patchiness because small prey frequently are 'useful' to predators only when they are concentrated (Golovkin, Shivokolobov & Garkavaya, 1972; Gaskin, 1976; Brodie, Sameoto & Sheldon, 1978; Lasker, 1979).

Steele (1978) proposed a simplified representation of the scale of patchiness in marine systems in which the temporal and spatial scale (ambit) at which organisms live increases with successively higher levels in the food chain (Fig. 2.2). In this conceptualization, life span is the critical feature of the time scale. Phytoplankton move over a distance of a few kilometers during periods of 1–10 days. Fish occupy or visit areas with bounds of the order of 1000 km over a period of 300–2000 days.

Although Steele (1978) restricts the use of the term ambit to the space used during an organism's lifetime, Haury *et al.* (1978) define ambits as 'the sphere of action...of individuals over days, weeks, lifetimes'. This introduction of a variable time scale is important, particularly for long-lived organisms, because the physical scale relevant to their ecology or population biology will vary greatly with temporal scale between daily feeding bouts, breeding or wintering seasons, and oceanographic events influencing lifetime distribution or survival. While lifetime ambits can be large for organisms at higher trophic levels, their distributions may be highly localized or patchy over short periods of time, such as during foraging. This increased short-term patchiness in high trophic-level prey is of consequence for top-level carnivores (Steele, 1980).

The control of phytoplankton patchiness is a complex function of growth, grazing and water mixing, interacting with factors that cause zooplankton to aggregate where barriers (fronts) limit dispersal. Vertical displacements in the water column have greater effect on photosynthesis than similar sized horizontal displacements, due to steeper gradients in light, nutrients and oxygen in the vertical (Owen, 1981). Thus phytoplankton and zooplankton have evolved structures and mechanisms for controlling their vertical position. In contrast, physical factors play a large role in structuring plankton patches horizontally and in time (Steele, 1976; Haury *et al.*, 1978). The minimum horizontal distance over which phytoplankton growth can maintain patch structure against horizontal turbulent diffusion is of the order of a few kilometers (Platt, 1972; Steele, 1976). At smaller scales, water turbulence (eddies) controls their distribution. Eddies smaller than a few kilometers in diameter and of less than a week's duration tend to break up biological patterning, while larger, longer-lived eddies may contribute to it (Owen, 1981).

The distribution of zooplankton relative to phytoplankton, and their interactions, are important because the chlorophyll content of water samples is frequently the only indicator marine ornithologists have of

Fig. 2.2. Temporal and spatial scales of lifetime ambits of various marine organisms: phytoplankton (P), zooplankton (Z) and fish (F) (after Steele, 1976, with permission of the author and Plenum Press).

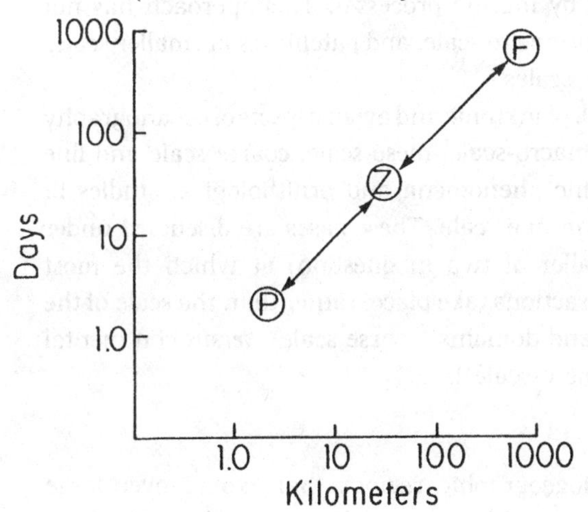

potential prey abundance. At fairly large scales there appears to be a correlation between phytoplankton standing stocks and herbivore standing stocks (Steele, 1974). At scales of 10 km or less (Riley, 1976), there may be an inverse relation between phytoplankton and herbivore abundance, with patches of the two being spatially separated (Riley & Bumpus, 1946; Steele, 1974).

Although marine ornithologists readily acknowledge the patchiness of pelagic bird distribution, patchiness has been largely ignored (Bailey & Bourne, 1972; Devillers, 1978). Devillers (1978) emphasizes the need for continuous observations to avoid missing small-scale patchiness. Averaging results of many short periods of observation or of separate trips will not compensate for incomplete coverage. Patchiness in bird distribution may indicate that food resources are patchy or that social interactions critical for successful foraging are occurring (Bailey & Bourne, 1972; Schneider, 1982; Wittenberger & Hunt, 1985). Small-scale patchiness in bird distribution results in high variance within regions and obscures patterns of bird use between regions.

Physical phenomena and pelagic bird distributions

Two methods have been used to examine the relationship between marine birds and the sea. The descriptive approach identifies patterns of bird distribution and compares them to other patterns in the ocean (e.g. water masses). This gives little or no information on the importance of the marine processes or features evoked as determinants of bird distribution. The second approach uses statistics to determine the amount of variation in bird numbers explained by marine processes. This approach has not been widely applied. It is sensitive to scale, and patchiness at smaller scales may mask pattern at larger scales.

Here we examine physical, planktonic and avian aspects of oceanography at five scales: mega-scale, macro-scale, meso-scale, coarse scale and fine scale. Various oceanographic phenomena and ornithological studies fit appropriately into more than one scale. These cases are discussed under the scale (usually the smaller of two in question) at which the most relevant ornithological interactions take place, rather than the scale of the system as a whole (fronts and domains (coarse scale), *versus* continental shelf systems as a whole (meso-scale)).

Mega-scale

Mega-scale systems are biogeographic regions that extend over large portions of the world's oceans, and are generally greater than 3000 km in extent (Haury *et al.*, 1978). The regions largely correspond to the major oceanic gyres and global circulation systems (region H in Fig. 2.1; see also Ashmole, 1971). These systems respond to climatic change but not to weather (McGowan, 1974); year-to-year changes in phytoplankton standing stocks reflect large-scale climatic shifts (Venrick, McGowan & Montyla, 1973). Patterns of biomass distribution have been established for most of these areas and the regions are marked by distinctly different species assemblages in the plankton (Fager & McGowan, 1963; Johnson & Brinton, 1963; Reid *et al.*, 1978). However, close examination of these systems reveals considerable overlap of plankton species, at least at the borders of the major divisions (McGowan, 1974; Geynrikh, 1976). Demersal fish also show distinct faunal changes at physical boundaries in the ocean (Fager & Longhurst, 1968), but it is rare for individual species to be restricted to a single region (Backus *et al.*, 1970). While Fager & McGowan (1963) found good evidence for plankton-group selection of water masses, their regression analysis of relative abundances against measured physical properties of the water showed little close relation. It is possible that small-scale plankton patchiness may have masked larger-scale associations.

Murphy (1914) described changes in the avifauna of the South Atlantic with latitude and was one of the first to recognize that major differences in marine avifaunas were due to latitudinal changes in oceanic conditions and productivity (Murphy, 1936). Shuntov (1968), van Oordt & Kruijt (1953) and Harris & Hansen (1974) related the distribution of several species of seabirds to sea surface temperature and to major current systems in the South Atlantic, as did Szijj (1967), Sanger (1972), Jehl (1973), King (1974*a*) and Gould (1983) in the Pacific.

The association between marine bird species and particular water masses has been summarized by Bourne (1963), Ashmole (1971), Watson *et al.* (1971), Shuntov (1974), Watson (1975), Brown (1980), Griffiths, Siegfried & Abrams (1982) and others. Shuntov (1974) noted the distribution of productivity and biomass within the oceans and elaborated several 'rules' relating the distribution of birds to the marine ecosystem.

These include: (1) that the distribution of birds is more closely related to major current systems and water masses than to latitudinal zones; (2) that overall there is a roughly symmetrical increase in bird density with latitude in both hemispheres (except for the area of the equatorial current systems with their large numbers of birds); and (3) that bird density is generally higher along the margins of oceans, particularly the eastern margin, with the difference between the western and eastern edges greatest in the tropics.

Within the Pacific Ocean, the avifaunas of the eastern and western sides differ considerably (Bourne, 1981*a*). These differences can be seen in the lack of symmetry in the mid-latitude observations of Szijj (1967) on a transect from New Zealand to Chile via the Antarctic. Several species were seen at one end of the transect, but not at the other. Bourne (1981*a*) suggests that differences between the eastern and western Pacific may be the result of isolation of temperate faunas on the two sides by the relatively impoverished waters of the central gyre. A similar, though less striking example, exists in the North Atlantic (Bourne, 1981*a*) in which there are breeding species in the eastern Atlantic that are absent in the west (see also Wynne-Edwards, 1935).

Ainley (1977) examined qualitative differences in the kinds of species and foraging behaviors used between high and low latitudes. He found that the foraging methods used were correlated with water clarity which, in turn, was negatively related to the standing stock of phytoplankton and food availability. Plunge-divers occurred in clear, tropical waters, while pursuit-divers and divers were confined to the rich, turbid polar and subpolar waters. Apparent exceptions to this pattern are the occurrence of a penguin and a cormorant on the Galapagos Islands, but these divers use rich, cold, frequently turbid upwelled water that surrounds the southern and western Galapagos (Boersma, 1978). Plunging by Peruvian Boobies *Sula variegata* in the turbid waters of the Peru Current is another exception, but their dives are shallow and directed at near-surface prey (Duffy, 1983*b*). Ainley's study was confined to the Pacific, and he did not consider the Northern Gannet *S. bassana* or the Cape Gannet *S. capensis*, species that regularly plunge in cold water (Nelson, 1978). Diving petrels, petrels and terns all, occasionally at least, plunge in turbid waters and fail to fit the generalizations of Ainley's hypothesis (Bourne, personal communication). Brown (personal communication) offers an alternative hypothesis that the absence of pursuit-divers in the tropics is because this mode of foraging requires structural adaptations that also require fast-flapping flight, which is energetically too expensive given the patchy distribution of prey in tropical seas.

Ainley & Boekelheide (1983) compared bird species distribution and abundance from the tropical Pacific to the ice pack of the Southern Ocean. They defined a series of water masses and climatic zones based on sea surface temperature and salinity. Seabird density and biomass were positively related to productivity and were higher in Antarctic than tropical waters. They also emphasized the importance of wind energy in shaping global patterns of bird distribution, as had Kuroda (1954), Harrington, Schreiber & Woolfenden (1972) and Manikowski (1975).

However, when Ainley & Boekelheide (1983) examined the distributions of individual species, the biogeographic picture was less clear. No oceanic seabird was confined to the sub-Antarctic waters and only 8% and 17% of the species in subtropical and Antarctic waters, respectively, were confined to these regions. In the low tropics 37% of the birds were restricted to a single oceanic region. The large numbers of species that use two or more mega-scale oceanic regions suggested that '...the major, classical oceanographic boundaries have few outstanding qualities as avifaunal barriers in the South Pacific' (Ainley & Boekelheide, 1983). Szijj (1967) found that 12 of 25 species covered a range of sea surface temperature exceeding five Celsius degrees, suggesting the use of more than one biogeographic region. Only seven species were restricted to zones with ranges of two Celsius degrees or less.

Macro-scale

Embedded within mega-scale systems are macro-scale features 1000–3000 km in size. They show less distinct changes in plankton species composition at 'boundaries' than mega-scale features (Haury *et al.*, 1978), but there are marked differences in plankton productivity or standing stocks. Examples of macro-scale features include the zonal band of high productivity associated with the equatorial circulation in the central Pacific, the Costa Rican Dome in the eastern tropical Pacific, and the central gyres of the North Atlantic and South Pacific (Haury *et al.*, 1978). In regions where water masses meet, fronts form in which the thermocline may be close to the surface. When this occurs, enhanced vertical flux of water and associated nutrients can result (Woods, Wiley & Briscoe, 1977).

In the tropical Pacific, the westward flowing North and South Equatorial Currents and the eastward flowing Equatorial Countercurrents are features of major biological importance (Murphy & Shomura, 1972). The two equatorial currents are driven by the Trade Winds, while the countercurrent represents the eastward return of this water. The cooler water at the equator is due to a divergence, driven by the Trade Winds and the Coriolis force (Cromwell, 1953; Ryther, 1963), which results in upwelling of cool nutrient-rich water. These diverging waters mix to the north and south. In the north there is a convergence at about 5° N near the southern boundary of the countercurrent (Murphy & Shomura, 1972).

This current system with its combination of upwelling and convergences results in an increased production of plankton, and in the production of small fish and invertebrates. Particularly when they aggregate near the convergence, these species provide food for tuna and birds, and it is here that surface-foraging tuna and birds were most frequently encountered (Murphy & Shomura, 1972; Seckel, 1972). Ashmole & Ashmole (1967) suggested that the patterns of abundance of prey, combined with the presence of large predatory tuna and other fish that force prey to the surface, were critical for supporting the large populations of breeding seabirds on mid-Pacific islands. Gould (1971, 1974) and King (1974*b*) demonstrated increased densities of Sooty Terns *Sterna fuscata*, Wedge-tailed Shearwaters *Puffinus pacificus* and other birds over the waters of the Equatorial Countercurrent. The smaller, planktivorous birds (storm petrels) were most common over the northern divergence and equatorial upwelling, in contrast to the larger nekton-foraging birds which were commonest near the convergence (Gould, 1971). These results with birds reflect the general pattern for plankton in macro-scale features – changes in density with little or no change in species composition between adjacent waters. The temporal scale of these changes in seabird abundance and in the physical features (Trade Winds) controlling upwelling was seasonal (Murphy & Shomura, 1972; Gould, 1974) or annual (Seckel, 1972). The patchiness of birds within the frontal areas of the countercurrent (Gould, 1974; King, 1974*b*) suggests that oceanic long waves (Legeckis, 1982) or meso-scale eddies were creating smaller-scale patches of prey distribution to which the tuna were responding (Murphy & Shomura, 1972; Uda, 1973).

In the western Indian Ocean, bird concentrations in areas of convergence at the northern edge of the Equatorial Countercurrent were correlated with flyingfish abundance but not with plankton abundance (Bailey, 1968). Forage fish were apparently responding to physical features which influenced plankton abundance, although sampling showed low volumes of plankton present. Within the waters of the Indian Ocean, Pocklington (1979) defined water masses on the basis of sea surface temperature and salinity relations and demonstrated that each water mass had a unique avifauna. He found major differences in abundance at macro-scales within the Tropical Indian Ocean, with concentrations of some species apparently greatest near the borders of water masses. Although changes in species composition at the macro-scale level are not usual, the two tropicbirds had non-overlapping ranges, with Red-billed Tropicbird *Phaethon aethereus* confined to water more saline than 35.2%, and Red-tailed Tropicbird *P. rubricauda* restricted to waters less saline than 35.0% (Pocklington, 1979).

Shuntov *et al.* (1981) described large-scale 'ornitho-geographical' regions within the Southern Ocean. These areas appear to be macro-scale in extent and avifaunal differences between these regions are largely changes in abundance (Shuntov *et al.*, 1982*a,b*), rather than in species composition. For example, Wandering Albatrosses *Diomedea exulans* are found over most of the region between Africa and Antarctica, and their occurrence is most frequent where wind speeds are between Beaufort force 5 and 6 (Abrams *et al.*, 1981).

Meso-scale

Meso-scale events in plankton patchiness are usually defined as having a spatial scale of 100–1000 km (Haury *et al.*, 1978). However, physical oceanographers often refer to meso-scale events as having a scale of tens to hundreds of kilometers. The physical forces driving meso-scale events frequently result from interactions of larger-scale systems with one another or with land masses, and occasionally cause the invasion of one large ecosystem by a community from another. This may allow observation of change in population and community structure as physical properties change simultaneously. Meso-scale events include the formation of rings, eddies, jets at the margins of eastern or western boundary currents, broad upwelling events in eastern boundary currents and fresh-water plumes from major river systems (Haury *et al.*, 1978). Wakes and eddies behind islands and cross-shelf frontal systems will be discussed under coarse scale.

Although events at the scale of 100–500 km are probably greater than the ambits of most larger nekton, they are within the foraging range of procellariiforms, pelecaniforms (other than cormorants), gulls, some terns and some alcids. Physical structures in this size range form habitat patches within which birds could organize their foraging on a daily basis. Eddies and jets on this scale lasting weeks to months could be large enough and stable enough to draw birds with them as these features invade nearby water masses. However, the short time scale of these phenomena will make it exceedingly difficult to obtain adequate ornithological sampling.

The productivity of ocean areas depends on the duration of the phytoplankton bloom which, in turn, depends on the period of nutrient availability in surface waters (Cushing, 1971, 1978). In temperate seas this period lasts for 1–2 months, while in upwelling systems in the tropics (Peru) production lasts for the better part of the year (Cushing, 1971).

However this productivity may be both patchy and episodic due to the sensitivity of upwelling to changes in wind strength (Huyer, 1976), thus emphasizing the importance of temporal as well as spatial scales.

The upwelling systems associated with eastern boundary currents result in marked local increases in biomass and changes in species composition that contrast with those of surrounding waters (Haury *et al.*, 1978). While our traditional view of these systems is of a broad, uniform sheet of upwelled water moving seaward, recent studies suggest that upwelling may be localized in a relatively small number of plumes (Faucheux & Arce, 1977; Andrews & Hutchings, 1980). If so, we might expect patchiness in prey and bird distributions.

Food chains may be shorter in upwelling systems and early workers believed this was due to forage fish feeding directly on phytoplankton (Ryther, 1969; Longhurst, 1971). However, the predominant fish of upwelling systems, sardines *Sardinops* and anchovies *Engraulis*, primarily filter zooplankton and in the course of foraging collect some larger diatoms and dinoflagellates (Cushing, 1978). The major biomass of their prey comes from large copepods and the hypothesis that food chains are shortened by herring-like fish feeding directly on phytoplankton has yet to be demonstrated.

Peru Current system The Peru Coastal Current system is the narrow belt of cold, upwelled water between Talcahuaho, Chile, and the Gulf of Guayaquil. In some years the coastal upwelling ceases, with catastrophic results to the marine ecosystem (Barber & Chavez, 1983). These El Niño years are characterized by northerly rather than southerly winds, high surface temperatures, and heavy rains (Smith, 1968; Cane, 1983). El Niño events occur infrequently and at irregular intervals: 1891, 1925, 1941, 1953, 1957–58, 1965–66, 1972, 1975–76 and 1982 (Smith, 1968, 1983), and are driven by geophysical events with extensive variation in amplitude and periodicity (Quinn *et al.*, 1978; Walsh, 1978). The effects of major El Niño events are not limited to coastal Peru; declines in bird reproductive success associated with El Niño events have been reported from the Galapagos Islands (Boersma, 1977, 1978), Christmas Island (Schreiber & Schreiber, 1984), and the California current system (see below). Oceanic and atmospheric effects of an El Niño may be felt in the Atlantic as well as the Pacific Ocean (Tourre & Rasmusson, 1984).

During El Niño in Peru, Anchovetas *Engraulis ringens* may retreat to pockets in a narrow zone close to the coast at depths below the diving capacity of birds, and may also shift their population center southward

(Jordan & Fuentes, 1966; Valdivia, 1978). More recent data suggest an absolute decrease in abundance following a 5- to 10-fold reduction of primary productivity (Barber & Chavez, 1983). Regardless of whether adult anchovies decrease in number or shift in distribution, the effect is a decrease in their availability to birds (Valdivia, 1978; Duffy, 1983*a*). During El Niño events, there are dramatic die-offs of adult breeding birds dependent upon Anchovetas and desertion of eggs and young on the offshore islands (Tovar, 1978; Duffy, 1983*a*).

Data on the distribution and abundance of seabirds off the west coast of South America are relatively scarce, which is surprising given the importance to birds of the upwelling system and Peru Coastal Current. Off the Chilean Coast, Jehl (1973) characterized bird species groups as associated with different sea surface temperatures, while Brown *et al.* (1975) suggested that a combination of surface temperature and salinity was best for defining seabird habitats. Murphy (1936) classified Peruvian bird communities by zones of water used (inshore turbid, offshore blue), and found the highest diversities in the cool waters of the Peru Current. The cool inshore waters support a larger diversity of bird species, many of which plunge for their prey, than the offshore waters where birds forage at the surface (Brown, 1981).

Benguela Current system The Benguela Current, the easternmost edge of the South Atlantic gyre, is a cool coastal current with an offshore flow due to wind forcing, similar to the Peru Coastal Current (Smith, 1968). As in Peru, there is also an offshore flow. Unlike the Peruvian upwelling system, in the Benguela system upwelling via wind forcing is apparently seasonal with a 100–170 km longitudinal shift in upwelling centers between summer and winter (Hart & Currie, 1960; Emery, Milliman & Uchupi, 1973).

Although the numbers of birds in the nearshore waters of the Benguela Current are high (Hockey, Cooper & Duffy, 1983), the most constant concentration of birds off the southwest coast of southern Africa occurs over the Offshore Divergence Belt near the edge of the continental shelf (Summerhayes, Hofmeyr & Rioux, 1974). Upwelling offshore occurs year-round, unlike the seasonal inshore upwelling of the Benguela Current. Over the continental shelf, bird concentrations were independent of surface temperature and discrete masses of upwelled water (Abrams & Griffiths, 1981), possibly because the productivity of the water drops rapidly as it leaves the coast (Summerhayes *et al.*, 1974). Much of the patchiness in seabird distribution in the Benguela occurs at scales of

0.3–10 km and is probably related to prey patchiness (Schneider & Duffy, 1985). Warm-water events in the Benguela system are apparently of less drastic consequence to bird populations than those in the Peru current system (Duffy *et al.*, 1984).

Canary Current–Senegal upwelling In the Senegal upwelling, bird densities are relatively low compared to other upwelling sites. Although an offshore front is the site of concentrations of planktivorous phalaropes (Emery *et al.*, 1974), the highest densities of birds in this upwelling system are close to shore (Brown, 1979; Cadee, 1981). Brown (1979) hypothesized that the relatively low numbers of birds in the Senegal upwelling result from the lack of large breeding populations nearby, and that the late occurrence of the upwelling precludes its use by wintering western hemisphere migrants.

California Current Changes in the strength of the California Current upwelling system can result in local warming of surface waters (Mooers & Robinson, 1984) and, in the case of major anomalies associated with El Niño events, the invasion of typically warm-water birds (Ainley, 1976, 1980), local reproductive failures (Ainley & Lewis, 1974; Boekelheide, McElroy & Carter, in press), and a major decrease in use by wintering northern species (Briggs, personal communication). Briggs *et al.* (1981 and Chapter 12) provide a detailed account of pelagic bird distribution in the Southern California Bight.

The distribution of colonies and birds at sea in the Southern California Bight reflect an ecotone between an ecosystem associated with the cold California Current and an ecosystem associated with the northward-flowing, warm, inshore California Countercurrent (Hunt, Pitman & Jones, 1980). Of the 13 species of marine birds breeding or formerly breeding in the Channel Islands, five have the southern and three have the northern limit of their nesting there. A similar mixing of faunas is found in other marine groups (Hubbs, 1967).

The California Current is constantly spawning eddies and sending high-velocity jets of water oceanward (Fig. 2.3) and significant changes can occur over a time scale of a week (Mooers & Robinson, 1984). These events transport prey assemblages into areas where they would not normally occur (Haury, 1983). At a longer time scale, interannual variations in the southward advection of nutrients in the California Current have a major influence on zooplankton volumes along the California coast (Chelton, 1981). The frequent occurrence of these structures and fluctuations requires ornithologists studying this and

Fig. 2.3. Satellite imagery of sea surface temperature in the California Current on 8 July 1981 showing meso-scale jets and eddies. Image was collected at the Scripps Institution of Oceanography Satellite Oceanography facility and processed by Mark Abbot (SIO/Jet Propulsion Laboratory) and Philip Zion (JPL).

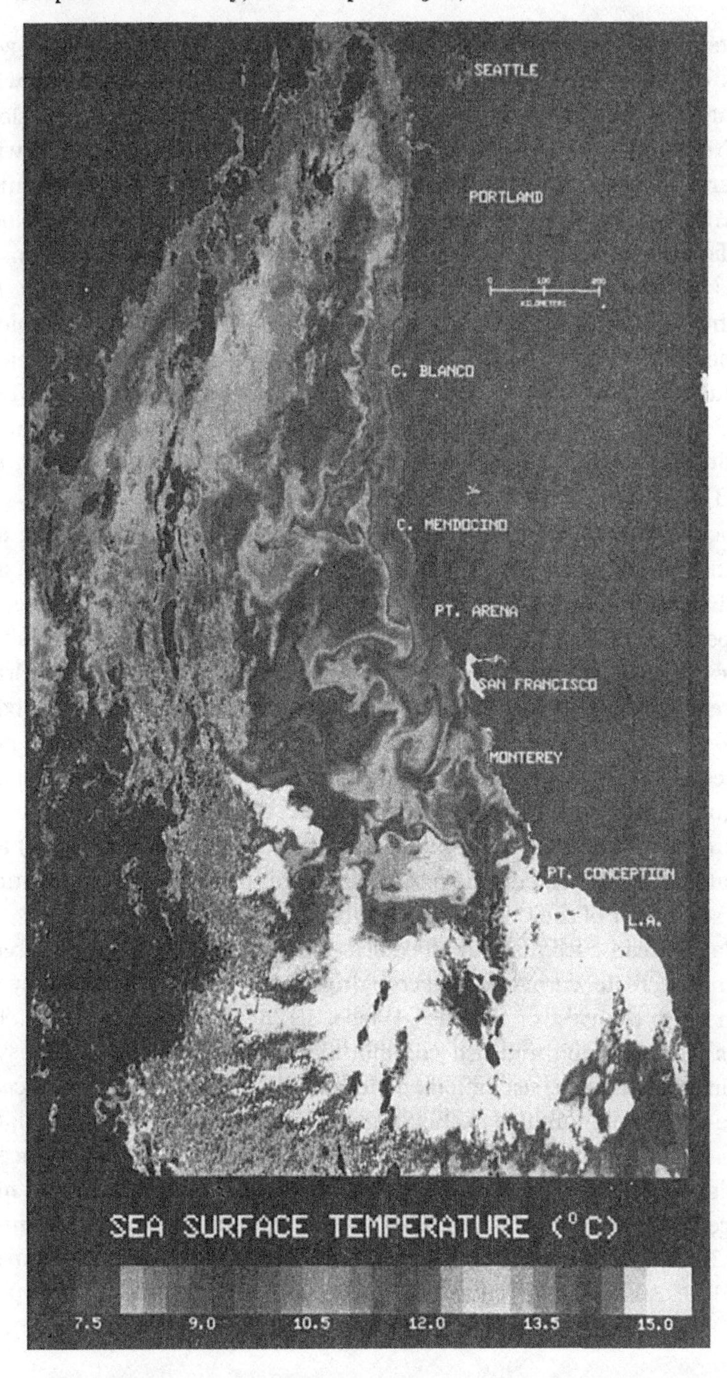

similar regions of fast-moving currents to know the origin of water masses and associated faunas when making bird observations.

Western Boundary Current systems Current systems also occur on the west sides of oceans. At high latitudes there are cold water masses flowing toward the equator (Labrador and Falkland in the Atlantic, the Oyasio in the North Pacific) and at lower latitudes there are warm currents flowing toward the poles (Florida Current/Gulf Stream and the Brazil Current in the Atlantic, Agulhas in the Indian Ocean, Kuroshio Current in the North Pacific) (Warnecke *et al.*, 1971; Robinson, 1976). There are also extensive areas of upwelling in the northwestern Arabian Sea driven by the southwest monsoon (Smith & Bottero, 1977). Eddies and fronts exist along the bounds of these currents and result in mixing of waters and increased productivity (Swallow, 1976); local upwelling may also be present (Uda, 1952; Han & Gong, 1970). These currents can dominate the adjacent continental shelves (e.g. Gulf Stream, see *J. Geophys. Res.*, **88**, No. C8, 1983).

Local assemblages of birds can be found in the areas of upwelling and mixing, such as at a front marking the confluence of the Gulf Stream and the Labrador Current (Lamb, 1964), or over water masses of different temperatures east of Newfoundland (Brown, 1968). In the South Atlantic, there are low densities of birds over the warmer shelf waters (Brazil Current) off Argentina, but many more in areas of upwelling and mixing (Jehl, 1974). Jehl also encountered large numbers of birds foraging near the edge of the continental shelf where sonar tracings revealed large schools of fish. In the North Pacific, Kuroda (1955) found concentrations of birds associated with 'irregular currents' (eddies and upwelling?) and elevated plankton catches near Attu and Agattu Islands in the Aleutians and southwest of the Commander Islands.

Major ocean currents spin off eddies as they meander. The Gulf Stream meanders create warm and cold core rings depending on whether they are spun off to the west or the east (Wiebe, 1976; Wiebe *et al.*, 1976). The rings trap plankton and fish communities typical of the water mass of origin which may persist for long periods in the rings. Cold core rings carry slope-water communities to the Sargasso Sea where they support higher densities of organisms than the surrounding Sargasso. The existence of similar eddies in the tropical Pacific can be deduced from patches of high concentrations of *Euphausia tenera* in the equatorial currents (Brinton, 1962). Rings are known to spin off from the Kuroshio, Agulhas, Somalia and East Australia Currents, and carry plankton and fish of one type of water into another (Nilsson, Andrews & Scully-Power, 1977; Haury *et al.*, 1978). Legeckis (1977) described wave-like fluctuations in the polar front in the Drake Passage with a scale of about 250 km over 19 days. He suggested that these waves could result in a water mass being transported across a frontal zone, and Hoffman, Nowlin & Whitworth (1983) describe rings detaching from the fronts and moving through Drake Passage.

Little or no ornithological work has investigated the possibility that birds would accompany or join a ring as it left one water mass and traveled through another. Particularly in areas of low productivity, rings and eddies may provide oases of greater forage for periods of days to weeks. Elevated densities of birds or predatory fish could form within, or at the edge of, these physical structures, at least soon after their formation when plankton and fish densities are highest (Wiebe, 1976).

Other meso-scale features that have received relatively little ornithological attention are the salinity fronts at the outflows of major river systems. Amos, Langseth & Markl (1972) describe convergent salinity fronts in the Indian Ocean 600 km southwest of Sumatra where sea surface salinity changed by 1.1% in less than 100 m. This front resulted from fresh water, entering the Bay of Bengal–Andaman Sea region, meeting the eastward-flowing Equatorial Countercurrent. Just seaward of the front there were concentrations of sharks, flyingfish, jellyfish and sea snakes. Plankton and debris of land origin were abundant in the turbulent zone of the front. Seabirds were observed foraging in the area, but quantitative data were not obtained. The observations of sea snakes, passively carried by weak currents, emphasize the importance of convergences for concentrating surface prey in tropical waters (Dunson & Ehlert, 1971; Kropach, 1975). Emery *et al.* (1974) also describe a low-salinity region where the combined waters of the Congo and Niger Rivers enter the Gulf of Guinea. High numbers of birds were associated with the region of low surface salinity in the Gulf. Based on their figures 6 and 15, areas of bird concentrations may have coincided with rapid changes in salinity, although they do not identify or discuss the presence of fronts.

Coarse scale

Coarse-scale features, 1–100 km in size, involve distinct aggregations of biomass and species (Haury *et al.*, 1978). They include plume-type upwelling events, widths of oceanic frontal zones, von Karmen vortex streets in the wake of islands, concentrations of prey near sea mounts, localized patches of plankton in open water and phenomena associated with the ice edge. Under some circumstances cyclonic storms may also cause mid-ocean upwelling (Beklemishev, 1961; Iverson, 1977). Coarse-scale features are probably within the ambits of larger nekton and, assuming that these organisms must find food every few days, coarse-scale structures provide the patches of prey upon which these populations depend. Patches of nekton or large zooplankton at this scale are undoubtedly important to foraging birds and may determine the patchiness of bird distribution within larger-scale features.

Island wakes enrich productivity and concentrate prey organisms (Uda & Ishino, 1958). Laboratory experiments on scale models of islands explained the location of major fishing and whaling grounds downstream from the Japanese islands as the result of eddies formed by currents. Flow around islands and resulting prey concentrations have been studied at the Great Barrier Reef of Australia (Hamner & Hauri, 1977, 1981; Alldredge & Hamner, 1980), where both plankton and planktivores aggregate in eddies. Ashmole & Ashmole (1967) suggested that tropical seabirds might depend on eddies with enriched productivity downstream from islands located in otherwise relatively unproductive waters. Murphy & Shomura (1972) and Uda (1973) observed concentrations of tuna and birds associated with equatorial Pacific islands. There are eddies downstream from the Galapagos Islands (White, 1973) and from Carribean islands that concentrate prey taken by birds (Ingham & Mahkan, 1966). In the Scotia Sea, the Weddell Gyre and the West Wind Drift create eddies with associated concentrations of krill and seabirds in the vicinity of both the South Orkney Islands and South Georgia Island (Bogdanov *et al.*, 1969; Everson, 1984). Stable eddy pairs may form around islands and Pacific islanders and seabirds alike take advantage of these eddies to locate concentrations of tuna and flyingfish (Johannes, 1981). Fay & Cade (1959) and Bédard (1969) suggested that eddies near St Lawrence Island, Alaska, were important for concentrating the prey of alcids.

Although bird concentrations have been found downstream from islands, no studies have quantified the relation between where the birds are seen and the position of either the island or eddies. Because different species of fish concentrate in different parts of eddy systems and different species of birds associate with different-sized fish, ecological segregation of birds may exist within eddy systems (Johannes, 1981).

Currents passing headlands can set up fronts and eddies similar to those near islands (Pingree, Bowman & Esaias, 1978). Plankton stocks are greater in the vicinity of Novaya Zemlya seabird colonies possibly due to fertilization of inshore waters by guano (Zelickman & Golovkin, 1972).

Alternatively, these plankton patches may result from eddies formed by currents passing headlands.

The importance of shallow continental shelves as areas of high productivity with large concentrations of birds was stressed by Ashmole (1971). Research activity, summarized in Bowman & Esaias (1978) and Allen *et al.* (1983), has increased in shelf waters, particularly in the North Sea and around North America. Shelf-break fronts may be expected when shelf water, diluted by fresh-water run-off meets oceanic water. Mooers *et al.* (1978) contrast the structure and function of these fronts with upwelling at shelf margins. Shelf-break fronts have been characterized for many shelf systems, e.g. Bering Sea (Kinder & Coachman, 1978; Iverson *et al.*, 1979), Celtic Sea (Pingree, 1979), Nova Scotian Shelf (Fournier *et al.*, 1979), Ross Sea (Ainley & Jacobs, 1981), and evidence for increased productivity and aggregations of birds has been found. Eddies associated with these fronts may enhance local productivity and small-scale patchiness in prey and predator populations.

Fronts occur in shallow water when there is an abrupt transition from a two-layer system to a well-mixed water column in which tidally stirred water near the bottom mixes with wind-stirred water at the surface (Fearnhead, 1975; Simpson, 1981), e.g. English Channel (Pingree, Forster & Harrison, 1974), Irish Sea (Pingree & Griffiths, 1978), Bering Sea (Schumacher *et al.*, 1979), around the Pribilof Islands (Kinder *et al.*, 1983). Enhanced vertical nutrient flux and productivity are associated with the fronts (Pingree, 1978). Concentrations of birds and their prey at these fronts were found by Pingree *et al.* (1974) in the English Channel, by Bourne (1981*b*) in the North Sea, and by Schneider (1982) and Kinder *et al.* (1983) in the Bering Sea.

In the southeastern Bering Sea, a series of three fronts partitions the shelf waters into separate domains (Fig. 2.4; see also Chapter 11, Fig. 11.2) (Coachman *et al.*, 1980). Carbon flux to the pelagic food web is greatest in the outer shelf domain (100–170 m depth), while carbon flux is greatest to the benthic food web in the middle shelf domain (50–100 m depth) and in the coastal domain (50 m depth) (Iverson *et al.*, 1979). Carbon flux to birds is primarily through surface foragers in the outer domain and to subsurface foragers in the middle and inner domains (Schneider, Hunt & Harrison, 1986). Harrison's (1982) observations of birds in the Gulf of Alaska suggest a similar pattern, with subsurface foragers primarily on the shelf and surface foragers at the shelf edge and beyond.

Similarities exist between the Bering Sea shelf and the North Sea (Fig.

Fig. 2.4. A comparison of the locations of fronts separating water masses in the Bering Sea and in the North Sea (after Coachman & Walsh, 1981, with permission of the authors and Deep-Sea Research © 1981 Pergamon Press).

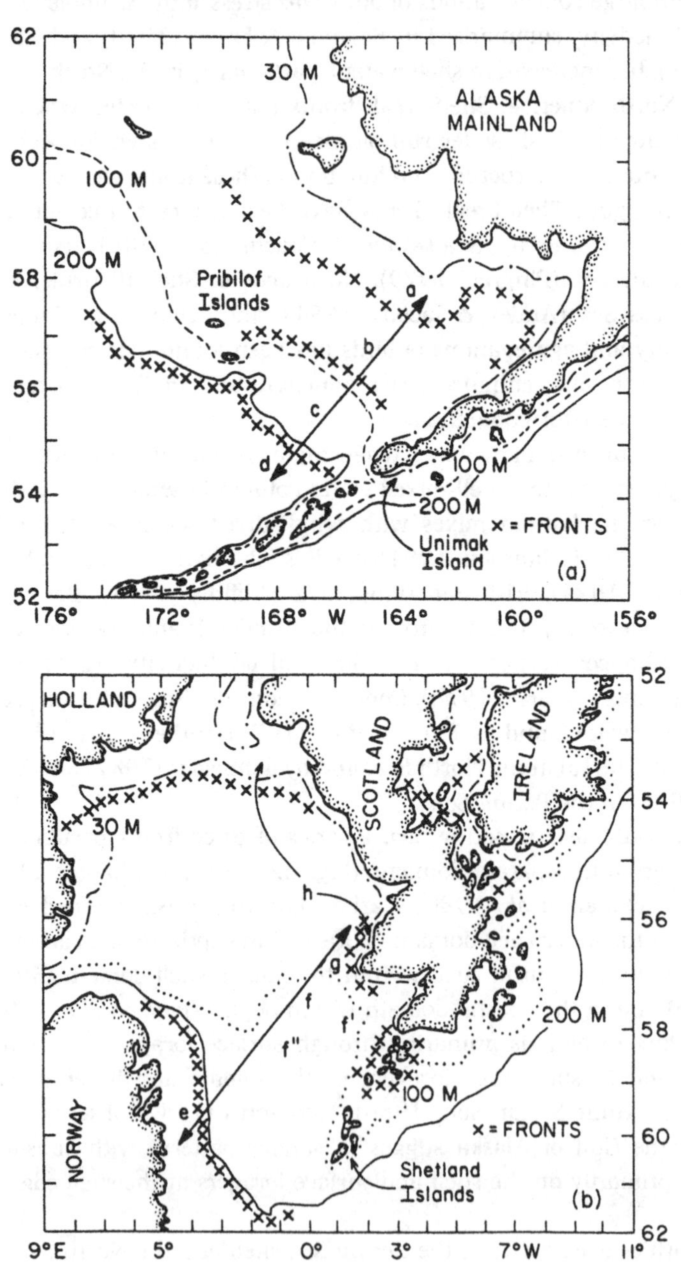

2.4) (Coachman & Walsh, 1981). Frontal systems similar to the inner front of the Bering Sea are found in the North Sea, as is a shelf edge front. Enhanced productivity and bird aggregations have been associated with these shallow fronts (Pingree, 1978; Bourne, 1981*a*, 1982; Joiris, 1982). Additionally, the waters of intermediate depths (50–200 m) on the shelf can be separated into two distinct water masses: high-salinity, low-temperature Atlantic water, and warmer, lower-salinity North Sea water (Joiris, 1978). Although primary production is similar in these two water masses, carbon flux in the Atlantic water is through zooplankton to pelagic fish, while in North Sea water heterotrophic bacteria apparently recycle much of the carbon (Joiris, 1978; Joiris *et al.*, 1982). The result is that North Sea water supports a smaller biomass of birds and Atlantic water supports the more pelagic species in both summer and winter (Joiris, 1978, 1983). Although the mechanisms structuring carbon pathways in the Bering Sea and the North Sea may be different, the similarities in domain structure, carbon flux to birds, and bird species distribution and abundance are noteworthy.

Plume upwellings, such as those off Baja California and South Africa, result in localized regions of increased productivity (Walsh *et al.*, 1974; Packard, Blasco & Barber, 1978; Andrews & Hutchings, 1980). They can be driven by steady longshore currents as well as by local wind forcing. Plume upwellings also occur when tidal or other currents cross underwater sills or seamounts thereby providing concentrated forage in surface waters (Gaskin, 1976; Brown, 1980).

Another important coarse-scale feature is the ice edge, an area of high productivity significant to high-latitude birds (Brown, 1980). Localized patches of enhanced productivity may be especially important in Arctic as compared to Antarctic systems. In the Antarctic there is widespread upwelling due to both the Antarctic Divergence and replacement of northward-flowing surface and bottom waters by nutrient-rich middle waters near the coast (Dunbar, 1968). In contrast, in the high Arctic the water column is generally stable and there is little flux of nutrients from deeper, dense water to lower-salinity surface water (Dunbar, 1968). Under these circumstances, localized upwelling and elevated productivity associated with the pack-ice edge, glaciers and icebergs may provide critical resources (Brown, 1980). The underside of pack ice provides a substrate for a rich community of algae, zooplankton and associated fish which, in turn, are important forage for planktivorous and piscivorous birds (Irving, McRoy & Burns, 1970; Meltofte, 1976; Bradstreet, 1982). Current research indicates significant differences between Antarctic and Arctic systems and extrapolation from one to the other should be done with caution.

Within the pack of the Beaufort Sea, Frame (1973) found the largest numbers of birds in areas with 50–90% ice cover. In the eastern Canadian Arctic, Northern Fulmars *Fulmarus glacialis*, Thick-billed Murres *Uria lomvia* and Black Guillemots *Cepphus grylle* occurred in higher densities in or near ice, while Black-legged Kittiwakes *Rissa tridactyla* preferred ice-free coasts (McLaren, 1982). The pack-ice edge may also constitute a barrier that concentrates birds waiting to migrate to nesting grounds in the spring (Shuntov, 1961).

Some species are restricted to the ice or its immediate vicinity (Meltofte *et al.*, 1981; Orr & Parsons, 1982; Griffiths, 1983; Ainley & Boekelheide, 1983). Emperor *Aptenodytes forsteri* and Adelie *Pygoscelis adeliae* Penguins and Snow Petrel *Pagodroma nivea* distributions are closely correlated with the Antarctic pack ice (Ainley *et al.*, 1984). In the ice-free central Ross Sea in summer birds may be scarce (Ainley *et al.*, 1984). Thus, in the Antarctic, ice has a strong influence on the community structure and density of birds (Kock & Reinsch, 1978; Zink, 1981; Thurston, 1982; Griffiths *et al.*, 1982). It is apparent that ice offers a great variety of habitats to birds; we need to learn more about habitats within ice and the mechanisms whereby they influence bird numbers.

Fine scale

Physical structures and patchiness at the scale of meters to hundreds of meters are classified as fine scale (Haury *et al.*, 1978). Until recently, variation at this scale was considered to be 'noise' generated by sampling error in plankton studies. However, the importance of patchiness at this scale is now realized for plankton, and it is over these distances that individual zooplankters interact. Haury *et al.* (1978) estimate that the time scale for events between copepods is of the order of hours to days and that in this time, particularly if the species migrate vertically, they could traverse distances of meters to several hundreds of meters or even several kilometers.

In the open ocean, vortices generated by vertical density gradients due to temperature and salinity differences may result in a series of parallel cells in calm surface waters with alternating divergence and convergence zones (Owen, 1966). Wind streaks, generated by alternating divergences and convergences parallel to the wind, result from Langmuir cells when floating particles and organisms are concentrated in the convergence (Barstow, 1983; Leibovich, 1983). Several sets of streaks may coexist, with 2–300 m between rows; the physical mechanism for driving these cells is not known although wind forcing is involved (Leibovich, 1983). Stavn (1971) demonstrated experimentally that the currents in Langmuir cells are capable of concentrating vertically migrating, photosensitive zooplankton, thus providing an explanation of Neess' (1949 in Stavn, 1971) finding that there was greater variability between plankton samples collected parallel to the wind than between those collected at right angles to wind direction. Brown (1980) observed phalaropes, gulls and terns off Peru feeding in convergences associated with Langmuir cells, but not between them. Brodie *et al.* (1978) suggested that Langmuir cells may also be important for concentrating the prey of baleen whales.

Fine-scale patchiness of plankton in the open ocean is also important (Cassie, 1963; Wiebe, 1970; Haury, 1976). Although physical processes are involved in some instances (e.g. Bailey, 1966; Packard *et al.*, 1978; Brown *et al.*, 1979), vertical migration (Brinton, 1967) or social interactions between plankton, e.g. the schooling of *Euphausia superba* (Marr, 1962), may predominate. Fenwick (1978) observed several species of birds foraging on small krill (*Nyctiphanes capensis* and *N. australis*) swarms near the surface off the Snares Islands, New Zealand. Off Peru, seabirds forage at small patches of crustaceans that may be available at the surface for periods as short as 5 min at a time (Duffy, 1983*b*), thus emphasizing that fine-scale features significant to birds may be of very short duration.

Concluding comments

Patterns of marine bird distribution and abundance fit remarkably well the scales of organization (Fig. 2.1) at which plankton are found. In the organizational scheme of Steele (1976), birds are rather like aerial fish (Fig. 2.2). Bird life spans and ambits are at the upper right of the diagram, events associated with breeding for many species fall nearer the middle, and critical foraging interactions are likely to be primarily in the lower left of the diagram.

Certainly the fine-scale patchiness within domains or regions can mask differences in bird distribution between water masses. For plankton, physical forces may predominantly control fine-scale interactions in the horizontal plane, while biological factors determine coarse-scale and larger patterns. For birds and nekton, fine-scale and coarse-scale patterns are probably controlled primarily by biological processes of social interaction and foraging behavior. At these smaller scales, information on prey availability is critical if we are to move from correlative studies to examination of causation. Patterns for marine birds at larger scales are most often a response to physical features of the environment. It is possible that large predatory fish behave in ways similar to birds and the possibility of using birds as models for the patterns of fish and *vice versa* should be investigated.

Because large-scale physical features have considerable inertia and therefore change over relatively long time scales (Owen, 1981), we can expect patterns of marine bird distribution at large scales to be more stable than those at small scales. This time linkage to spatial scale for birds has not been investigated directly. The role of weather, in particular wind speed and cyclonic storms, has been largely ignored but the available evidence suggests weather may be an important physical variable. We have yet to develop sufficient descriptive measures that summarize the influence of weather on bird distribution and movement.

The greatest difficulty in documenting differences in patterns of bird activity comes at the level of meso-scale events. This difficulty is the result of large variances in bird abundance at coarse and fine scales. If we are to understand the role of meso-scale events we will have to design with care our observation programs to contrast water masses and features. We will need continuous observations of birds at sea to provide adequate samples to resolve sources of variation and we will have to obtain real-time data on a variety of physical and biological properties such as sea surface temperature and salinity, prey taken and prey availability. As we shift from studies describing the distribution of birds to those seeking to determine the environmental variables controlling variance in bird numbers, we will become increasingly aware of the observational and statistical problems caused by patterns at different spatial and temporal scales.

Acknowledgements Useful suggestions for improving the manuscript were provided by W.R.P. Bourne, K. Briggs, R.G.B. Brown, J.P Croxall, L. Haury, T. Kinder, A. Longhurst, J. McGowan, B. Obst and J. Steele. We thank Zoe Eppley for invaluable bibliographic aid and for suggestions on improving an earlier draft. Karen Christensen drew the figures and Dianne Christianson typed the several drafts of the manuscript. The authors' research on pelagic bird distribution has been supported in part by the Division of Polar Programs, National Science Foundation (Grants DPP 79-10386, DPP 83-08232, DPP 83-18469) and by the Bureau of Land Management through interagency agreement with the National Oceanographic and Atmospheric Administration, under which a multi-year program responding to needs of petroleum development of the Alaska continental shelf is managed by the Outer Continental Shelf Environmental Assessment Program (OCSEAP).

References

Abrams, R.W. & Griffiths, A.M. (1981). Ecological structure of the pelagic seabird community in the Benguela Current region. *Mar. Ecol.-Prog. Ser.*, **5**, 269–77.

Abrams, R.W., Griffiths, A.M., Hajee, Y. & Schoeppe, E. (1981). A computer-assisted plotting program for analysing the dispersion of pelagic seabirds and environmental features. *P.S.Z.N.I. Mar. Ecol.*, **2**, 363–8.

Ainley, D.G. (1976). The occurrence of seabirds in the coastal region of California. *Western Birds*, **7**, 33–68.

Ainley, D.G. (1977). Feeding methods in seabirds: a comparison of polar and tropical nesting communities in the eastern Pacific Ocean. In *Adaptations within Antarctic ecosystems*, ed. G.A Llano, pp. 669–85. Washington D.C.: Smithsonian Institution.

Ainley, D.G. (1980). Birds as marine organisms: a review. *Calif. Coop. Ocean. Fish. Invest. Rep.*, **21**, 48–53.

Ainley, D.G. & Boekelheide, R.L. (1983). An ecological comparison of oceanic seabird communities of the south Pacific Ocean. In *Tropical seabird biology*, ed. R.W. Schreiber. *Studies in avian biology*, **8**, 2–23.

Ainley, D.G. & Jacobs, S.S. (1981). Sea-bird affinities for ocean and ice boundaries in the Antarctic. *Deep-Sea Res.*, **28A**, 1173–85.

Ainley, D.G. & Lewis, T.J. (1974). The history of the Farallon Islands marine bird populations 1854–1972. *Condor*, **76**, 432–46.

Ainley, D.G., O'Connor, E.F. & Boekelheide, R.J. (1984). *The marine ecology of birds in the Ross Sea, Antarctic*. New York: American Ornithological Union, Ornithological Monographs No. 32, 97 pp.

Alldredge, A.L. & Hamner, W.M. (1980). Recurring aggregation of zooplankton by a tidal current. *Est. Coast. Mar. Sci.*, **10**, 31–57.

Allen, J.S., Beardsley, R.C., Blanton, J.O., Boicourt, W.C., Butman, B., Coachman, L.K., Huyer, A., Kinder, T.H., Royer, T.C., Schumacher, J.D., Smith, R.L., Sturges, W. & Winant, C.D. (1983). Physical oceanography of continental shelves. *Rev. Geophys. Space Phys.*, **21**, 1149–81.

Amos, A.F., Langseth, M.G., Jr & Markl, R.G. (1972). Visible oceanic saline fronts. In *Studies in physical oceanography*, vol. 1, ed. A.L. Gordon, pp. 49–62. New York: Gordon & Breach.

Andrews, W.R.H. & Hutchings, L. (1980). Upwelling in the southern Benguela Current. *Prog. Oceanogr.*, **9**, 1–81.

Ashmole, M.J. & Ashmole, N.P. (1968). The use of food samples from seabirds in the study of seasonal variation in the surface fauna of tropical oceanic areas. *Pacif. Sci.*, **22**, 1–10.

Ashmole, N.P. (1971). Sea bird ecology and the marine environment. In *Avian biology*, vol. 1. ed. D.S. Farner & J.R. King, pp. 223–86. New York: Academic Press.

Ashmole, N.P. & Ashmole, M.J. (1967). Comparative feeding ecology of seabirds of a tropical oceanic island. *Bull. Peabody Mus. Nat. Hist.*, **24**, 1–131.

Backus, R.H., Craddock, J.E., Haedrich, R.L. & Shores, D.L. (1970). The distribution of mesopelagic fishes in the equatorial and western North Atlantic Ocean. *J. Mar. Res.*, **28**, 179–201.

Bailey, R.S. (1966). The seabirds of the southeast coast of Arabia. *Ibis*, **108**, 224–64.

Bailey, R.S. (1968). The pelagic distribution of sea-birds in the western Indian Ocean. *Ibis*, **110**, 493–519.

Bailey, R.S. & Bourne, W.R.P. (1972). Notes on sea-birds 36: counting birds at sea. *Ardea*, **60**, 124–7.

Barker, R.T. & Chavez, F.P. (1983). Biological consequences of El Nio. *Science*, **222**, 1203–10.

Barstow, S.F. (1983). The ecology of Langmuir circulation: a review. *Mar. Environ. Res.*, **9**, 211–36.

Bédard, J. (1969). Feeding of the Least, Crested and Parakeet Auklets around St. Lawrence Island, Alaska. *Can. J. Zool.*, **47**, 1025–50.

Beklemishev, K.V. (1961). Influence of atmospheric cyclone on feeding grounds of whales in Antarctic. *Trudy Inst. Okeanol. Akad. Nauk SSSR*, **51**, 121–41.

Boekelheide, R.J., McElroy, T. & Carter, H.R. (In press). Farallon Island seabirds and the 1983 El Niño. *Pacif. Seabird Gp. Bull.*

Boersma, P.D. (1977). An ecological and behavioral study of the Galapagos penguin. *Living Bird*, **15**, 43–93.

Boersma, P.D. (1978). Breeding patterns of Galapagos penguins as an indicator of oceanographic conditions. *Science*, **200**, 1481–3.

Bogdanov, M.A., Oradovskiy, S.G., Solyankin, V. & Khvatskiy, N.V. (1969). On the frontal zone of the Scotia Sea. *Oceanology*, **9**, 777–83.

Bourne, W.R.P. (1963). A review of oceanic studies of the biology of seabirds. *Proc. XIII Int. Ornithol. Congr. 1962*, 831–54.

Bourne, W.R.P. (1981*a*). The habits, distribution and numbers of northern seabirds. *Trans. Linn. Soc. N.Y.*, **9**, 1–14.

Bourne, W.R.P. (1981*b*). Some factors underlying the distribution of seabirds. In *Proceedings of the symposium on birds of the sea and shore, 1979*, ed. J. Cooper, pp. 110–34. Cape Town, South Africa: African Seabird Group.

Bourne, W.R.P. (1982). Concentrations of Scottish seabirds vulnerable to oil pollution. *Mar. Poll. Bull.*, **13**, 270–3.

Bowman, M.J. & Esaias, W.E. (1978). *Oceanic fronts in coastal processes*. Berlin: Springer-Verlag.

Bradstreet, M.S.W. (1982). Occurrence, habitat use, and behavior of seabirds, marine animals, and Arctic Cod at the Pond Inlet ice edge. *Arctic*, **35**, 28–40.

Briggs, K.T., Chu, E.W., Lewis, D.B., Tyler, W.B., Pitman, R.L. & Hunt, G.L., Jr (1981). Distribution, numbers and seasonal status of seabirds of the southern California Bight area, 1975–1978. *Summary of marine mammal and seabird surveys of the southern California Bight area, 1975–1978. Vol. III. Investigator's reports, Part III – Seabirds, Book I, Chapter I.* BLM/YM/SR-81/03.

Brinton, E. (1962). Variable factors affecting the apparent range and estimated concentration of euphausiids in the North Pacific. *Pacif. Sci.*,**14**, 374–408.

Brinton, E. (1967). Vertical migration and avoidance capability of euphausiids in the California Current. *Limnol. Oceanogr.*, **23**, 451–83.

Brodie, P.F., Sameoto, D.D. & Sheldon, R.W. (1978). Population densities of euphausiids off Nova Scotia as indicated by net samples, whale stomach contents, and sonar. *Limnol. Oceanogr.*, **23**, 1264–7.

Brown, R.G.B. (1968). Seabirds in Newfoundland and Greenland waters, April–May 1966. *Can. Field-Nat.*, **82**, 88–102.

Brown, R.G.B. (1979). Seabirds of the Senegal upwelling and adjacent waters. *Ibis*, **121**, 283–92.

Brown, R.G.B. (1980). Seabirds as marine animals. In *Behavior of marine animals, Vol. 4 Marine birds*, ed. J. Burger, B. Olla & H.E. Winn, pp. 1–39. New York: Plenum Press.

Brown, R.G.B. (1981). Seabirds in northern Peruvian waters, November–December 1977. *Bol. Inst. Mar. Peru*, vol. extraordinario, 34–42.

Brown, R.G.B., Barker, S.P. & Gaskin, D.E. (1979). Daytime surface-swarming in *Meganyctiphanes norvegica* (M. Sars.) (Crustacea, Euphausiacea) off Brier Island, Bay of Fundy. *Can. J. Zool.*, **51**, 2285–91.

Brown, R.G.B., Cooke, F., Kinnear, P.K. & Mills, E.L. (1975). Summer seabird distributions in Drake Passage, the Chilean fjords and off southern South America. *Ibis*, **117**, 339–56.

Cadee, G.C. (1981). Seabird observations between Rotterdam and the equatorial Atlantic. *Ardea*, **69**, 211–16.

Cane, M.A. (1983). Oceanographic events during El Niño. *Science*, **222**, 1189–95.

Cassie, R.M. (1959). Micro-distribution of plankton. *N.Z. J. Sci.*, **2**, 398–409.

Cassie, R.M. (1963). Microdistribution of plankton. *Oceanogr. Mar. Biol. Ann. Rev.*, **1**, 223–52.

Chelton, D.B. (1981). Interannual variability of the California current – physical factors. *Calif. Coop. Ocean. Fish. Invest. Rep.*, **22**, 34–48.

Coachman, L.K., Kinder, T.H., Schumacher, J.D. & Tripp, R.B. (1980). Frontal systems of the southeastern Bering Sea shelf. In *Stratified flows*, ed. T. Carstens & T. McClimans, pp. 917–93. Trondheim: TAPIR Pub.

Coachman, L.K. & Walsh, J.J (1981). A diffusion model of cross-shelf exchange of nutrients in the southeastern Bering Sea. *Deep-Sea Res.*, **28A**, 819–46.

Cromwell, T. (1953). Circulation in a meridianal plane in the central equatorial Pacific. *J. Mar. Res.*, **12**, 196–213.

Crowley, P.H. (1977). Spatially distributed stochasticity and the constancy of ecosystems. *Bull. Math. Biol.*, **39**, 157–66.

Cushing, D. (1971). A comparison of production in temperate seas and the upwelling areas. *Trans. R. Soc. S. Afr.*, **40**, 17–33.

Cushing, D. (1978). Upper trophic levels in upwelling areas. In *Upwelling ecosystems*, ed. R. Boje & M. Tomczak, pp. 261–81. New York: Springer-Verlag.

Dayton, P.K. & Tegnor, M.J. (1984). The importance of scale in community ecology: a kelp forest example with terrestrial analogs. In *A new ecology: novel approaches to interactive systems*, ed. P.W. Price, C.N. Slobodchikoff & W.S. Gaud, pp. 457–81. New York: Wiley.

Devillers, P. (1978). Patchiness of seabird distribution and shortcomings of sampling methods. *Ibis*, **120**, 125.

Duffy, D.C. (1983*a*). Environmental uncertainty and commercial fishing: effects on Peruvian guano birds. *Biol. Conserv.*, **26**, 227–38.

Duffy, D.C. (1983*b*). The foraging ecology of Peruvian seabirds. *Auk*, **100**, 800–10.

Duffy, D.C., Berruti, A., Randall, R.M. & Cooper, J. (1984). Effects of the 1982–3 warm water event on the breeding of South African seabirds. *S. Afr. J. Sci.*, **80**, 65–9.

Dunbar, M.J. (1968). *Ecological development in polar regions*. Englewood Cliffs, N.J.: Prentice-Hall.

Dunson, W.A. & Ehlert, G.W. (1971). Effects of temperature, salinity and surface water flow on the distribution of the sea snake *Pelamis*. *Limnol. Oceanogr.*, **16**, 845–53.

Emery, K.O., Lepple, F., Toner, L., Uchupi, E., Rioux, R.H., Pople, W. & Hulbert,

E.M. (1974). Suspended matter and other properties of surface waters of the northeastern Atlantic Ocean. *J. Sedim. Petrol.*, **44**, 1087–110.

Emery, K.O., Milliman, J.D. & Uchupi, E. (1973). Physical properties and suspended matter of surface waters in the southeastern Atlantic Ocean. *J. Sedim. Petrol.*, **43**, 822–37.

Everson, I. (1984). Marine interactions. In *Antarctic ecology*, ed. R.M. Laws, pp. 783–819. London: Academic Press.

Fager, E.W. & Longhurst, A.R. (1968). Recurrent group analysis of species assemblages of demersal fish in the Gulf of Guinea. *J. Fish. Res. Bd Can.*, **25**, 1405–21.

Fager, E.W. & McGowan, J.A. (1963). Zooplankton species groups in the North Pacific. *Science*, **140**, 453–60.

Faucheux, C.V. & Arce, J.Q. (1977). *Variaciones termicas superficiales del Pacifico Sur Oriental (Junio 75–Septiembre 79)*. La Punta, Callao: Minist. Marina, Republ. Peru.

Fay, F.H. & Cade, T.J. (1959). An ecological analysis of the avifauna of St. Lawrence Island, Alaska. *Univ. Calif. Publs Zool.*, **63**, 73–150.

Fearnhead, P.G. (1975). On the formation of fronts by tidal mixing around the British Isles. *Deep-Sea Res.*, **22**, 311–21.

Fenwick, G.D. (1978). Plankton swarms and their predators at the Snares Islands. *N.Z. J. Mar. Freshw. Res.*, **12**, 223–4.

Fournier, R.O., van Det, M., Wilson, J.S. & Hargraves, N.B. (1979). Influence of the shelf-break front off Nova Scotia on phytoplankton standing stock in winter. *J. Fish. Res. Bd Can.*, **36**, 1228–37.

Frame, G.W. (1973). Occurrence of birds in the Beaufort Sea, summer 1969. *Auk*, **90**, 552–63.

Gaskin, D.E. (1976). The evolution, zoogeography and ecology of the Cetacea. *Oceanogr. Mar. Biol. Ann. Rev.*, **14**, 247–346.

Geynrikh, A.K. (1976). Role of displaced species in the structure of Pacific tropical plankton communities. *Oceanology*, **15**, 492–5.

Golovkin, A.N., Shivokolobov, V.N. & Garkavaya, G.P. (1972). Peculiarities of biogenic element distribution near birds' bazaars in the north of Novaya Zemlya. In *Peculiarities of biological productivity of waters near birds' bazaars in the north of Novaya Zemlya*, pp. 33–46. Leningrad: Nauka. (In Russian.)

Gould, P.J. (1971). *Interactions of seabirds over the open ocean*. Ph.D diss., Univ. Arizona, Tucson.

Gould, P.J. (1974). Sooty tern (*Sterna fuscata*). In *Pelagic studies of seabirds in the central and eastern Pacific Ocean*, ed. W.B. King, pp. 6–52. Washington D.C.: Smithsonian Institution.

Gould, P.J. (1983). Seabirds between Alaska and Hawaii. *Condor*, **85**, 286–91.

Griffiths, A.M. (1983). Factors affecting the distribution of the Snow Petrel (*Pagodroma nivea*) and the Antarctic Petrel (*Thalassoica antarctica*). *Ardea*, **71**, 145–50.

Griffiths, A.M., Siegfried, W.R. & Abrams, R.W. (1982). Ecological structure of a pelagic seabird community in the Southern Ocean. *Polar Biol.*, **1**, 39–46.

Gudkov, V.M. (1962). Relationship between the distribution of zooplankton, seabirds and baleen whales. *Trudy Inst. Okeanol., Akad. Nauk SSSR*, **58**, 298–313. Translation 1974, Dept. of Navy NOO T-7.

Hamner, W.M. & Hauri, I.R. (1977). Fine-scale surface currents in the Whitsunday Islands, Queensland, Australia: effect of tide and topography. *Aust. J. Mar. Freshw. Res.*, **28**, 333–59.

Hamner, W.M. & Hauri, I.R. (1981). Effect of island mass: water flow and plankton pattern around a reef in the Great Barrier Reef lagoon, Australia. *Limnol. Oceanogr.*, **26**, 1084–102.

Han, H.S. & Gong, Y. (1970). Relation between oceanographical conditions and catch of saury in the eastern Sea of Korea. In *The Kuroshio-A symposium on the Japan Current*, ed. J. Marr, pp. 585–92. Honolulu: East–West Center Press.

Harrington, B.A., Schreiber, R.W. & Woolfenden, G. (1972). The distribution of the male and female Magnificent Frigatebirds, *Fregata magnificens*, along the gulf coast of Florida. *Am. Birds*, **26**, 927–31.

Harris, M.P. & Hansen, L. (1974). Sea-bird transects between Europe and Rio Plate, South America, in autumn 1973. *Dansk Orn. Foren Tidsskr.*, **68**, 117–37.

Harrison, C.S. (1982). Spring distribution of marine birds in the Gulf of Alaska. *Condor*, **84**, 245–54.

Hart, T.J. & Currie, R.I. (1969). The Benguela Current. *Discovery Rep.*, **31**, 123–93.

Haury, L.R. (1976). A comparison of zooplankton patterns in the California Current and North Pacific central gyre. *Mar. Biol.*,**37**, 159–67.

Haury, L.R. (1983). Biological consequence of offshore eddies in the California Current. *CalCOFI Annual Conference, 1983, Program and Abstracts*, p. 10. (Abstract.)

Haury, L.R., McGowan, J.A. & Wiebe, P.H. (1978). Patterns and processes in the time-space scales of plankton distributions. In *Spatial pattern in plankton communities*, ed. J.H. Steele, pp. 277–327. New York: Plenum Press.

Hockey, P.A.R., Cooper, J. & Duffy, D.C. (1983). The roles of coastal birds in the functioning of marine ecosystems in southern Africa. *S. Afr. J. Sci.*, **79**, 130–4.

Hoffman, E.E., Nowlin, W.D., Jr & Whitworth, T., III (1983). Effects of frontal exchange processes in the Antarctic circumpolar current at Drake Passage. Chemical Variability in Ocean Frontal Areas, Sept. 1983. *N.O.R.D.A.* (Abstract.)

Hubbs, C.L. (1967). A discussion of the geochronology and archeology of the California Islands. In *Proceedings of the symposium on the biology of the California Islands*, ed. R.N. Philbrick, pp. 337–41. Santa Barbara, CA: Santa Barbara Botanic Garden.

Hunt, G.L., Pitman, R.L. & Jones, H.L. (1980). Distribution and abundance of seabirds breeding on the California Channel Islands. In *The California Islands: proceedings of a multidisciplinary symposium*, ed. D.M. Power, pp. 443–59. Santa Barbara, CA: Santa Barbara Museum of Natural History.

Hutchinson, G.E. (1953). The concept of pattern in ecology. *Proc. Philadelphia Acad. Nat. Sci.*, **105**, 1–12.

Huyer, A. (1976). A comparison of upwelling events in two locations: Oregon and northwest Africa. *J. Mar. Res.*, **34**, 531–46.

Ingham, M.C. & Mahken, C.V.W. (1966). Turbulence and productivity near St. Vincent Is., B.W.I. A preliminary report. *Carib. J. Sci.*, **6 (3–4)**, 83–92.

Irving, L., McRoy, C.P. & Burns, J.J. (1970). Birds observed during a cruise in the ice-covered Bering Sea in March 1968. *Condor*, **72**, 110–12.

Iverson, R.L. (1977). Mesoscale oceanic phytoplankton patchiness caused by hurricane effects on nutrient distribution in the Gulf of Mexico. In *Oceanic sound scattering prediction*, ed. N.R. Anderson & B.J. Zahuranec, pp. 767–78. New York: Plenum Press.

Iverson, R.L., Coachman, L.K., Cooney, R.T., English, T.S., Goering, J.J., Hunt, G.L., Jr, MacCauley, M.C., McRoy, C.P., Reeburg, W.S. & Whitledge, T.E. (1979). Ecological significance of fronts in the southeastern Bering Sea. In *Ecological processes in coastal and marine systems*, ed. R.L. Livingston, pp. 437–65. New York: Plenum Press.

Jehl, J.R., Jr (1973). The distribution of marine birds in Chilean waters in winter. *Auk*, **90**, 114–35.

Jehl, J.R., Jr (1974). The distribution and ecology of marine birds over the continental shelf of Argentina in winter. *Trans. San Diego Soc. Nat. Hist.*, **17**, 217–34.

Jesperson, P. (1930). Ornithological observations in the north Atlantic Ocean. *Oceanogr. Rep. Dana Exped.*, **7**, 1–36.

Johannes, R.E. (1981). *Words of the lagoon: fishing and marine lore in the Palau District of Micronesia.* Berkeley, CA: Univ. California Press.

Johnson, M.W. & Brinton, E. (1963). Biological species, water-masses and current. In *The sea, vol II. The composition of sea-water – comparative and descriptive oceanography*, ed. M.N. Hill, pp. 381–414. New York: Wiley.

Joiris, C. (1978). Seabirds recorded in the northern North Sea in July: the ecological implications of their distribution. *Le Gerfaut*, **68**, 419–40.

Joiris, C. (1983). Winter distribution of seabirds in the North Sea: an oceanographical interpretation. *Le Gerfaut*, **73**, 107–22.

Joiris, C., Billen, G., Lancelot, C., Daro, M.H., Mommaerts, J.P., Bertels, A., Boissicort, M., Nijs, J. & Hecq, J.H. (1982). A budget of carbon cycling in the Belgian coastal zone: relative roles of zooplankton, bacterioplankton and benthos in the utilization of primary production. *Neth. J. Sea. Res.*, **16**, 260–75.

Jordan, R. & Fuentes, H. (1966). Las poblaciones de aves guaneras y su situacion actual. *Inf. Inst. Mar Peru-Callao*, **10**, 1–31.

Kelley, J.C. (1976). Sampling the sea. In *Ecology of the seas*, ed. D.H. Cushing & J.J. Walsh, pp. 361–78. Philadelphia: Saunders.

Kinder, T.H. & Coachman, L.K. (1978). The front overlaying the continental slope of the eastern Bering Sea. *J. Geophys. Res.*, **83**, 4551–9.

Kinder, T.H., Hunt, G.L., Jr, Schneider, D. & Schumacher, J.D. (1983). Correlations between seabirds and oceanic fronts around the Pribilof Islands, Alaska. *Est. Coast. Shelf Sci.*, **16**, 309–19.

King, W.B. (1974*a*). *Pelagic studies of seabirds in the central and eastern Pacific Ocean.* Washington D.C.: Smithsonian Institution.

King, W.B. (1974*b*). Wedge-tailed Shearwater (*Puffinus pacificus*). In *Pelagic studies of seabirds in the central and eastern Pacific Ocean*, ed. W.B. King, pp. 53–95. Washington D.C.: Smithsonian Institution.

Kock, K.H. & Reinsch, H.H. (1978). Ornithological observations during the German Antarctic Expedition 1975/76. *Beitr. Vogelkd.*, **24**, 305–28.

Kropach, C. (1975). The yellow-bellied sea snake, *Pelamis*, in the eastern Pacific. In *The biology of sea snakes*, ed. W.A. Dunson, pp. 185–213. Baltimore: University Park Press.

Kuroda, N. (1954). *On the classification and phylogeny of the order Tubinares, particularly the shearwaters* Puffinus. Tokyo: Herald Co.

Kuroda, N. (1955). Observations on pelagic birds of the northwest Pacific. *Condor*, **57**, 290–300.

Lamb, K.D.A. (1964). Sea birds at the confluence of the Gulf Stream and the Labrador Current east of New York. *Sea Swallow*, **16**, 65.

Lasker, R. (1979). Factors contributing to variable recruitment of the Northern Anchovy (*Engraulis mordax*) in the California Current: contrasting years 1975 through 1978. *Rapp. P.-v. Reun. Cons. Int. Explor. Mer*, **178**, 375–88.

Legeckis, R. (1977). Oceanic polar front in the Drake Passage-satellite observations during 1976. *Deep-Sea Res.*, **24**, 701–4.

Legeckis, R. (1982). Satellite infrared observations of oceanic long waves in the equatorial Pacific, 1975 to 1981. *NOAA-TR-NESS-92. National Environmental Satellite Service, Washington, D.C.*, 113 pp.

Leibovich, S. (1983). The form and dynamics of Langmuir circulations. *Ann. Rev. Fluid Mech.*, **15**, 391–427.

Longhurst, A.R. (1971). The clupeoid resources of tropical seas. *Oceanogr. Mar. Biol. Ann. Rev.*, **9**, 349–85.

McGowan, J.A. (1974). The nature of oceanic ecosystems. In *The biology of the oceanic Pacific*, ed. C.B. Miller, pp. 9–28. Corvallis: Oregon State University Press.

McLaren, P.L. (1982). Spring migration and habitat use by seabirds in eastern Lancaster Sound and western Baffin Bay. *Arctic*, **35**, 88–111.

Manikowski, S. (1975). The effect of weather on the distribution of kittiwakes and fulmars in the north Atlantic. *Acta Zool. Cracoviensia*, **13**, 489–98.

Marr, J.W.S. (1962). The natural history and geography of the Antarctic Krill (*Euphausia superba* Dana). *Discovery Rep.*, **32**, 33–464.

Meltofte, H. (1976). Ornithologiske observationer i Scoresbysundomradet, Ostgronland, 1974. *Dansk Orn. Foren Tidsskr.*, **70**, 107–22.

Meltofte, H., Edelstam, C., Granstrom, G., Hammer, J. & Hjort, C. (1981). Ross's gulls in the arctic pack-ice. *Br. Birds*, **74**, 316–20.

Mooers, C.N.K., Flagg, C.N. & Boicourt, W.C. (1978). Prograde and retrograde fronts. In *Oceanic fronts in coastal processes*, ed. M.J. Bowman & W.E. Esaias, pp. 43–58. Berlin: Springer-Verlag.

Mooers, C.N.K. & Robinson, A.R. (1984). Turbulent jets and eddies in the California Current and inferred cross-shelf transports. *Science*, **223**, 51–3.

Murphy, G.I. & Shomura, R.A. (1972). Pre-exploitation abundance of tunas in the equatorial central Pacific. *Fish. Bull.*, **72**, 875–913.

Murphy, R.C. (1914). Observations on birds of the South Atlantic. *Auk*, **31**, 439–55.

Murphy, R.C. (1936). *Oceanic birds of South America*. New York: Am. Mus. Nat. Hist.

Neess, J.C (1949). *A contribution to aquatic population dynamics*. Ph.D. diss. Madison, WI: Univ. Wisconsin.

Nelson, J.B. (1978). *The Sulidae: gannets and boobies*. Oxford: Oxford University Press.

Nilsson, C.S., Andrews, J.C. & Scully-Power, P. (1977). Observations of eddy formation off east Australia. *J. Phys. Oceanogr.*, **7**, 659–69.

Obst, B.S. (1985). Densities of Antarctic seabirds at sea and the presence of the *krill Euphausia superba*. *Auk*, **102**, 540–9.

Orr, C.D. & Parsons, J.L. (1982). Ivory Gulls, *Pagophila eburnea*, and ice edges in Davis Strait and the Labrador Sea. *Can. Field-Nat.*, **96**, 323–8.

Owen, R.W., Jr (1966). Small-scale, horizontal vortices in the surface layer of the sea. *J. Mar. Res.*, **24**, 56–66.

Owen, R.W., Jr (1981). Fronts and eddies in the sea: mechanisms, interactions and biological effects. In *Analysis of marine ecosystems*, ed. A.R. Longhurst, pp. 197–233. New York: Academic Press.

Packard, T.T., Blasco, D. & Barber, R.T. (1978). *Mesodinium rubrum* in the Baja California upwelling system. In *Upwelling ecosystems*, ed. R. Roje & M. Tomczak, pp. 73–89. New York: Springer-Verlag.

Pennycuick, C.J. (1982). The flight of petrels and albatrosses (Procellariiformes), observed in South Georgia and its vicinity. *Phil. Trans. R. Soc. Lond. B*, **300**, 75–106.

Pielou, E.C. (1960). A single mechanism to account for regular, random and aggregated populations. *J. Ecol.*, **48**, 575–84.

Pingree, R.D. (1978). Mixing and stabilization of phytoplankton distributions on the northwest European continental shelf. In *Spatial pattern in plankton communities*, ed. J.H. Steele, pp. 181–220. New York: Plenum Press.

Pingree, R.D. (1979). Baroclinic eddies bordering the Celtic Sea in late summer. *J. Mar. Biol. Assoc. U.K.*, **59**, 689–98.

Pingree, R.D., Bowman, M.J. & Esaias, W.E. (1978). Headland fronts. In *Oceanic fronts in coastal processes*, ed. M.J. Bowman & W.E. Esaias, pp. 78–86. New York: Springer-Verlag.

Pingree, R.D., Forster, G.R. & Harrison, G.K. (1974). Turbulent convergent tidal fronts. *J. Mar. Biol. Assoc. U.K.*, **54**, 469–79.

Pingree, R.D. & Griffiths, D.K. (1978). Tidal fronts on the shelf seas around the British Isles. *J. Geophys. Res.*, **83**, 4615–22.

Platt, T. (1972). Local phytoplankton abundance and turbulence. *Deep-Sea Res.*, **19**, 183–7.

Platt, T. & Denman, K.L. (1975). Spectral analysis in ecology. *Ann. Rev. Ecol. Syst.*, **6**, 189–210.

Pocklington, R. (1979). An oceanographic interpretation of seabird distributions in the Indian Ocean. *Mar. Biol.*, **51**, 9–21.

Quinn, W.H., Zopf, D.O., Short, K.S. & Kuo Yang, R.T.W. (1978). Historical trends and statistics of the southern oscillation, El Niño, and Indonesian droughts. *Fish. Bull.*,**76**, 663–78.

Reid, J.L., Brinton, E., Fleminger, A., Venrick, E.L. & McGowan, J.A. (1978). Ocean circulation and marine life. In *Advances in oceanography*, ed. H. Charnock & G. Deacon, pp. 65–130. New York: Plenum Press.

Ricklefs, R.E. (1973). *Ecology*. Newton, Mass.: Chiron Press.

Rijke, A.M. (1968). The water repellancy and feather structure of cormorants, Phalacrocoracidae. *J. Exp. Biol.*,**48**, 185–9.

Riley, G.A. (1976). A model of plankton patchiness. *Limnol. Oceanogr.*, **21**, 873–80.

Riley, G.A. & Bumpus, D.F. (1946). Phytoplankton–zooplankton relationships on George's Bank. *J. Mar. Res.*, **6**, 33–47.

Robinson, A.R. (1976). Eddies and ocean circulation. *Oceanus*, **19**, 2–17.

Ryther, J.H. (1963). Geographical variations in productivity. In *The sea, vol. II. The composition of sea-water – comparative and descriptive oceanography*, ed. M.N. Hill, pp. 347–80. New York: John Wiley & Sons.

Ryther, J.H. (1969). Photosynthesis and fish production in the sea. *Science*, **166**, 72–6.

Sanger, G.A. (1972). Preliminary standing stock and biomass estimates of seabirds in the subarctic Pacific Region. In *Biological oceanography of the northern North Pacific*, ed. A.Y. Takenouti, pp. 589–622. Tokyo: Idemitsu Shoten.

Schneider, D.C. (1982). Fronts and seabird aggregations in the southeastern Bering Sea. *Mar. Ecol.-Progr. Ser.*, **10**, 101–3.

Schneider, D.C. & Duffy, D.C. (1985). Scale-dependent variability in seabird abundance. *Mar. Ecol.-Progr. Ser.*, **25**, 211–18.

Schneider, D.C., Hunt, G.L., Jr & Harrison, N.M. (1986). Mass and energy transfer to seabirds in the southeastern Bering Sea. *Coast. Shelf Processes.*

Schreiber, R.W. & Schreiber, E.A. (1984). Central Pacific seabirds and the El Niño Southern Oscillation: 1982 to 1983 perspectives. *Science*, **225**, 713–15.

Schumacher, J.D., Kinder, T.H., Pashinski, D.J. & Charnell, R.L. (1979). A structural front over the continental shelf of the eastern Bering Sea. *J. Phys. Oceanogr.*, **9**, 79–87.

Seckel, G. (1972). Hawaiian-caught skipjack tuna and their physical environment. *Fish. Bull.*, **72**, 763–87.

Shuntov, V.P. (1961). Migration and distribution of marine birds in the southeastern Bering Sea during spring–summer season. *Zool. Zh.*, **40**, 1058–69. (In Russian.)

Shuntov, V.P. (1968). Some regularities in distribution of albatrosses (Tubinares, Diomedeidae) in the northern Pacific. *Zool. Zh.*, **47**, 1054–64. (In Russian.)

Shuntov, V.P. (1974). *Sea birds and the biological structure of the ocean.* Washington D.C.: U.S. Dept. Commerce, Nat. Tech. Inf. Serv. (NTIS-TT-74-55032).

Shuntov, T.P., Kirlan, D.F., Batytskaya, L.V., Glebova, S.V. & Kolesova, N.G. (1981). Geographical distribution of sea birds in connection with zonality of oceanological environment in the Southern Ocean. *Biologia Morya 1981*, No. 6, 16–26. (In Russian.)

Shuntov, V.P., Kirlan, D.F., Batytskaya, L.V., Glebova, S.V. & Kolesova, N.G. (1982*a*). Areas of concentrations of different bird groups, their interannual variability and abundance in the Southern Ocean. *Biologia Morya 1982*, No. 3, 3–11. (In Russian.)

Shuntov, V.P., Kirlan, D.F., Batytskaya, L.V., Glebova, S.Y. & Kolesova, N.G. (1982*b*). General regularities of quantitative distribution of sea birds in the Southern Ocean. *Biologia Morya 1982*, No. 2, 3–11. (In Russian.)

Simpson, J.H. (1981). The shelf-sea fronts: implications of their existence and behaviour. *Phil. Trans. R. Soc. Lond.*, **302A**, 531–46.

Smith, R.L. (1968). Upwelling. *Oceanogr. Mar. Biol. Ann. Rev.*, **6**, 11–46.

Smith, R.L. (1983). Peru coastal currents during El Niño: 1976–1982. *Science*, **221**, 1397–8.

Smith, R.L. & Bottero, J.S. (1977). On upwelling in the Arabian Sea. In *A voyage of discovery*, ed. M. Angel, pp. 291–304. Oxford: Pergamon Press.

Stavn, R.H. (1971). The horizontal-vertical distribution hypothesis: Langmuir circulations and *Daphnia* distribution. *Limnol. Oceanogr.*, **16**, 453–66.

Steele, J.H. (1974). *The structure of marine ecosystems*. Cambridge, Mass.: Harvard Press.

Steele, J.H. (1976). Patchiness. In *Ecology of the sea*, ed. D.H. Cushing & J.J. Walsh, pp. 98–115. Philadelphia: Saunders.

Steele, J.H. (1978). Some comments on plankton patches. In *Spatial pattern in plankton communities*, ed. J.H. Steele, pp. 1–20. New York: Plenum Press.

Steele, J.H. (1980). Patterns in plankton. *Oceanus*, **23**, 3–8.

Stommel, H. (1963). Varieties of oceanographic experience. *Science*, **139**, 572–6.

Summerhayes, C.P., Hofmeyr, P.K. & Rioux, R.H. (1974). Seabirds off the southwestern coast of Africa. *Ostrich*, **45**, 83–109.

Swallow, J.C. (1976). Variable currents in mid-ocean. *Oceanus*, **19**, 19–25.
Szijj, L.J. (1967). Notes on the winter distribution of birds in the western Antarctic and adjacent Pacific waters. *Auk*, **84**, 366–78.
Thurston, M.H. (1982). Ornithological observations in the South Atlantic Ocean and Weddell Sea, 1959–1964. *Bull. Br. Antarct. Surv.*, **55**, 77–103.
Tourre, Y. & Rasmusson, E. (1984). The tropical Atlantic region during the 1982–83 equatorial Pacific warm event. *Tropical Ocean-Atmosphere Newsletter*, **25**, 1–2.
Tovar, H. (1978). Las poblaciones de aves guaneras en las ciclos reproductivos de 1969/70 a 1973–74. *Inf. Inst. Mar Peru-Callao*, **45**, 1–13.
Uda, M. (1952). On the relation between the variation of important fisheries conditions and the oceanographical conditions in the adjacent waters of Japan I. *J. Tokyo Univ. Fish.*, **38**, 363–89.
Uda, M. (1973). Pulsative fluctuations of oceanic fronts in association with the tuna fishing grounds and fisheries. *J. Fac. Mar. Sci. Technol. Tokai Univ.*, **7**, 245–65.
Uda, M. & Ishino, M. (1958). Enrichment pattern resulting from eddy systems in relation to fishing ground. *J. Tokyo Univ. Fish.*, **44**, 105–29.
Valdivia, G.J.E. (1978). The Anchoveta and El Niño. *Rapp. P-v. Reun. Cons. Perm. Int. Explor. Mer*, **173**, 196–202.
van Oordt, G.J. & Kruijt, J.P. (1953). On the pelagic distribution of some Procellariiformes in the Atlantic and Southern Ocean. *Ibis*, **95**, 615–37.
Venrick, E.L., McGowan, J.A. & Montyla, A.W. (1973). Deep maxima of photosynthetic chlorophyll in the Pacific Ocean. *Fish. Bull.*, **71**, 41–52.
Visser, G.A., Krager, I., Coetzee, D.J. & Cram, D.L. (1973). *The environment. Cape Cross Programme (Phase III)*. Internal Report. Cape Town, South Africa: Sea Fisheries Research Institute.
Walsh, J.J (1978). The biological consequences of the interaction of the climatic, El Niño, and event scales of variability in the eastern tropical Pacific. *Rapp. P-v. Reun. Cons. Perm. Int. Explor. Mer*, **173**, 182–92.
Walsh, J.J., Kelley, J.C., Whitledge, T.E., MacIssac, J.J. & Huntsman, S.A (1974). Spin-up of the Baja California upwelling ecosystem. *Limnol. Oceanogr.*, **19**, 553–72.
Warnecke, G., Allison, L.J., McMillin, L.M.& Szekielda, K.-H. (1971). Remote sensing of ocean currents and sea surface temperature changes derived from the Nimbus II satellite. *J. Phys. Oceanogr.*, **1**, 45–60.
Watson, G.E. (1975). *Birds of the Antarctic and Subantarctic*. Washington D.C.: Am. Geophys. Union.
Watson, G.E., Angle, J.P., Harper, P.C., Bridge, M.A., Schlatter, R.P., Tickell, W.L.N., Boyd, J.C. & Boyd, M.M. (1971). *Birds of the Antarctic and Subantarctic*. Antarctic Map Folio Ser. Folio 14. Washington D.C.: Am. Geophys. Union.
White, W.B. (1973). An oceanic wake in the equatorial undercurrent downstream from the Galapagos Archipelago. *J. Phys. Oceanogr.*, **3**, 156–61.
Wiebe, P.H. (1970). Small-scale spatial distribution in oceanic zooplankton. *Limnol. Oceanogr.*, **15**, 205–17.
Wiebe, P.H. (1971). A computer model study of zooplankton patchiness and its effects on sampling error. *Limnol. Oceanogr.*, **16**, 29–38.
Wiebe, P.H. (1972). A field investigation of the relationship between length of tow, size of net and sampling error. *J. Cons. Perm. Int. Explor. Mer*, **34**, 268–75.
Wiebe, P.H. (1976). The biology of cold-core rings. *Oceanus*, **19**, 69–76.

Wiebe, P.H., Hulbert, E.M., Carpenter, E.J., Jahn, A.E., Knapp, G.P., III, Boyd, S.H., Ortner, P.B. & Cox, J.L. (1976). Gulf Stream cold core rings: large-scale interaction sites for open ocean plankton communities. *Deep-Sea Res.*, **23**, 695–710.

Wittenberger, J.F. & Hunt, G.L., Jr (1985). The adaptive significance of coloniality in birds. In *Avian biology*, vol. 8, ed. D.S. Farner, J.R. King & K.C. Parkes, pp. 1–78. New York: Academic Press.

Woodby, D. (1984). The April distribution of murres and prey patches in the southeastern Bering Sea. *Limnol. Oceanogr.*, **29**, 181–8.

Woods, J.D., Wiley, R.L. & Briscoe, M.G. (1977). Vertical circulation of fronts in the upper ocean. In *A voyage of discovery*, ed. M. Angel, pp. 253–75. Oxford: Pergamon Press.

Wynne-Edwards, V.C. (1935). On the habits and distribution of birds on the North Atlantic. *Proc. Boston Soc. Nat. Hist.*, **40**, 233–346.

Zelickman, E.A. & Golovkin, A.N. (1972). Composition, structure and productivity of neritic plankton communities near the bird colonies of the northern shore of Novaya Zemlya. *Mar. Biol.*, **17**, 265–74.

Zink, R.M. (1981). Observations of seabirds during a cruise from Ross Island to Anvers Island, Antarctica. *Wilson Bull.*, **93**, 1–20.

3

Flight of seabirds

C. J. PENNYCUICK

Department of Biology, P.O. Box 249118, University of Miami, Coral Gables, Florida 33124, USA

Introduction

The uses to which seabirds put the power of flight may be divided into four broad categories, which impose different and sometimes conflicting requirements. These are:

1. Catching food
2. Commuting to and from feeding grounds when incubating, or when bringing food for the young
3. Access to nest sites
4. Migration.

As a device for classifying the great variety of wing forms and styles of flight seen in seabirds, we can begin with a 'standard seabird'. From this arbitrarily chosen starting point, various types of modification can be followed, to see how they allow their owners to specialize in one or other of the above respects, and impose restrictions in other ways. The 'standard seabird' is not primitive in the phylogenetic sense, and the sequences of modification considered below are not proposed as likely paths of evolution.

The 'standard seabird'

The 'standard seabird' is defined in terms of three primary variables. It has a **body mass** (denoted by m) of 700 g, a **wing span** (b) of 1.09 m, and a **wing area** (S) of 0.103 m². The meaning of these variables is illustrated in Fig. 3.1. Also shown is the **disc area** (S_d), which is the area of a circle whose diameter is the same as the wing span. It is derived from the wing span, thus:

$$S_d = \pi b^2/4 \qquad (3.1)$$

The following further variables are also derived from the three primary variables:

The **wing loading** (Q), is defined as

$$Q = mg/S \qquad (3.2)$$

where g is the acceleration due to gravity, and Q is the weight (not mass) supported by unit area of the wing. The 'standard seabird' has a wing loading of 66.6 N m^{-2}. The **disc loading** (Q_d) is

$$Q_d = mg/S_d \qquad (3.3)$$

It is the weight supported by unit disc area. The standard seabird has a disc loading of 7.36 N m^{-2} The **aspect ratio** (R) is

$$R = b^2/S \qquad (3.4)$$

It is a dimensionless index of the shape of the wing, being high for a long narrow wing, and low for a short broad one. The 'standard seabird' has an aspect ratio of 11.5.

Fig. 3.1. Definitions of wing area (S), wing span (b), disc area (S_d) and cross-sectional area of the body (A).

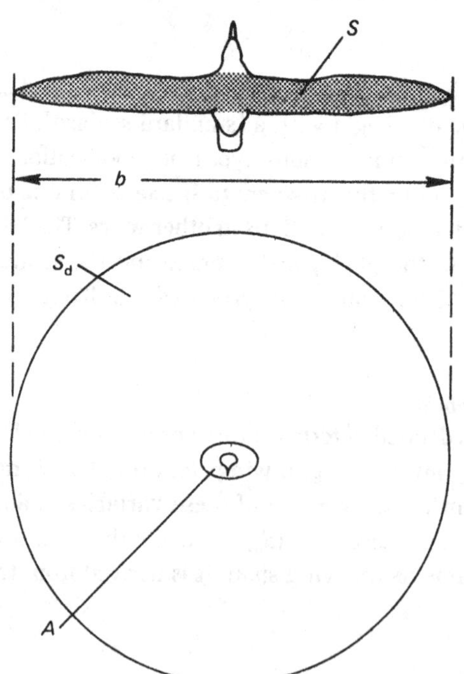

Variations from the 'standard seabird'

Most seabirds differ from this 'standard seabird' in one or several respects. The effects on flight performance of changing the above variables, either singly or in combination, can be deduced in a general way from considerations of flight mechanics. A simplified flight model has been given by Pennycuick (1975), and some references are included below to equation numbers in that publication in order to avoid going into mathematical detail here. The primary effects of changing the variables just mentioned may be summarized as follows:

The **wing loading** determines the gliding speed – the higher it is, the faster the bird must glide. A high **disc loading** calls for a high power output from the muscles in horizontal flight, and also a high speed for economical cruising. The **wing length** is the main factor determining the wingbeat frequency – the shorter the wing, the higher the frequency at which it has to flap. Increasing the **wing area** (by making the wing broader) reduces the wingbeat frequency at low speeds, and to a lesser extent at cruising speeds. A high **aspect ratio** is conducive to efficient gliding; that is, it allows the bird to glide at a shallow angle to the horizontal. The ratio of the disc area to the cross-sectional area of the body (also shown in Fig. 3.1) has a major influence on the effective lift: drag ratio, which, when multiplied by the weight, yields directly the energy consumed per unit distance flown in cruising flight.

It may be noted that the wing loading and disc loading change if the size of the bird is changed, keeping all dimensions in proportion. If the 'standard seabird' is scaled to a larger size, the mass increases by a bigger factor than the wing area or disc area, so that both the wing loading and the disc loading increase. The aspect ratio remains the same, however, as does the ratio of the disc area to the cross-sectional area of the body. Variation of size in this way, keeping the shape constant, generates a series of 'geometrically similar' species. Exact geometrical similarity probably does not occur in nature over a wide range of body mass, but quite a close approximation to it does, and this is the first type of variation from the 'standard seabird' to be considered.

Variation of mass

The procellariiform sequence

The Procellariiformes, excluding the diving petrels Pelecanoididae, represent a tolerably close approximation to a geometrically similar series. They are a relatively homogeneous group covering an unusually wide range of body mass, which includes both the largest and the smallest truly pelagic flying birds. The middle-sized members of this group closely resemble the 'standard seabird' defined above, although it may be noted that the rudimentary description given so far would also fit a tropicbird Phaethontidae or skua (jaeger) Stercorariidae.

If the 'standard seabird' is understood to be a medium-sized petrel or shearwater Procellariidae, then some more details can be added. It has webbed feet, placed far back, which makes it clumsy on land. It takes off from land by launching downwards from a slope or cliff, and from the water by launching from the windward slope of a wave, pattering with its feet until it is airborne. Once in the air, it glides along the windward slopes of waves at speeds in the region of 10–14 m s^{-1}, pulling up to cross from one wave to another. In powered flight it never flaps its wings continuously but proceeds by **flap-gliding**, in which short glides alternate with short bursts of flapping. At the breeding grounds it slope-soars along windward cliffs and slopes, but lands clumsily, owing to difficulty in manoeuvering at low speeds, especially in light winds. When feeding young, it forages out to a few hundred kilometers from its nest site, bringing back food loads around 15% of its own body mass at intervals of 2–3 days. Its long range gives it access to a large foraging area, but the low rate at which it can supply food precludes raising more than one nestling in a brood.

The main procellariiform sequence comprises storm petrels, with a minimum body mass of around 20 g, through petrels and shearwaters, up to albatrosses, of which the largest exceed 9 kg. The larger species, of course, have larger wings. If all species were geometrically similar to the 'standard seabird', one would expect the wing area to be proportional to the two-thirds power of the mass. In this case, if the logarithm of the wing area were plotted against the logarithm of the mass for the complete series of species, a straight line should result, with a slope equal to two-thirds. Fig. 3.2 is such a plot, showing 10 species from the author's field observations (Pennycuick, 1982), and a further 37 species from an extensive survey of the Procellariiformes by Warham (1977). The latter includes data from a number of sources, some using inappropriate or doubtful methods of measurement. As in all very large surveys of this type, the data are somewhat heterogeneous, and statistics drawn from them have to be treated with some reserve. These problems undoubtedly contribute to the scatter of points about the line shown in Fig. 3.2, and may also have biased its slope. The line shown is not strictly a regression line, but is a fitted line whose slope is that of the major axis of an ellipse enclosing the scatter of points. This avoids a known bias, due to the asymmetric nature of the usual regression calculation (Rayner, 1985). The slope of the line so fitted is 0.63, a little below the expected value of two-thirds. This means that wing loading increases with mass a little more steeply than it would if all the birds were geometrically similar. From eqn 3.2, the wing loading varies in this group of birds as the 0.37 power of the mass. Gliding speed is proportional to the square root of the wing loading, and therefore would vary with the one-sixth power of the mass in geometrically similar birds, and actually varies with the 0.19 power of the mass. The speed for any particular type of gliding, such as the minimum gliding speed, the speed for minimum sink, or that for best glide ratio is thus greater in large species than in small ones. If the mass of the largest albatross were 450 times that of the smallest storm petrel, then the ratio of gliding speeds would be 450 raised to the 0.19 power, which is 3.1. Estimated minimum gliding speeds range from about 4 m s^{-1} in small storm petrels to about 12 m s^{-1} in large albatrosses. The gliding speeds of the former are low enough to permit their unique 'sea-anchor' method of soaring, described by Withers (1979) (see Fig. 3.3). Storm petrels feed from the water surface whilst soaring in this manner, and are the only members of this group in which flight is directly involved in feeding behaviour.

Fig. 3.2. Wing area related to body mass 'for normal' Procellariiformes, that is, Diomedeidae, Procellariidae and Hydrobatidae, but excluding Pelecanoididae. The cross represents the standard seabird'.

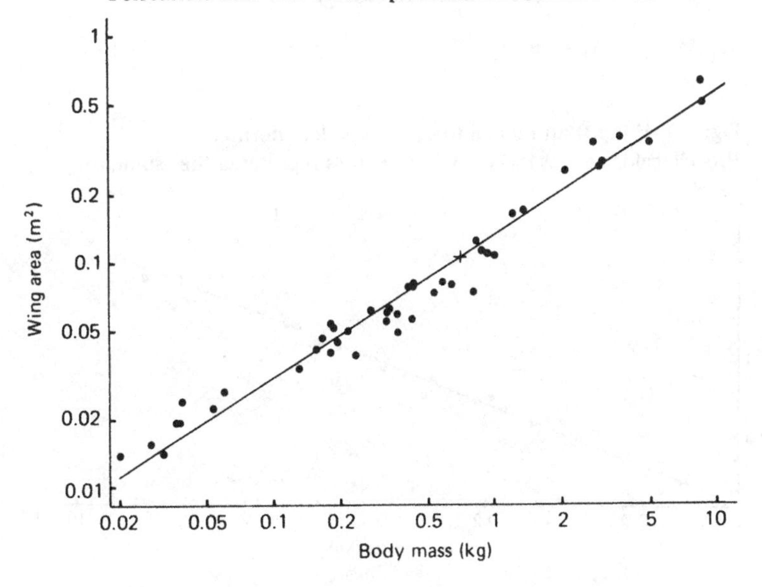

Fig. 3.4 is a plot of wing span *versus* body mass for the main procellariiform sequence. Only 22 species are represented in this plot, since Warham (1977) did not record the span for all the species in his set. If all species were geometrically similar, the slope of this line should be one-third, but it is in fact considerably steeper than that, at 0.39. This means that the proportions change so that the larger species have relatively longer wings than the smaller ones. Combined with the 0.63 slope for wing area, it results in a marked increase of aspect ratio in the larger species. The progressive change in wing shape can be seen in Fig.

Fig. 3.3. 'Sea-anchor soaring', as described by Withers (1979) for Wilson's Storm Petrel. The bird's feet move backwards through the water at a speed V_w which is represented by a forward flow of water, relative to the feet. This produces a drag force D, directed forwards, which balances the backward aerodynamic drag on the wings, produced by the relative airspeed V_a.

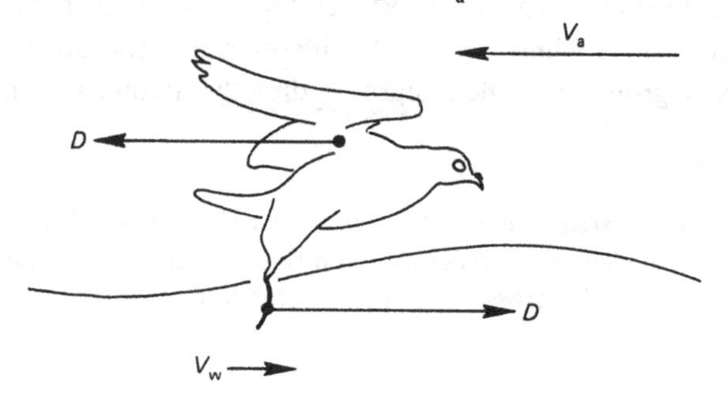

Fig. 3.4. Wing span related to body mass for 'normal' Procellariiformes, as in Fig. 3.1. The cross represents the 'standard seabird'.

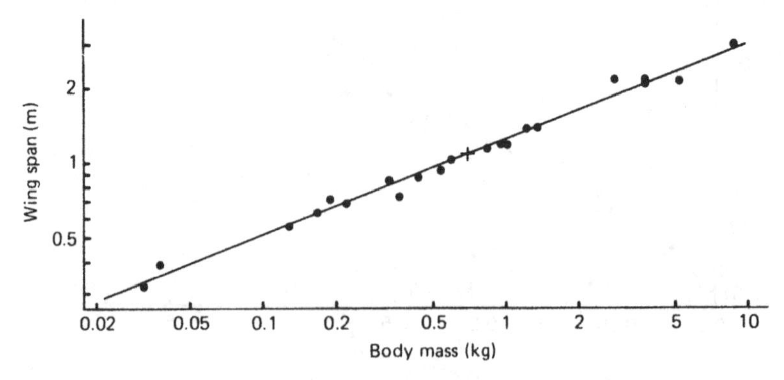

3.5, which shows silhouettes of nine procellariiform species of different size, scaled to the same apparent wing span.

The equations given for speed in flapping flight by Pennycuick (1975: equations 10 and 12) can be used to estimate how these should vary in different-sized members of the sequence. They suggest that for geometrically similar birds, the speed for maximum range (for example) should vary in the same way as gliding speeds, with the one-sixth power of the mass. However, if the disc area is allowed to vary allometrically according to the slope of Fig. 3.4, then the speed for maximum range would vary with the 0.14 rather than the 0.17 power of the mass. Over

Fig. 3.5. Allometry of wing shape in Procellariiformes. The span of the largest species (Wandering Albatross *Diomedea exulans*, top left), is 7.7 times that of the smallest (Wilson's Storm Petrel *Oceanites oceanicus*, bottom right), and the respective aspect ratios are 15 and 8.0. From top left to bottom right: *Diomedea exulans* (8.7 kg), Black-browed Albatross *D. melanophris* (3.8 kg), Grey-headed Albatross *D. chrysostoma* (3.8 kg), Light-mantled Sooty Albatross *Phoebetria palpebrata* (2.8 kg), giant petrel *Macronectes* sp. (5.2 kg), White-chinned Petrel *Procellaria aequinoctialis* (1.4 kg), Cape Pigeon *Daption capense* (0.43 kg), Antarctic Prion *Pachyptila desolata* (0.17 kg), *Oceanites oceanicus* (0.032 kg). From Pennycuick (1982).

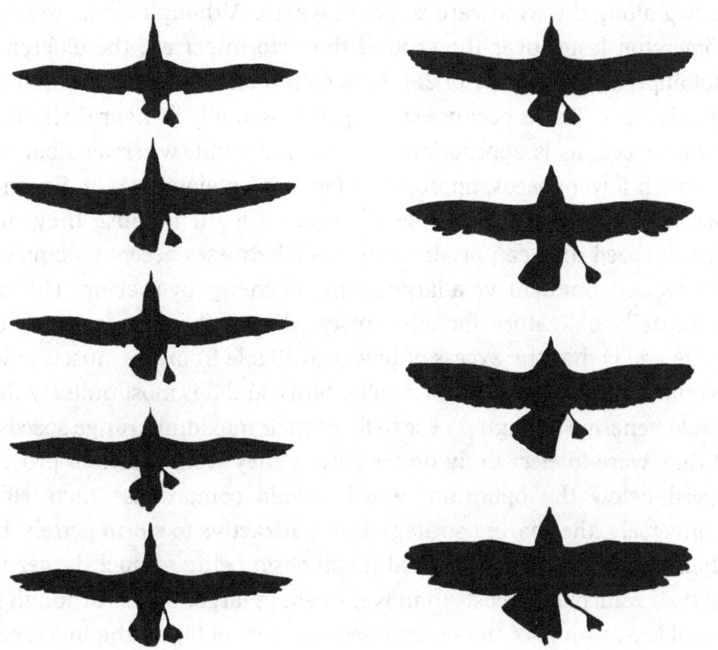

a mass range of 450:1, the maximum range speed would vary by a factor of 2.4 rather than 3.1, as estimated for gliding speeds. Flap-gliding remains the characteristic style of powered flight in all members of the sequence, but the speeds for both flapping and gliding are lower, relative to the bird's power curve, in the larger species. Storm petrels flap-glide at speeds near their maximum range speeds, whereas albatrosses do so at speeds near their minimum power speeds. Much greater divergence of the speeds for flapping and gliding would make flap-gliding impracticable at one end or other of the scale.

It will be clear that simply scaling a given body form to larger or smaller sizes does not result in similar styles of flight throughout the range. Gross departures from geometrical similarity would be required to keep speeds constant. Such departures have other consequences and also can only be maintained over a limited range of size. The Procellariiformes maintain a remarkably close approximation to geometrical similarity over a wider range of mass than any other group of birds, but they do it by accepting a progressive change of flight style. Storm petrels and albatrosses fly in very different ways, which are connected by a spectrum of intermediate styles in the intervening species. The difference is most apparent in foraging. Storm petrels proceed on foraging expeditions in steady flap-gliding flight, whereas albatrosses are heavily dependent on soaring, and zigzag along the windward slopes of waves. Although albatrosses proceed somewhat faster over the ground than storm petrels, the difference does not approach the factor of 2.4, by which their respective maximum range speeds differ. This is because storm petrels actually fly near their maximum range speed, as is appropriate for powered flight, whereas albatrosses fly at much lower speeds, appropriate for soaring along waves. Storm petrels accept the metabolic expense of powered flight because they need the fastest speed they can attain, whereas albatrosses accept a comparatively slow speed, but achieve a large saving of energy by soaring. This solution is actually obligatory for albatrosses. One of the consequences of their large size is that the excess of power available from the muscles over that required to fly is less than in smaller birds, and it is most unlikely that they could generate enough power to fly at their maximum range speeds. Thus, if they were to elect to fly under power, they would have to proceed at a speed below the optimum, which would compromise their efficiency. Conversely, the soaring strategy is not attractive to storm petrels, because their small size results in basal metabolism being a much larger fraction of their total energy costs than is the case in larger birds. Although soaring would save some of the direct energetic cost of flight, the increased flight time, caused by flying at the low speeds appropriate to soaring, would lead to a more-than-compensating loss of energy on basal metabolism.

A point is ultimately reached where further increase or decrease of body mass is impracticable without some discontinuous change, which is not, of course, allowed by evolution. It appears that further reduction in the size of storm petrels would make their foraging speeds too low, and metabolic costs too high, to allow them to supply food to the nestling at a rate sufficient to sustain its growth. Albatrosses bigger than a wandering albatross would be limited by the same problem, but for a different reason. Their foraging speeds are high, and metabolic costs low, but the amount of food that can be lifted declines if the body mass is too great, and this reduces the rate of food supply to the young (Pennycuick, Croxall & Prince, 1984).

Changes of wing size and shape

We can now consider the effects of four types of variation from the type of wing seen in normal Procellariiformes. These are illustrated in Fig. 3.6, in which the White-chinned Petrel *Procellaria aequinoctialis* is shown in the

Fig. 3.6. Seabird silhouettes, adjusted so that the bodies appear about the same size. Centre: White-chinned Petrel *Procellaria aequinoctialis*, representing the 'standard seabird', modified as follows: 1. Shortening the wings with little change of aspect ratio leads first to Razorbill *Alca torda*, then to Macaroni Penguin *Eudyptes chrysolophus*. 2. Lengthening the wings with little change of aspect ratio leads to magnificent Frigatebird *Fregata magnificens*. 3. Shortening the wings without reducing the wing area leads to Blue-eyed Shag *Phalacrocorax atriceps*. 4. Increasing the wing area with only slight increase of span leads to Brown Pelican *Pelecanus occidentalis* .

centre, representing an approximation to the 'standard seabird'. The silhouettes of the six species shown in Fig. 3.6 have been enlarged by different amounts, such that the linear size of each image is proportional to the minus one-third power of the body mass. The effect of this is that the bodies are about the same size. The frigatebird's body (top left) looks bigger than the others, but this is an illusion caused by its long tail and prominent scapular feathers. Fig. 3.6 gives, as closely as possible, a direct comparison of the sizes and shapes of the wings of the six species, if all were adjusted to the same body mass. The four types of deviation from the standard procellariiform shape are numbered in the Figure as follows.
1. Reduction of wing length, keeping the aspect ratio roughly constant. This is associated with the use of the wings for swimming under water. Two stages of reduction are shown, representing auks Alcidae and penguins Spheniscidae. 2. Increase of wing length, keeping the aspect ratio roughly constant. This is seen in frigatebirds Fregatidae. 3. Reduction of

Fig. 3.7. Wing area plotted against body mass for various seabirds. The solid line is the fitted line for Procellariiformes, transferred from Fig. 3.2. Large circles: Pelecanidae. Small circles: Laridae, Stercorariidae. Large solid square: Fregatidae. Small solid squares: Phalacrocoracidae. Small open squares: Sulidae, Phaethontidae. Solid stars: Alcidae. Small open stars: Pelecanoididae. Large open star: Spheniscidae. The dashed line is fitted through the five points for Pelecanoididae and Alcidae.

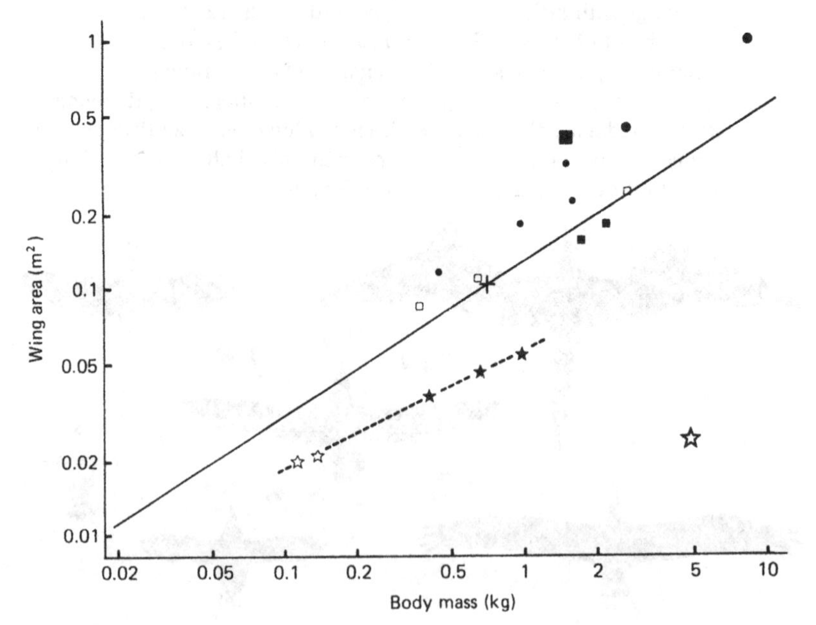

wing length, keeping the wing area roughly constant. This results in a drastically reduced aspect ratio, and is seen in cormorants Phalacrocoracidae. 4. Increase of wing area, keeping the wing span roughly constant. This also results in a reduced aspect ratio, and is seen in pelicans Pelecanidae.

The wing span and area of procellariiform wings change progressively with changing body size, and the fitted lines of Figs 3.2 and 3.4 can now be taken as a moving standard, against which the wing measurements of other seabirds can be compared. The procellariiform line has been transferred from Fig. 3.2 to Fig. 3.7, where the wing areas of various other seabirds are plotted against their masses. Species which fall above the line have more wing area than the 'procellariiform standard', whereas those which fall below have less. Similarly, in Fig. 3.8, the wing spans of other seabirds are compared with the fitted line for Procellariiformes, transferred from Fig. 3.4. The effects of the four types of deviations from one or both lines, listed above, can now be examined.

Reduced wing length: constant aspect ratio

Birds with dual-medium wings

The petrels and albatrosses considered in the previous section are essentially surface feeders, and do not use their wings for propulsion under water. Their wings have to be adapted to get them to the feeding grounds and back again, but not for the feeding process itself. Other seabirds, notably auks Alcidae and diving petrels Pelecanoididae, use their wings also to swim under water in pursuit of their prey, using a flapping motion which

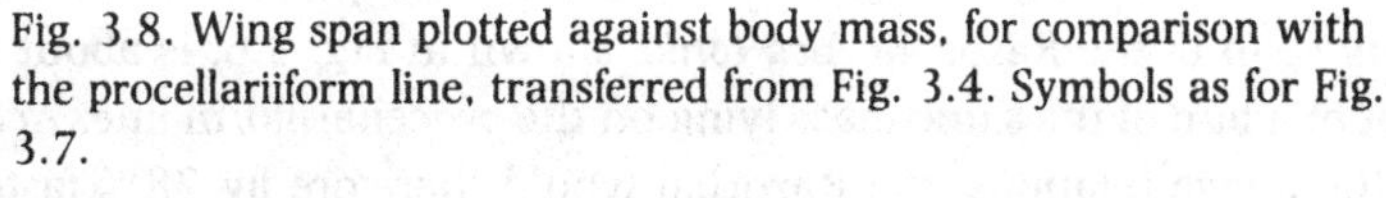

Fig. 3.8. Wing span plotted against body mass, for comparison with the procellariiform line, transferred from Fig. 3.4. Symbols as for Fig. 3.7.

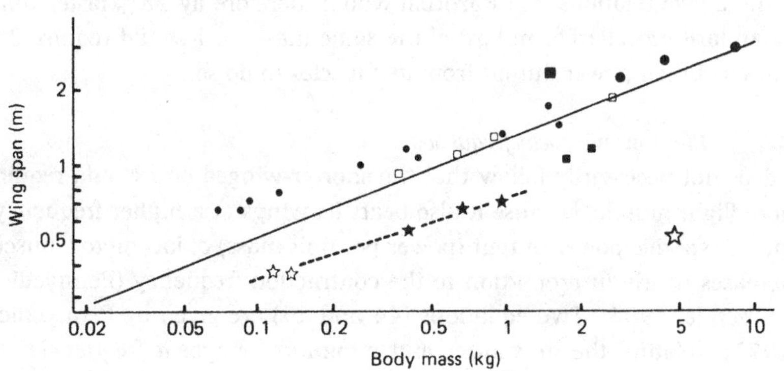

looks not unlike flapping flight in air. As the density of sea water is over 800 times that of air at sea level, there is some difficulty in using the same propulsive organ in both media. The wing is held with the wrist and elbow joints flexed, to minimize its span, and flapped at a much lower frequency than in air. Even so, a normal procellariiform wing would be far too large to operate effectively under water. To modify a 'standard seabird' into one which can swim with its wings under water, the wing has to be shortened, and this has effects in turn on its ability to fly in air.

Three species of auks and two of diving petrels are plotted in Figs 3.7 and 3.8, with their own fitted lines. These slope much less steeply than the procellariiform lines, perhaps indicating that auks and diving petrels are not sufficiently convergent to be combined in this way. Be that as it may, all of these species show a large reduction of both wing span and area, relative to the procellariiform lines. The net result is that the wing shape of an auk or diving petrel is not very different from that of a petrel or shearwater of similar mass, but the whole wing is smaller. The aspect ratio is much the same, but the wing loading and disc loading are higher.

Effect on speed and power

The effect of the reduced wing span can be seen from equations given by Pennycuick (1975: equations 12 and 13) which relate the maximum range speed, and the power to fly at that speed to various body measurements and other variables. Of these variables, the only one affected by shortening the span is the disc area. The maximum range speed depends on the -0.25 power of the disc area, and therefore on the -0.5 power of the span, if the aspect ratio is held constant. The power required to fly at the maximum range speed varies with the -1.5 power of the span. The wing span of the Razorbill *Alca torda*, shown in Fig. 3.6, is about 60% of that of a bird of the same mass lying on the procellariiform line. According to the above relations, the Razorbill would therefore fly 28% faster than a standard procellariiform bird of the same mass, and would require 2.1 times as much power output from its muscles to do so.

Effect on wingbeat frequency

It does not necessarily follow that the shorter-winged bird would require more flight muscle, because it also beats its wings at a higher frequency, and the specific power output (power per unit mass) of locomotor muscle increases nearly in proportion to the contraction frequency (Pennycuick & Rezende, 1984). Two equations (44 and 49) are given by Pennycuick (1975), relating the maximum and minimum wingbeat frequencies to various quantities. If only the span were varied, as proposed here, the minimum frequency should vary with the -2 power of the span, and the maximum frequency with the -1 power. Thus, shortening the wing does not cause difficulty in meeting the power required to fly, but it does cause the minimum and maximum frequencies to converge. The higher level of power output also requires a greater proportion of the total body mass to be devoted to the heart and lungs, and this in turn must reduce the amount of food which can be lifted on foraging flights.

The convergence of the maximum and minimum wingbeat frequencies, just noted, also occurs without shortening the wings in a series of geometrically similar birds, as the body mass is increased. For instance, in the main procellariiform sequence, the large albatrosses have only a narrow range of wingbeat frequencies available to them, whilst smaller species have a wider choice. The minimum frequency is set by the need to generate enough relative air flow over the wing for lift and propulsion. The maximum is set by the inertial properties of the wing, and the maximum angular acceleration which can be imparted to it without breaking the insertion of the pectoralis muscle. In very large birds, the range of available flapping frequencies becomes narrow, a fact which has been documented from field observations by Scholey (1982). At some level of body mass, probably a little above that of the largest extant flying birds, a limit would be reached where the bird is just able to fly at its minimum power speed, whilst flapping at a frequency just within the limit set by the mechanical strength of its muscles and bones. This limit is reached at a lower body mass if the wings are shortened relative to other body dimensions. The mass of the largest auks extends only up to about 1 kg, as compared to about 12 kg for flying birds in general, and is most probably limited by this factor. If the power of flight is relinquished, as in penguins and the Great Auk *Pinguinus impennis*, then the upper mass limit no longer applies, and larger birds become possible.

Extreme shortening of the wing

The difficulties caused by shortening the wings can be understood by comparing a normal procellariiform bird with a penguin. In Fig. 3.6 the Macaroni Penguin *Eudyptes chrysolophus* is shown as a further stage in wing reduction, along the same path as the Razorbill. Figs 3.7 and 3.8 show that its wing area and span are about 7% and 22% of the values for a procellariiform bird of the same mass (about 4.8 kg). This mass is well within the range of flying birds, and similar to that of a giant petrel *Macronectes*. The penguin's aspect ratio is 10.7, which compares quite favourably with the giant petrel's 12.0. Why then, exactly, should the penguin not be able to fly?

The penguin's wing span is 0.51 m, which is less than the giant petrel's by a factor of about 4. If the penguin flew, its maximum range speed would therefore exceed that of the giant petrel by a factor of $4^{0.5}$, i.e. 2, which would put its cruising speed at about 40 m s^{-1}. In order to achieve this impressive speed, however, its wingbeat frequency would have to be 4^2, or 16, times that of the giant petrel, which would be somewhere around 50 Hz. Frequencies in this range are satisfactory in the smaller hummingbirds, but the penguin wing is too big to withstand the forces needed to accelerate it at such a high frequency. This is what is meant by the minimum frequency for flight being higher than the maximum frequency which can be tolerated by the structure. Because it does not fly, the penguin need only flap its wings at the much lower frequency required for propulsion in water. It is also free to adapt its muscles to work efficiently at this lower frequency. An auk, on the other hand, has to work efficiently at the high frequency required for flight, and its muscles cannot therefore also be efficient in water.

Effects on foraging flight

Presumably auks are able to accept low efficiency in underwater swimming, provided that only a small fraction of their total energy expenditure is used for this. However, their efficiency in powered flight is also reduced. As compared to a 'standard seabird' of the same mass, an auk flies faster but uses more energy per unit distance flown. The fuel economy in this sense is determined by the effective lift:drag ratio, which depends very strongly on the ratio of the disc area to the cross-sectional area of the body. As noted above, the amount of food that can be lifted is a smaller fraction of the body mass than in the 'standard seabird'. For example, Harris & Hislop (1978) found that the mean feed mass in puffins was about 2% of the adult's body mass, and never exceeded 7%. This compares with a typical **mean** value of 15% in medium-sized petrels (Croxall & Prince, 1980). Auks and diving petrels are therefore not well adapted for long-range foraging. They typically forage a few tens of kilometers from the nest, and bring small loads of food several times per day, whereas the 'standard seabird' brings larger loads, at longer intervals, from further away. The auk method may be safer in one way, in that one food delivery intercepted by a skua represents less of a setback for the chick than it would in a typical petrel.

Gliding and access to nests

Another interesting characteristic of the hypothetical flying penguin is that its gliding speed would be even higher than its speed in powered flight. Minimum gliding speed is proportional to the square root of the wing loading. When the wing span alone is varied, as assumed in this comparison, the minimum gliding speed is inversely proportional to the span. The giant petrel's minimum gliding speed is about 13 m s^{-1}, so the penguin's would be four times this, or 52 m s^{-1}. Flap-gliding would therefore be impossible when cruising at the maximum range speed of about 40m s^{-1}. This separation of gliding and flapping speeds is already apparent in auks, which never flap-glide in cruising flight. They can and do glide, but only at very high speeds. In gale-force winds, guillemots, razorbills and puffins slope-soar along cliffs and slopes like other seabirds.

It has been noted that petrels and shearwaters, resembling the 'standard seabird', tend to be clumsy in take-off and landing, but auks have even more difficulty, because they cannot glide at low speeds. The larger auks prefer to nest on ledges or in crevices on vertical cliffs, and have evolved a characteristic 'ballistic' method of landing in such places, which was described by Pennycuick (1956). To initiate its approach, the bird begins some hundreds of meters out from the cliff, and somewhat above its intended landing place. It approaches the cliff in a shallow dive, building up a high speed well below the landing place, and then pulls up into a steep climb. As the speed drops, the bird's wings suffice only for a limited amount of steering. If the manoeuvre is correctly judged, the speed drops to zero at the top of the ballistic trajectory, just above the landing spot, and the bird drops lightly on to the ledge. If it misses, it dives away to regain speed, then flies out to sea, climbs up, and repeats the whole process. In very strong winds, gliding becomes possible, and auks can then manoeuver in front of a cliff face like other seabirds, albeit somewhat unsteadily, because of the turbulence usually associated with gale-force winds.

Increase of wing length: constant aspect ratio

Reverting to variations from the 'standard seabird', frigatebirds show the opposite modification to that seen in auks. Figs 3.7 and 3.8 show that the Magnificent Frigatebird *Fregata magnificens* has a wing span and area, respectively, 1.55 and 2.42 times those for a typical petrel or shearwater of similar mass. This results in the opposite of the trends noted above. Speeds in flapping flight are reduced, and those in gliding flight are reduced even more. Frigatebirds can glide at very low speeds, down to about 6 m

s^{-1}, which also means that they can make turns of very small radius. They are highly manoeuverable, and can soar in small thermals, and even exploit patches of turbulence containing irregular vertical motion. Their distribution coincides with the Trade Wind zones, and it has been argued that they are adapted to soar in thermals over the sea, which are characteristic of those zones (Pennycuick, 1983).

Feeding adaptations

Unlike most seabirds, frigatebirds are not able to alight on the water. According to Nelson (1975) their main food is flyingfish and jumping squid, which they snatch just above the water surface. Typically they patrol by soaring at a low altitude, and gain speed by diving (in air, not water), twisting and turning with great agility to catch their prey. Frigatebirds can keep up 9–10 m s^{-1} in steady flapping flight, but this would not be sufficient for their other main feeding method, klepto-parasitism of such birds as boobies, terns and tropicbirds (see Chapter 5). It is not clear whether frigatebirds and other kleptoparasites such as skuas are capable of short bursts of sufficiently high speed to enable them to overtake their faster victims in horizontal flight. The main requirement for this would be a massive amount of white (anaerobic) flight muscle, which has not been recorded. The main method of catching fast-flying birds seems to be to patrol well above the victim's usual flying height, and then catch up with it in a short burst of speed, gained by diving.

Foraging and dispersal

Although frigatebirds are capable of powered horizontal flight, their main method of travelling about is by soaring in thermals, particularly under cumulus clouds over the Trade Wind zones of the oceans. It is still often stated that frigatebirds seldom venture far from land, long after Sibley & Clapp (1967) demonstrated by banding studies that they actually travel vast distances over the open ocean. The reason why frigatebirds are not often seen at sea is probably that they spend most of their time soaring near the bases of Trade Wind cumulus clouds, where they are difficult to see. Unlike cumulus clouds over land, Trade Wind cumuli, and their associated thermals, continue by day and night, and would allow frigate-birds to soar over the long distances they are known to cover.

Thermal soaring is an economical method of travel in terms of energy expenditure, but slow. It is not entirely clear whether this is a contributory factor to the unusually low growth rate of nestlings noted by Nelson (1975) as characteristic of the family. Anecdotal accounts indicate that frigatebirds forage close to their nests, and feed their young at least once per day, but this may not be so throughout the fledging period.

Access to nests

Frigatebirds nest on the tops of trees or bushes, which present no problems for landing. They are able to slope-soar in tiny areas of lift created by a single tree or a line of mangroves. The long, pointed wings are not, however, the best shape for take-off from a level surface. Frigatebirds are not well adapted for powered flight at very low speeds, and need to dive from an elevated perch to pick up speed.

Reduced wing length: constant area

Another type of departure from the standard seabird is to reduce the wing span but not the area, so producing a short wing of very low aspect ratio. This type of wing is characteristic of cormorants Phalacrocoracidae. The two cormorants plotted in Fig. 3.8 have wing spans almost as far below the procellariiform line as the auks, but Fig. 3.7 shows their wing areas only just below the line. The reduction of span produces an increase in disc loading, as in auks and diving petrels, and such effects as follow directly from this are also seen in cormorants. In particular, the cruising speed is higher than in a 'standard seabird', and the amount of food that can be lifted is less. Cormorants do not, however, show the excessive increase in gliding speeds, seen in birds with dual-medium wings. Their normal mode of powered flight is continuous flapping, but this is because their high speed calls for more power output than a 'standard seabird', not because they cannot glide at a low enough speed. Cormorants slope-soar readily along sea cliffs, and occasionally soar in thermals.

Feeding, foraging and access to nests

Cormorants have much the same combination of high cruising speed and low load-lifting ability as auks. They forage at shorter ranges than petrels, and bring smaller loads of food for the young, at more frequent intervals. The wings are too large to be used with a flapping action under water. Propulsion under water comes mainly from the feet, with limited use of the wings for steering. Cormorants do not have the problems in take-off and landing seen in auks. Their broad wings allow them to exert power, and to glide and manoeuver at low speeds. They are able to land on trees, posts or even wires. Some species nest on cliffs like the larger auks, but others build nests in trees. Cormorants take off without difficulty from a level water surface. Like pelicans, they paddle with both feet together when taking off, instead of using the feet alternately like most other water birds.

Increased wing area: constant span

The two species of pelicans plotted in Fig. 3.7 show wing areas nearly double the values on the procellariiform line, but their wing spans are only slightly above the line in Fig. 3.8. This means that their wing loading and aspect ratios are much lower than those of procellariiform birds, but their disc loadings are only a little lower. Consequently they glide at low speeds, with little sacrifice of speed in powered flight. White pelicans such as *Pelecanus onocrotalus* and *P. erythrorhynchus* are mainly inland birds, which make long migrations by thermal soaring. Brown pelicans *P. occidentalis* and *P. thagus* take any opportunity to soar in slope lift and thermals, and will make detours in order to do so, but are also able to flap or flap-glide steadily over the open sea. Pelicans have emarginated primaries and slotted wing tips, resembling those of such birds as vultures Aegypiinae and Cathartidae and storks Ciconiidae, but they have higher aspect ratios than these land soaring birds. Brown pelicans plunge-dive, whereas white pelicans do not, but this is not reflected in any obvious differences of wing size or shape.

Minor variations from the 'standard seabird'

A number of seabirds are plotted in Figs 3.7 and 3.8 which are distinct taxonomically and in other ways but show only minor departures from the wing form chosen as the 'standard seabird', some of which may be briefly characterized as follows.

Plunge-divers

Boobies and gannets Sulidae, tropicbirds Phaethontidae and terns Sternidae all use plunge-diving to various degrees as a feeding technique. This does not result in any easily recognizable adaptations of the wings, and all of these groups have wing spans and areas close to those of procellariiformes. Indeed, errant albatrosses in the northern hemisphere are frequently mistaken for immature gannets, so close is the resemblance in general shape and size.

Tropicbirds, like typical petrels, forage far afield, and carry large loads of food back for their young at infrequent intervals (F.C. Schaffner, personal communication). The data on sulids reviewed by Nelson (1978) indicate that they too carry large loads, comparable with those of Procellariiformes, but feed their young more frequently, at least once per day. Although they slope-soar readily at the breeding grounds, they characteristically proceed in direct powered flight on foraging expeditions, flapping more than their procellariiform counterparts, and not soaring along waves. Evidently fast progress is more important to them than the energy savings to be gained (at the cost of reduced speed) by soaring. Tropicbirds also prefer flapping flight, and are seldom seen gliding.

Other surface feeders

A number of species of gulls Laridae and skuas Stercorariidae are plotted in Figs 3.7 and 3.8. Their wing areas tend to fall somewhat above the procellariiform line, but not so far above as those of pelicans. Most gulls do much of their foraging along coastlines or even over land. They use slope lift and thermals when patroling for food, and their low wing loadings adapt them for this. Some gulls, particularly Herring Gulls *Larus argentatus*, soar a great deal in thermals, over both sea and land.

Conclusions

One might suppose that the great variety of feeding methods used by seabirds, as reviewed and classified by Ashmole (1971), would be reflected in a corresponding variety of styles of flight. In fact, only the use of the wings under water has a drastic effect on flight adaptations. Other feeding methods, which do not call for this, can be accommodated with little effect on flight. Migration also is not a major factor in determining flight adaptations, since migration in seabirds is really a negative phenomenon, consisting of the relaxation of the need to return to a nest. Once breeding is at an end, seabirds can simply remain at sea wherever the feeding is best, and if they follow seasonally shifting feeding areas, 'migration' results.

Flight adaptations in seabirds seem to be determined mainly by the requirements of foraging and access to nests. The many variations of wing shape and flight style may be understood in terms of gliding speed, cruising speed in powered flight, and load-lifting ability. Low gliding speed is required for soaring and manoeuverability at nesting sites. High cruising speed and heavy load are needed to provision the young at a high rate. As has been seen, these requirements cannot be maximized simultaneously. The compromise adopted by any particular type of seabird reflects a particular combination of type of nest site, range and speed in foraging flights, and load-lifting ability, which in turn limits the number of young that can be raised, and their growth rates.

References

Ashmole, N.P. (1971). Sea bird ecology and the marine environment. In *Avian biology*, vol. 1, ed. D.S. Farner & J.R. King, pp. 224–86. New York: Academic Press.

Croxall, J.P. & Prince, P.A. (1980). Food, feeding ecology and ecological segregation of seabirds at South Georgia. *Biol. J. Linn. Soc.*, **14**, 103–131.

Harris, M.P. & Hislop, J.R.G.(1978). The food of young Puffins, *Fratercula arctica*. *J. Zool., Lond.*, **185**, 213–36.

Nelson, J.B. (1975). The breeding biology of frigatebirds – a comprehensive review. *Living Bird*, **14**, 113–56.

Nelson, J.B. (1978). *The Sulidae*. Oxford: Oxford University Press.

Pennycuick, C.J. (1956). Observations on a colony of Brunnich's Guillemot *Uria lomvia* in Spitsbergen. *Ibis*, **98**, 80–99.

Pennycuick, C.J.(1975). Mechanics of flight. In *Avian biology*, vol. 5, ed. D.S. Farner & J.R. King, pp. 1–75. New York: Academic Press.

Pennycuick, C.J.(1982). The flight of petrels and albatrosses (Procellariiformes), observed in South Georgia and its vicinity. *Phil. Trans. R. Soc. Lond. B*, **300**, 75–106.

Pennycuick, C.J. (1983). Thermal soaring compared in three dissimilar tropical bird species, *Fregata magnificens*, *Pelecanus occidentalis* and *Coragyps atratus*. *J. Exp. Biol.*, **102**, 307–25.

Pennycuick, C.J., Croxall, J.P. & Prince, P.A. (1984). Scaling of foraging radius and growth rate in petrels and albatrosses *Ornis Scand.*, **15**, 145–54.

Pennycuick, C.J. & Rezende, M.A. (1984). The specific power output of aerobic muscle, related to the power density of mitochondria. *J. Exp. Biol.*, **108**, 377–92.

Rayner, J.M.V. (1985). Linear relations in biomechanics: the statistics of scaling functions. *J. Zool., Lond.* (A) 206, 415–39.

Scholey, K.D. (1982). *Developments in vertebrate flight: climbing and gliding of mammals, and the flapping flight of birds*. Ph.D. thesis, University of Bristol.

Sibley, F.C. & Clapp, R.B. (1967). Distribution and dispersal of central Pacific Lesser Frigatebirds *Fregata ariel*. *Ibis*, **109**, 328–37.

Warham, J. (1977). Wing loadings, wing shapes and flight capabilities of Procellariiformes. *N.Z. J. Zool.*, **4**, 73–83.

Withers, P.C.(1979). Aerodynamics and hydrodynamics of the hovering' flight of Wilson's Storm Petrel. *J. Exp. Biol.*, **80**, 83–91.

4

Diving behavior and performance, with special reference to penguins

G. L. KOOYMAN and R. W. DAVIS
Physiological Research Laboratory, Scripps Institution of Oceanography, University of California, San Diego, La Jolla, California 92093, USA

Introduction

There have been few studies of the diving behavior or swimming performance of seabirds and data for some species are available only from laboratory experiments. This is partly because the present dive-recording instruments are too large for the small size of most aquatic birds. The recent development of small and cheap velocity recorders and depth recorders (Wilson & Bain, 1984*a,b*) may, however, substantially improve the situation.

The main features of aquatic behavior and performance relate to dive depth and duration and swimming speeds. These are influenced by hydrodynamic properties, body oxygen stores and metabolic rates. We shall review what is known on these topics. Most work has been done on penguins, but comparisons with other aquatic birds will be made where information is available.

Dive capacities

Most dive depths of seabirds have been obtained indirectly from individuals caught in fishing nets (Table 4.1). Despite the potential biases and uncertainties of this method, some remarkable values should be noted. Comparisons with data obtained from other sources should be made with caution and with due allowance for often very different sample sizes, e.g. with the 180 m dive depth of the Common Murre *Uria aalge* and that of 265 m for the Emperor Penguin *Aptenodytes forsteri*. In general, the underwater flyers (alcids) and kickers (cormorants and loons) are not as large as the totally aquatic penguins, and they may not dive consistently as deep. However, the data collected so far would not support this hypothesis except for the largest penguins. Only penguin data have been

Table 4.1. *Maximum reported dive depths of birds*

Species	Maximum depth (m)	Method of observation	Sample size[a]	Reference
Common Loon *Gavia immer*	60	Trammel net	—	Schorger, 1947
Great Cormorant *Phalacrocorax carbo*	37	Net	—	Kooyman, 1975
Common Eider *Somateria mollissima*	55	Net	—	Kooyman, 1975
Oldsquaw *Clangula hyemalis*	60	Gill net	—	Schorger, 1947
Common Murre *Uria aalge*	180	Gill net	12243	Piatt & Nettleship, 1985
Razorbill *Alca torda*	120	Gill net	9	Piatt & Nettleship, 1985
Atlantic Puffin *Fratercula arctica*	60	Gill net	875	Piatt & Nettleship, 1985
Black Guillemot *Cepphus grylle*	50	Gill net	36	Piatt & Nettleship, 1985
Chinstrap Penguin *Pygoscelis antarctica*	70	Recorder	1110	Lishman & Croxall, 1983
Adélie Penguin *P. adeliae*	68	Compression chamber	—	Kooyman *et al.*, 1973
Gentoo Penguin *P. papua*	100	Net	1	Conroy & Twelves, 1972
Gentoo Penguin	70	Recorder	19	Adams & Brown, 1983
King Penguin *Aptenodytes patagonicus*	240+	Recorder	2500	Kooyman *et al.*, 1982
Emperor Penguin *A. forsteri*	265	Recorder	5	Kooyman *et al.*, 1971

[a] Number of dives recorded or dead birds caught in nets.

obtained using recorders (chiefly because penguins are generally large and easy to recapture).

Without the use of recorders, the reported duration of dives is most likely to be conservative because of the reluctance of observers to report exceptionally long dives as a single breath-hold. This would have been true of the 18 min dive of an Emperor Penguin, the longest unrestrained dive reported, were it not that the bird was observed from an under-ice chamber. Most observations show that birds dive for about 1–2.5 min (Dewar, 1924). Laboratory studies show that medium-sized penguins have a higher tolerance of about 5–7 min (Scholander, 1940; Kooyman *et al.*, 1973), but these durations are seldom achieved in the wild under unrestrained conditions except for the largest, and deepest diver, the Emperor Penguin (Kooyman *et al.*, 1971) and probably also the King Penguin *A. patagonicus*.

Field estimates of swimming speeds are also potentially unreliable as observers tend to over-estimate because the apparent velocity appears much higher than the real velocity. The field data in Table 4.2 were cruising speeds estimated by reliable methods. Emperor and Adélie *Pygoscelis adeliae* Penguins were timed while swimming a measured course (Kooyman *et al.*, 1971); Jackass Penguin *Spheniscus demersus* values were averaged over several hours (Nagy, Siegfried & Wilson, 1985). The Little Penguin *Eudyptula minor* was swimming at a known metabolic rate in a water mill (Baudinette & Gill, 1985); the other data are from birds swimming in aquaria where the size of the tank and bird density may influence speeds.

Hydrodynamics

The drag coefficient is a dimensionless number related to the total drag of the animal as follows:

$$C_d = 2D/pAV^2 \quad (4.1)$$

Table 4.2. *Normal cruise/glide velocities of penguins underwater*

Species	Mass	Velocity (m s⁻¹)	Velocity (km h⁻¹)	Reference
Emperor	~ 25	2·1	7·5	Kooyman *et al.*, 1971
Adélie	~ 5	2·0	7·2	Kooyman, 1975
Jackass	3	1·8	6·5	Nagy *et al.*, 1984
Little	1·2	0·7	2·5	Baudinette & Gill, 1985

where p = density of the medium, A = the frontal, or surface area, or volume of the body, V = the velocity, and D = the drag. Because it is dimensionless, bodies of different size can be compared for their relative degree of streamlining. It is difficult to measure the drag in most animals, and so far it has not been done during active swimming. In birds (and to our knowledge only in penguins) drag has been measured only during gliding, or from cast models (Table 4.3). The lowest C_d was for an Emperor Penguin, with the Gentoo Penguin *Pygoscelis papua* about 1.5 times greater, and this species in turn was about 1.8 times greater than an ideal, streamlined spindle. The Emperor Penguin, therefore, is very close to the ideal shape and design when gliding at 2 m s^{-1}. The problem of cast models is indicated by the C_d of 2.5 times that of the similar-sized Gentoo Penguin. There may be several reasons for this; two of which are inaccurate body shape and lack of feathers. Feathers may dampen turbulent eddies in the boundary layer. The advantage of the cast model is that velocity can be varied to assess its affects on C_d. From eqn 4.1 it can be seen that once the C_d is known, the minimum power requirements of the bird (P) can be estimated assuming that C_d changes little over the velocity ranges considered. In this case:

$$P = (\rho A V^3/2).C_d \qquad (4.2)$$

Therefore, if a bird's velocity is increased from 1 to 2 m s^{-1} the power requirement will increase by about eight times. This is discussed later in relation to dive durations.

The above C_d values are all for submerged bodies. When the bird comes to the surface to breathe, the greatly increased drag due to surface interactions with the penguin's body makes surface swimming energetically costly. As speed increases, the surface drag increases exponentially and, in theory, when a critical velocity is reached, it will become energetically more economical to porpoise, or leap clear of the water for those brief breathing periods (Au & Weihs, 1980). The equation derived for the critical velocity is:

$$Y = 2.54X^{0.167}$$

where Y = velocity (m s^{-1}), and X = body mass (kg). In Table 4.4 the porpoise velocity of several other species of penguin has been calculated based on this equation. Later equations incorporated drag augmentation due to swimming motion (Blake, 1983). In this work it was predicted that penguins may leap at a lower speed, for example the critical velocity for the Gentoo Penguin would be 2.5 m s^{-1} instead of 3.4 m s^{-1}, and for the Emperor Penguin 3.6 m s^{-1} instead of 4.4 m s^{-1} (Table 4.4). None of these

Table 4.3. *Drag coefficients of penguins*

Species	Wetted surface area (C_{d_A})	Frontal area (C_{d_F})	Mean velocity (m s^{-1})	Reynolds number (*Re*)	Method	Reference
Ideal spindle	—	0·04	—	5×10^6		Nachtigall & Bilo, 1980
Humboldt (cast model)	0·01	—	1·0	10^6	Pushed carcass	Hui, 1983
Gentoo	0·0044	0·07	—	10^6	Glide	Nachtigall & Bilo, 1980
Emperor	0·003	—	2	$1{\cdot}6 \times 10^6$	Glide	Clark & Bemis, 1979

theoretical manipulations has any supportive experimental data, but the point is that animals may porpoise at predictable speeds. Once some of these values have been determined it might make field observations of swimming penguins more reliable. Based on our own experience, Macaroni *Eudyptes chrysolophus*, Gentoo, Adélie and Chinstrap Penguins routinely porpoise and King Penguins occasionally do so, but so far we have never seen Emperor Penguins porpoise.

The optimum swimming speed may depend upon the activities of the bird. When they are traveling between food patches, Norberg (1981) argues that for flying birds minimum flight speed should be at least their maximum range speed (V_{mr}). If the increased cost of travel for flying faster than V_{mr} can be more than compensated for by the additional foraging time and prey captured, then the optimum speed should be higher than V_{mr}. If the same criteria apply to swimming birds, then penguins with chicks might swim at higher rates than V_{mr}. The only simultaneous measurements of swimming velocity and metabolic rate are for Humboldt *Spheniscus humboldti* (Hui, 1983) and Little Penguins (Baudinette & Gill, 1985) and there was little or no change over most of the range of speeds measured. However it is likely that the measurements were obtained over too narrow a velocity range and assessments of optimum travel speed must await data collected over a broad range of speeds. If penguins porpoise *en route* to feeding areas, then their optimum velocities might be the minimum porpoise speeds of, for example, 3.2 and 2.6 m s^{-1}, respectively, for Humboldt and Little Penguins. This speed may be different from diving

Table 4.4. *Calculated crossover above which penguins should porpoise to save energy. Based on the equation* $Y = 65{\cdot}3X^{0{\cdot}33}$ *where* Y = *velocity in* m s^{-1} *and* X = *body volume in* m^{3}, *assuming body density* = 1 kg l^{-1} *(from Au & Weihs, 1980)*

Species	Mass (kg)	Velocity (m s^{-1})
Little	1·2	2·6
Jackass		
Macaroni	3·2–3·8	3·2
Humboldt		
Gentoo	6·2	3·4
King	13	3·9
Emperor	25	4·4

speed because while diving a bird may need to increase its breath-hold limit (which determines search time) depending upon how densely and reliably distributed the prey is.

Since oxygen consumption is directly related to swim speed, the faster a bird swims, the shorter the aerobic breath-hold duration. Specifically, in the Little Penguin, oxygen consumption when swimming submerged at 0.8 m s^{-1} is 50% higher than in resting controls (Baudinette & Gill, 1985). This would cause a depletion of body oxygen stores at about a 50% higher rate. If the bird should swim even faster, then drag would increase as V^2 (eqn 4.1), and power requirements would increase as V^3 (eqn 4.2). The result would be a sharp decline in breath-hold duration.

Where some underwater travel is needed to locate the prey, and then some pursuit required after finding it, a compromise must be achieved between the maximum aerobic dive limit and the most productive swim speed. Such a swim speed is probably near the lowest cost of transport velocity. At the lowest cost of transport, the oxygen consumption would be the least for the distance swum, and probably only about 50% greater than the maximum aerobic breath-hold limit. This may be the speed penguins select when routinely swimming and gliding in aquaria. If so, then the values in Table 4.2 may be close to that speed. If this lowest cost of transport speed is the optimum dive velocity, then once the metabolic rate is known for that swim speed, the upper limit to the duration of foraging dives can be predicted.

Oxygen stores, swimming metabolism and dive durations

When a bird dives, the duration of the submersion is dictated by how the body oxygen stores are utilized. The management of these oxygen stores is influenced by variable distribution of blood flow to the organs and the metabolic rate of the organs. In a swimming bird the major consumer of oxygen is the muscle and, as we indicated above, the level of power output of the muscles is the primary determinant of the rate of depletion of the oxygen store, which is found in three main compartments: (1) bound to myoglobin in the muscle; (2) bound to haemoglobin in the blood; and (3) stored in the air sacs. If we assume that the bird remains within the limits of this oxygen store, then the dive limit at which rapid recovery is possible can be estimated if the available body oxygen stores, and the metabolic rate (MR) at the swimming speed normally occurring during the dive are known. Aerobic metabolism is assumed to provide the major source of energy, because only then is rapid recovery from dives possible. Under aerobic conditions only oxygen stores have to be replenished, and the restoration of blood and muscle chemistry to pre-dive concentrations of glucose and lactic acid is not required. The latter would require a substantially longer recovery in order for the metabolites to return to pre-dive levels, and this might make it impossible for the bird to make a series of hunting dives in a short period of time.

This aerobic dive limit (ADL) is useful in predicting the maximum dive durations that animals will rarely exceed. For Weddell Seals *Leptonychotes weddelli*, a diving mammal for which the ADL has been studied in detail, this limit is exceeded in less than 5% of dives (Kooyman *et al.*, 1980). Unfortunately, data which make possible the calculation of the body oxygen stores of penguins and other diving birds are only available for a few species (Table 4.5). Metabolic rates and swimming speeds have been measured on only two species of penguins and to our knowledge in no other diving birds (Table 4.6). Swimming speeds of only three species of penguins have been measured in the wild (Table 4.6). Therefore, calculations are made only for a small species (Little Penguin) and a medium-sized species (Humboldt Penguin) in which metabolic rate relative to swimming speed has been measured, but not total oxygen stores. We assume the measured total oxygen store value for Adélies and Gentoos to be the same for other penguins and we use this estimate. If we assume an MR of 2.1 times for the Little Penguin and 3.0 times for the Humboldt Penguin above resting metabolism (Table 4.6), the ADLs are 1.7 and 2.5 min, respectively. If we assume rates as high as those estimated for the Jackass Penguin of 8.0 times MR (Table 4.6), which might represent only a doubling in swimming speed above minimum cost of transport speed (see Hydrodynamics), then the ADLs are 0.4 and 0.6 min, respectively. To permit comparison with the largest penguin, the Emperor Penguin, we assume twice the Standard Metabolic Rate (SMR) and an equal mass-specific oxygen store. The calculated ADL is 4.1 min at a swim velocity of 2 m s^{-1} (Table 4.3). At 8 times resting MR the ADL is 1.0 min. At the above calculated dive durations and the assumed most economical swimming speeds of 0.7, 1.0 (Table 4.6) and 2.0 m s^{-1} (Table 4.3) for Little, Humboldt and Emperor Penguins, their maximum economical foraging depths would be 35 m, 75 m and 250 m, respectively. These estimates are based on direct descents and ascents with few options in the dive for hunting strategies.

Hunting economics

Dives to the birds' maximum economical limit represents one strategy in a continuum from dives near the surface to dives attaining that limit.

Table 4.5. *Body oxygen stores of diving birds*

Species	Haemoglobin (g per 100 g blood)	Oxygen capacity (ml O_2 per 100 ml blood)	Blood volume (ml kg^{-1})	Myoglobin (wet weight) (g per 100 g)	Available O_2 store[a] (ml kg^{-1})	Reference
Pigeon *Columba livia*	14·2	—	92	0·25	—	Lawrie, 1950; Bond & Gilbert, 1958
Adélie Penguin	16·3	22	93	2·9	43	Lenfant *et al.*, 1969; Weber *et al.*, 1974
Gentoo Penguin	16·4	22[b]	—	4·6	46	Kooyman *et al.*, 1973; Milsom *et al.*, 1973
Macaroni Penguin	—	23	93	—	—	Scholander, 1940
Chinstrap Penguin	19·6	26[b]	—	3·2	—	Weber *et al.*, 1974; Milsom *et al.*, 1973
King Penguin	19·7	26[b]	—	—	—	Kooyman, unpublished observation
Emperor Penguin	16·8	23[b]	—	4·3	—	Lenfant *et al.*, 1969 Panganis, unpublished observation
South Georgia Diving Petrel *Pelecanoides georgicus*	19·7	26[b]	—	—	—	Kooyman, unpublished observation

[a] Calculation is based on: arterial blood volume is 1·3 total blood volume; O_2 available is from 95 to 20% saturation; venous O_2 content is 5 vol % < arterial to 0% saturation; muscle mass is 30% of body weight.
[b] Calculated from haemoglobin binding capacity of 1·34 ml O_2 g^{-1}.

Table 4.6. *Metabolic rate of penguins resting and swimming, expressed as a multiple of the calculated standard metabolic rate (MR) of non-passerine birds from the regression equation* $Y = 4{\cdot}41\ Mb^{0{\cdot}729}$, *Y in watts; Mb in kg, derived by Aschoff & Pohl, 1970*

Species	Mass (kg)	Resting MR	Brooding MR	Overall MR at sea & shore	Swim MR	Velocity at min. cost of transport ($m\ s^{-1}$)	Cost of transport transport ($ml\ O_2$ per kg per m)	Reference
Little	1·2	1·2	—	—	2·1	0·72	0·6	Baudinette & Gill, 1985
Jackass	3·2	—	1·4	2·9[a]	8·2	1·8	—	Nagy *et al.*, 1984
Macaroni	3·6	—	1·8	2·9	—	—	—	Davis *et al.*, 1983
Humboldt	3·8	1·1	—	—	3·0	1·0	0·69	Hui, 1983
Gentoo	6·2	—	—	2·6	—	—	—	Davis *et al.*, 1983
King	13·0	—	—	2·8[b]	—	—	—	Kooyman *et al.*, 1982

[a] Time on land 67%; time at sea 33%.
[b] About 90% of the time at sea.

Before making a shallow dive the prey may be in sight, and thus the dive becomes simply one of pursuit and capture. In this case the likelihood of capturing prey is presumably high compared with deeper dives where the prey is not found until sometime during the dive. Because search time is absent or reduced, the energy commitment of the shallow dive is low, and the prey sought can be small and dispersed. As long as the average result is a net energy gain over the energy expended if no hunting were done, it is profitable to continue hunting in this fashion. This kind of hunting may be in accord with the suggestion of Norberg (1977) that predators should shift to less energy-consuming foraging procedures when prey availability is minimal. Norberg's example was for terrestrial birds and mammals foraging during the winter, but in marine divers this may be the case either in the winter or during daylight hours at any time of year, when most prey is deep. During this dispersed-prey condition of daylight hours it may not be much more expensive to keep swimming and searching for prey than to rest. In summary, the advantages of continuous activity while birds are at sea, especially for those with chicks, would be that, as long as there were any net energy gain in foraging, it would shorten the period at sea and make possible a quicker return to chicks than if they rested for the hours when prey was dispersed.

Exploitation of deeper-living prey should normally be done when the prey is large, aggregated, or both, or prey at the surface is absent or at such low density that the cost of maintenance and search exceeds the energy intake. During dives to depth the prey initially is not in sight so there is a search component to the dive. This might even involve most of the descent time and parts of the ascent, assuming that immediately after encounter the descent ends, pursuit of the prey begins, and pursuit is normally from below the prey (Hobson, 1966). Because the prey is not in sight at the beginning of the dive, there is a greater chance of failure compared with shallow dives. As mentioned earlier, in order to extend search time the swim velocity may be near the minimum power speed (Table 4.2) until prey is detected. Pursuit speeds may be fast or slow depending upon the best strategy for the type of prey and its capabilities.

The dive depth ability for King Penguins (Table 4.1) may enable them to hunt for prey even during the daytime when it has descended to depths of low light and, consequently, reflect some of the problems of deep feeding just discussed. If we assume that they feed solely on squid of the size, energy content and assimilation efficiency proposed in Chapter 6, then based on the measured energy expenditure of the birds (Table 4.6), they would require about 30 squid for a 4-day period at sea. If the average number of dives per day is about 150 (Kooyman *et al.*, 1982), then for a 4-day trip they would catch a squid on one dive out of every 20. Such a low success rate is suggested by the earlier comments on deep diving where much of the dive is dedicated to travel to and search for the prey. This success rate is much lower than that estimated for the smaller penguins (Chapter 6) which feed on krill, a smaller prey that may be within sight from the surface.

Conclusions

Most information on dive capacities is for penguins, but for all diving birds it is scarce. Only recently have swimming speeds been recorded. This information, together with drag coefficients, will make it possible to calculate the minimum power requirements of birds while diving. The power estimates may make it possible to predict the optimum swimming speeds, dive durations and most likely depth range of a particular species of bird when the oxygen stores are better known, and when more measurements of swimming metabolism have been obtained to justify the theoretical calculations. This collection of variables should provide a sound basis for estimating the quantity of prey consumed when the type of prey is known.

Acknowledgement Supported by USPHS Grant No. HL 17731 and NSF, Division of Polar Programs Grant No. 78-22999.

References

Adams, N.S. & Brown, C.R. (1983). Diving depths of the Gentoo Penguin *Pygoscelis papua*. *Condor*, **85**, 503–4.

Aschoff, S. & Pohl, H. (1970). Rhythmic variation in energy metabolism. *Fedn Proc. Fedn Am. Socs Exp. Biol.*, **29**, 1541–52.

Au, D. & Weihs, D. (1980). At high speeds dolphins save energy by leaping. *Nature, Lond.*, **284**, 548–50.

Baudinette, R.V. & Gill, P. (1985). The energetics of 'flying' and 'paddling' in water: locomotion in penguins and ducks. *J. Comp. Physiol.*, **155**, 373–80.

Blake, R.W. (1983). Energetics of leaping in dolphins and other aquatic animals. *J. Mar. Biol. Assoc. U.K.*, **63**, 61–70.

Bond, C.F. & Gilbert, P.W. (1958). Comparative study of blood volume in representative aquatic and nonaquatic birds. *Am. J. Physiol.*, **194**, 519–21.

Clark, B.D. & Bemis, W.(1979). Kinematics of swimming of penguins at the Detroit Zoo. *J. Zool., Lond.*, **188**, 411–28.

Conroy, J.W.H. & Twelves, E.L. (1972). Diving depths of the Gentoo Penguin and Blue-eyed Shag from the South Orkney Islands. *Bull. Br. Antarct. Surv.*, **30**, 106–8.

Davis, R.W., Kooyman, G.L. & Croxall, J.P. (1983). Water flux and estimated metabolism of free-ranging Gentoo and Macaroni Penguins at South Georgia. *Polar Biol.*, **2**, 41–6.

Dewar, J.M. (1924). *The bird as a diver.* London: Witherby.

Hobson, E.S. (1966). Visual orientation and feeding in seals and sea lions. *Nature, Lond.*, **210**, 326–7.

Hui, C.A. (1983). *Swimming in penguins.* Ph.D. thesis, Univ. California, Los Angeles.

Kooyman, G.L. (1975). Behaviour and physiology of diving. In *The biology of penguins*, ed. B. Stonehouse, pp. 115–37. London: Macmillan.

Kooyman, G.L., Davis, R.W., Croxall, J.P. & Costa, D.P. (1982). Diving depths and energy requirements of King Penguins. *Science*, **217**, 726–7.

Kooyman, G.L., Drabek, C.M., Elsner, R. & Campbell, W.B. (1971). Diving behavior of the Emperor Penguin, *Aptenodytes forsteri. Auk*, **88**, 775–95.

Kooyman, G.L., Schroeder, J.P., Greene, D.G. & Smith, V.A. (1973). Gas exchange in penguins during simulated dives to 30 and 68 m. *Am. J. Physiol.*, **225**, 1467–71.

Kooyman, G.L., Wahrenbrock, E.A., Castellini, M.A., Davis, R.W. & Sinnett, E.E. (1980). Aerobic and anaerobic metabolism during voluntary diving in Weddell Seals: evidence of preferred pathways from blood chemistry and behavior. *J. Comp. Physiol. B*, **138**, 335–46.

Lawrie, R.A. (1950). Some observations on factors affecting myoglobin concentrations in muscle. *J. Agric. Sci.*, **40**, 356–66.

Lenfant, C., Kooyman, G.L., Elsner, R. & Drabek, C.M. (1969). Respiratory function of the blood of the Adélie Penguin, *Pygoscelis adeliae. Am. J. Physiol.*, **216**, 1598–600.

Lishman, G.S. & Croxall, J.P. (1983). Diving depths of the Chinstrap Penguin *Pygoscelis antarctica. Bull. Br. Antarct. Surv.*, **61**, 21–5.

Milsom, W.K., Johansen, K. & Millard, R.W. (1973). Blood respiratory properties in some antarctic birds. *Condor*, **75**, 472–4.

Nachtigall, W. & Bilo, D. (1980). Stromungsanpassung des Pinguins beim Schwimmen unter Wasser. *J. Comp. Physiol. A*, **137**, 17–26.

Nagy, K.A., Siegfried, W.R. & Wilson, R.P. (1985). Energy utilization of free-ranging Jackass Penguins, *Spheniscus demersus. Ecology*, **65**, 1648–55.

Norberg, R.A. (1977). An ecological theory on foraging time and energetics and choice of optimal food-searching method. *J. Anim. Ecol.*, **46**, 511–29.

Norberg, R.A. (1981). Optimal flight speed in birds when feeding young. *J. Anim. Ecol.*, **50**, 473–7.

Piatt, J.F. & Nettleship, D.N. (1985). Diving depths of the Alcidae in Newfoundland. *Auk*, **102**, 293–7.

Scholander, P.F. (1940). Experimental investigations on the respiratory function in diving mammals and birds. *Hvalrad. Skr.*, **22**, 1–131.

Schorger, A.W. (1947). The deep diving of the loon and the old-squaw and its mechanism. *Wilson Bull.*, **59**, 151–9.

Weber, R.E., Hemmingsen, E.A. & Johansen, K.(1974). Functional and biochemical studies of penguin myoglobins. *Comp. Biochem. Physiol.*, **49**, 197–214.

Wilson, R.P. & Bain, C.A.R. (1984*a*). An inexpensive speed meter for penguins at sea. *J. Wildl. Mgmt*, **48**, 1360–4.

Wilson, R.P. & Bain, C.A.R. (1984*b*). An inexpensive depth gauge for penguins. *J. Wildl. Mgmt*, **48**, 1077–84.

5

Kleptoparasitism in seabirds

R. W. FURNESS
Zoology Department, Glasgow University, Glasgow G12 8QQ, UK

Introduction

Kleptoparasitism is the deliberate stealing by one animal of food which has already been captured by another. The same behaviour has also been called 'food parasitism' (Hopkins & Wiley, 1972), 'robbing' (Dunn, 1973; Hulsman, 1976; Rockwell, 1982) or 'piracy' (Meinertzhagen, 1959; Ashmole, 1971; Hatch, 1975; Andersson, 1976; Duffy, 1980). It may be an interspecific or intraspecific interaction, and occurs widely throughout the vertebrates, but is particularly well documented for birds. While most documented cases of robbing behaviour in birds are ecologically and evolutionarily trivial, kleptoparasitism may be an important feeding technique of Fregatidae (frigatebirds), Chionididae (sheathbills), Stercorariidae (skuas) and Laridae (gulls and terns), the only bird families where the habit is recorded for more than 25% of the species (Brockman & Barnard, 1979). It is probably more than coincidental that all these are families of seabirds.

Brockman & Barnard (1979) tried to explain why some bird groups tend to employ kleptoparasitism while others do not. They suggested that kleptoparasitism would only become more than an occasional opportunistic event in species which occur where there are large concentrations of potential hosts, where large quantities of food are carried, and where the food is carried in large, discrete masses, is available in a predictable manner, and is visible to the kleptoparasite. Finally, the kleptoparasite should itself be short of food so that robbing would be more profitable than capturing food directly. Duffy (1980, 1982) further suggested that seabirds would feed by kleptoparasitism only if their victims had access to prey at greater depths in the sea, and he also considered the possibility that, as Kushlan (1978) showed for an ardeid community, larger species may tend to pirate from smaller ones.

The aims of this chapter are:
1. To review data on kleptoparasitism among seabirds. 2. To examine these data in the light of the general rules of kleptoparasite behaviour developed by Brockman & Barnard (1979), as more than half the studies of kleptoparasitic relationships were published since that review. 3. To examine why certain seabirds adopt kleptoparasitic habits, how such habits may evolve within populations, what responses may be induced from victims and, thus, the nature of the resulting evolutionary arms race.

Methods of kleptoparasitism

Kleptoparasitism can occur on the ground. Sheathbills throw themselves against penguins which are bringing up food for the chick and steal boluses spilt as a result of their interference (Burger, 1981*a*); terns may steal fish being presented to mates or chicks (Dunn, 1973). Most kleptoparasitism, however, involves aerial interceptions or chases. In some relationships the kleptoparasite can fly faster than its victim (e.g. Great Skua *Catharacta skua* – Northern Gannet *Sula bassana*; Arctic Skua *Stercorarius parasiticus* – Kittiwake *Rissa tridactyla*). In these interactions the victim is usually harried for a period of several seconds or even minutes, until it either gives up food or the skua gives up the chase. Such chases may result in physical attack and in convoluted aerobatic manoeuvres, and may lead to several other kleptoparasites joining in (Hatch, 1970, 1975; Andersson, 1976). In other cases, the victim is capable of flying as fast, or faster, than the kleptoparasite in continuous horizontal flight, so the attack involves either head-on confrontation causing the victim to brake (e.g. some attacks of Atlantic Puffins *Fratercula arctica* by Arctic Skuas (Grant, 1971)) or the kleptoparasite stoops onto the victim from above. Most chases of auks by skuas are of the latter type. If the victim notices the stooping skua in time it can accelerate to maximum speed and avoid being caught. If it is slow to react then the skua is able to catch up with it before it has time to take evasive action and in this situation it is more likely to drop its food (Furness, 1978).

General descriptions of the kleptoparasitic feeding behaviour of frigatebirds are given by Stonehouse & Stonehouse (1963), Diamond (1973, 1975), Nelson (1976) and Schreiber & Hensley (1976), but no detailed studies of their interactions with hosts have been made. Detailed studies have been published giving accounts of the kleptoparasitic interactions with their victims by skuas (references in Table 5.1), gulls and terns (Table 5.2) and Lesser Sheathbills *Chionis minor* (Burger, 1981*a,b*).

Brockman & Barnard (1979) reiterated a generally held view that frigatebirds and skuas are 'specialised kleptoparasites' and implied that these birds have evolved structural adaptations to enhance their success as kleptoparasites. They cited the small unwebbed feet, vestigial uropygial gland and exceptionally large wings and small body mass of frigatebirds as examples of such adaptations. In contrast, they considered gulls and terns to be 'opportunistic kleptoparasites'. This distinction will be considered further in the following sections.

Patterns of kleptoparasitism within seabird communities

Because authors have examined kleptoparasitic interactions with different ideas in mind it is often difficult to compare results between studies. For example, Dunn (1973) reported only the outcome of chases of Common Terns *Sterna hirundo* by Roseate Terns *Sterna dougallii*, although other interactions also occurred, while Duffy (1980, 1982) recorded piracy by 23 species of seabirds in three different seabird communities, but did not record the outcome of chases. Verbeek (1977*a*) tabulated data for success rates of chases by single Herring Gulls *Larus argentatus* and Lesser Black-backed Gulls *L. fuscus*, but presented only graphical results of chases by more than one gull and did not distinguish between the two species in multiple chases. Most authors present data lumped for all age classes, weather conditions, stages of tide, distances from breeding colonies, stages of the breeding season, sizes of prey carried by hosts, and year of study. All of these factors may influence the probability of a chase being successful (i.e. resulting in the host dropping food).

In order to examine patterns of kleptoparasitism on a broad scale, I have listed all the published data I could locate in which the percentage of successful chases of one host species by one kleptoparasite species have been given for a sample of at least 10 chases (Table 5.1, skuas and frigatebirds; Table 5.2, gulls and terns). Unless specified in the tables, data refer to all the observed instances of kleptoparasitism for these species pairs – i.e. they reflect the situation in the field and are not standardised for any of the parameters known to affect success. Results derived from Tables 5.1 and 5.2 allow us to examine a number of generally held concepts or predictions' such as those mentioned earlier.

It is generally considered that skuas and frigatebirds are more highly adapted for kleptoparasitism and, as a consequence, that skuas and frigatebirds would have a higher rate of success in kleptoparasitic chases

Table 5.1. *The success rate of skuas and frigatebirds in forcing hosts to drop food*

Author and locality	Kleptoparasite	Host	Total chases	Percent where food dropped
Andersson (1976)				
Unst, Shetland	Great Skua	Gannet	93	12
	Great Skua	Puffin	32	19
	Arctic Skua	Puffin	88	22
Arnason (1978)				
Iceland	Arctic Skua	Puffin	1130	69
Arnason & Grant (1978)				
Iceland	Great Skua	Puffin	32	34
Diamond (1975)				
Aldabra	Frigatebird	Red-f. Booby	55	18
B. L. Furness (1980)				
Orkney	Arctic Skua	Guillemot	108	10
	Arctic Skua	Arctic Tern	19	37
S. Iceland	Arctic Skua	Arctic Tern	50	22
Fetlar, Shetland	Arctic Skua	Puffin	68	22
Noss, Shetland	Arctic Skua	Guillemot	153	16
Foula, Shetland	Arctic Skua	Guillemot	26	8
	Arctic Skua	Puffin	144	22
	Arctic Skua	Arctic Tern	17	6
	Arctic Skua	Guillemot	36	22
	Arctic Skua	Puffin	289	26
	Arctic Skua	Kittiwake	30	33
	Arctic Skua	Arctic Tern	374	24
	Arctic Skua	Guillemot	41	22
	Arctic Skua	Puffin	250	31
	Arctic Skua	Arctic Tern	23	17

Source / Location	Kleptoparasite	Host		
B. L. Furness (1983)				
South Africa	Arctic Skua	Sabine's Gull	16	25
	Arctic Skua	Hartlaub's Gull	144	25
	Arctic Skua	Common Tern	19	17
	Arctic Skua	Sandwich Tern	11	0
R. W. Furness (1978)				
Foula	Arctic Skua	Arctic Tern	220	48
	Arctic Skua	Kittiwake	16	25
	Arctic Skua	Puffin	13	15
R. W. Furness (1978)				
Foula	Great Skua	Kittiwake	13	15
	Great Skua	Puffin	223	38
	Great Skua	Razorbill	38	18
	Great Skua	Guillemot	113	29
	Great Skua	Gannet	69	30
	Arctic Skua	Puffin	110	21
	Arctic Skua	Razorbill	18	11
	Arctic Skua	Guillemot	27	11
	Arctic Skua	Kittiwake	33	33
	Arctic Skua	Arctic Tern	87	44
Grant (1971)				
S. Iceland	Arctic Skua	Puffin	140	51
Maxson & Bernstein (1982)	Antarctic Skua	Blue-eyed Shag	50	2
Palmer Station, Antarctica	Antarctic Skua	Blue-eyed Shag	280	5
Nelson (1976)				
Galapagos	Frigatebird	Red-f. Booby	—	12
R. W. Schreiber (unpublished)				
Christmas Is.	Frigatebird	Red-f. Booby	—	63
Taylor (1979)				
Aberdeenshire	Arctic Skua	Arctic Tern	70	40
	Arctic Skua	Common Tern	100	9
	Arctic Skua	Sandwich Tern	254	23

than gulls or terns. Great Skuas, being less dependent on kleptoparasitism than are Arctic Skuas (Furness & Hislop, 1981), might perhaps be intermediate between these 'specialists' and 'opportunists'. This prediction is not borne out by the data. For Arctic Skuas and frigatebirds, chase success rates vary from 0% to 69% with a median of 22% (mean 23%; 95% confidence interval from 18% to 29%; n = 36; arcsine transformations). For *Catharacta* skuas (Great Skua and South Polar Skua *C. maccormicki*), success rates vary from 2% to 38%, median 19% (mean 19%; 95% confidence interval from 11% to 27%; n = 10). For gulls and terns the success rates vary from 1% to 85%, median 23% (mean 26%; 95% confidence interval from 18% to 34%; n = 28).

The mean success rates do not differ significantly between groups of kleptoparasite. However, the success rate is very variable between different localities and species pairs, and one possible explanation for the unexpectedly poor performance of 'specialist' kleptoparasitic species may be that they chase a wider spectrum of hosts, many of which the opportunistic kleptoparasites (e.g. gulls) would not attempt to rob because they could not hope to be successful. Let us then compare the performances of 'specialist'

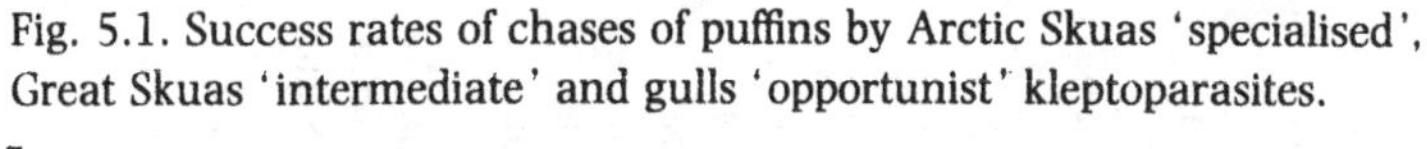
Fig. 5.1. Success rates of chases of puffins by Arctic Skuas 'specialised', Great Skuas 'intermediate' and gulls 'opportunist' kleptoparasites.

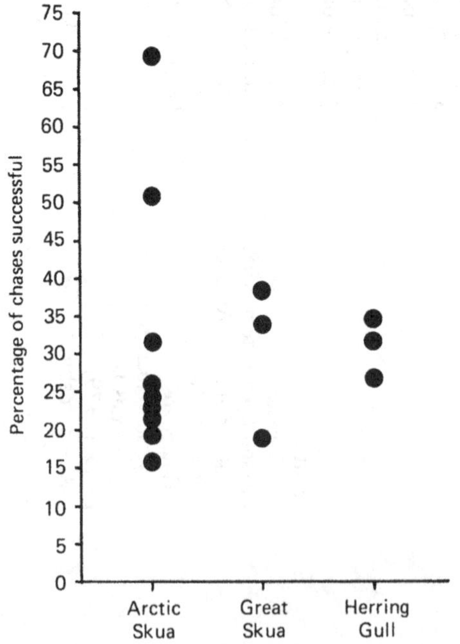

and 'opportunistic' kleptoparasites in robbing from the same host species.

There is no evidence that Arctic Skuas are generally more successful in robbing Puffins than are Great Skuas or Herring Gulls (Fig. 5.1). Nor do Arctic Skuas do better against Arctic Terns *Sterna paradisaea* or Common Terns than do Common Terns or Roseate Terns; and all these kleptoparasites are distinctly less successful per chase than are Laughing Gulls *Larus atricilla* (Fig. 5.2).

Another possible explanation for the lack of superiority of 'specialist' kleptoparasites may be that kleptoparasitism only occurs as a regular feeding method if its chances of success are above a certain threshold. In this case there should be few documented instances of low success rates and these should only be found for relatively uncommon species interactions. There is some support for this idea in Tables 5.1 and 5.2. Very few interactions have a success rate of less than 10%. However, several

Fig. 5.2. Success rates of chases of Arctic Terns (solid symbols), Common Terns (open symbols) and 'Comic' Terns (hatched symbols) by Arctic Skuas 'specialised kleptoparasites' and by terns and Laughing Gulls 'opportunist kleptoparasites'.

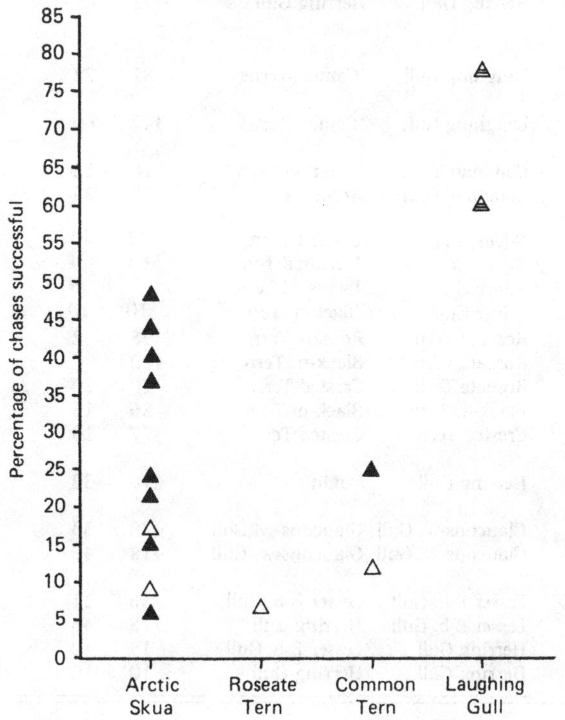

relationships with very low chase success rates are both frequent and apparently important to the host. Roseate Terns force Common Terns at Coquet Island to drop fish in only 7% of attacks, but Dunn (1973) observed 2358 such chases. Hulsman (1976) recorded numerous attempts at

Table 5.2. *The success rates of gulls and terns in forcing hosts to drop food*

Author and locality	Kleptoparasite	Host	Total chases	Percent where food dropped
Arnason & Grant (1978)				
Iceland	Herring Gull	Puffin	29	34
Barnard *et al.* (1982)	Black-h. Gull	Lapwing	108	74
Midlands, England	Black-h. Gull	Golden Plover	46	36
Burger & Gochfeld (1979)				
New York	Ring-b. Gull	Starling	206	41
Corkhill (1983)				
Skomer, Wales	Herring Gull	Puffin	15	27
Dunn (1973)				
Northumberland	Roseate Tern	Common Tern	2358	7
Fuchs (1977)				
Aberdeenshire	Black-h. Gull	Sandwich Tern	238	22
Greig *et al.* (1983)	Herring Gull	Herring Gull	112	85
Durham				
Hatch (1970)				
Maine	Laughing Gull	'Comic' Terns	87	78
Hatch (1975)				
Maine	Laughing Gull	'Comic' Terns	342	60
Hopkins & Wiley (1972)				
Maine	Common Tern	Common Tern	16	12
	Common Tern	Arctic Tern	20	25
Hulsman (1976) One Tree				
Island, Australia	Silver Gull	Crested Tern	3932	13
(at colony)	Silver Gull	L-crested Tern	563	4
(at sea)	Silver Gull	L-crested Tern	—	9
	Silver Gull	Black-n. Tern	10	20
	Roseate Tern	Roseate Tern	98	2
	Roseate Tern	Black-n. Tern	420	1
	Roseate Tern	Crested Tern	10	20
	Black-n. Tern	Black-n. Tern	86	15
	Crested Tern	Crested Tern	377	13
Nettleship (1972)				
Newfoundland	Herring Gull	Puffin	416	32
Rockwell (1982)				
Washington	Glaucous-w. Gull	Glaucous-w. Gull	428	33
Washington	Glaucous-w. Gull	Glaucous-w. Gull	418	49
Verbeek (1977*a*)				
Lancashire	Lesser B-b. Gull	Lesser B-b. Gull	86	21
	Lesser B-b. Gull	Herring Gull	118	43
	Herring Gull	Lesser B-b. Gull	19	10
	Herring Gull	Herring Gull	10	10

kleptoparasitism of Lesser-crested Terns *Sterna bengalensis* by Silver Gulls *Larus novaehollandiae*, and Black-naped Terns *Sterna sumatrana* by Roseate Terns, where success rates were only 4% and 1%, respectively. Maxson & Bernstein (1982) recorded several hundred chases of Blue-eyed Shags *Phalacrocorax atriceps* by South Polar Skuas, although success rates averaged only 2% in one season and 5% in the next. They suggested that many of these chases were exploratory on the part of the skuas. Although generally unsuccessful, chasing could be profitable under particular circumstances and the skuas may have been monitoring profitability in order to be able to take advantage of periods when success rates became higher.

The available evidence suggests that 'specialist' kleptoparasites are not very highly evolved to be successful in their piratical activities. It is perhaps surprising that in most instances chase success rate is as low as 20–25%, but a similar situation is found with most predators, where prey capture occurs in only a small proportion of attacks (Curio, 1976).

Intraspecific kleptoparasitism is often considered to be somewhat different from interspecific kleptoparasitism. Nine cases in Table 5.2 have success rates comparable to those for interspecific kleptoparasitism (intra: median 15%, mean 25%; inter: median 23%, mean 25%). Nor is there any obvious difference between inter- and intraspecific robbery in terms of the proportion of potential hosts which are attacked. In some studies attack rates are high, in some lower (Table 5.3). However, only the 'opportunistic' kleptoparasites indulge regularly in intraspecific kleptoparasitism. Skuas, sheathbills and frigatebirds apparently only steal food from other species.

Barnard & Sibly (1981) provide a general model which describes the changes in profitability of kleptoparasitic feeding within groups. Their model assumed that the pay-off is purely a function of group composition, but in several cases of intraspecific kleptoparasitism, and all interspecific cases, there are clear differences in foraging ability of individuals, which complicate the situation. For example, Greig, Coulson & Monaghan (1983) showed that, within flocks on refuse tips in winter, adult Herring Gulls expend less energy and feed faster than juvenile Herring Gulls. This asymmetry leads to a food shortage for juveniles while surrounded by adults with food. As a result, juveniles rely to a much greater extent on kleptoparasitism than do adults. Verbeek (1977*b*,*c*) found the same where Herring Gulls were feeding intertidally or on refuse tips in Lancashire.

Table 5.3. *Proportions of hosts which suffer attacks from Kleptoparasites*

Kleptoparasite	Victim	Percent: Chased	Percent: Losing food	Reference
Silver Gull	L-crested Tern	100	28	Hulsman, 1976
Silver Gull	Crested Tern	100	12	Hulsman, 1976
Glaucous-w. Gull	Glaucous-w. Gull	72	35	Rockwell, 1982
Laughing Gull	Brown Pelican	67	—	Schnell *et al.*, 1983
Lesser B-b. Gull	Herring Gull	65	32	Verbeek, 1977*a*
Lesser B-b. Gull	Lesser B-b. Gull	36	18	Verbeek, 1977*a*
Herring Gull	Puffin	30	10	Nettleship, 1972
Black-h. Gull	Sandwich Tern	27	6	Fuchs, 1977
Herring Gull	Herring Gull	23	20	Greig *et al.*, 1983
Glaucous-w. Gull	Glaucous-w. Gull	21	5	Rockwell, 1982
Roseate Tern	Black-n. Tern	10	0	Hulsman, 1976
Antarctic Skua	Blue-eyed Shag	10	0	Maxson & Bernstein, 1982
Great Skua	Gannet	9	3	Furness, 1978
Jackdaw	Puffin	7	1	Corkhill, 1973
Arctic Skua	Puffin	6	4	Arnason & Grant, 1978
Great Skua	Great B-b. Gull	6	2	Furness, 1978
Great Skua	Herring Gull	4	2	Furness, 1978
Black-h. Gull	Lapwing	2	1	Barnard & Stephens, 1982
Black-h. Gull	Golden Plover	2	1	Thompson, 1983
Arctic Skua	Arctic Tern	2	1	Furness, 1978
Great Skua	Guillemot	2	< 1	Furness, 1978
Arctic Skua	Puffin	1	< 1	Furness, 1978
Great Skua	Puffin	1	< 1	Furness, 1978
L. Sheathbill	Rockhopper Penguin	1	1	Burger, 1981*a*
Herring Gull	Puffin	1	0	Corkhill, 1973
Black-n. Tern	Black-n. Tern	< 1	0	Hulsman, 1976
Crested Tern	Crested Tern	< 1	0	Hulsman, 1976
Roseate Tern	Black-n. Tern	< 1	0	Hulsman, 1976

Context of kleptoparasitism

All 74 interactions listed in Tables 5.1 and 5.2 involve hosts which occur in large and predictable aggregations, which Brockman & Barnard (1979) suggested would be a precondition for the evolution of kleptoparasitism as a regular feeding method. As seabirds are generally colonial nesters and many are also highly aggregated when feeding (Brown, 1980), it is not surprising that most of the well-known instances of kleptoparasitic feeding involve seabirds as hosts or parasites, or both. With a low chase success rate the prospects would be bleak for a kleptoparasite attempting to make a living by chasing dispersed hosts. In fact, only the Arctic Skua seems to make use of kleptoparasitism at all times of year, and even in winter, when conditions are likely to be least conducive, it can find and associate with large aggregations of hosts (B.L. Furness, 1983).

Large concentrations of hosts are not alone sufficient to allow kleptoparasitic feeding. In all of the documented interactions, kleptoparasitism occurs either where birds are feeding socially at highly aggregated food sources, such as refuse tips or in flocks on pasture fields, or when hosts are carrying food to breeding colonies. In the latter case, most authors note that kleptoparasitic behaviour is rarely, if ever, seen early in the breeding season.

Kleptoparasitism by Lesser Sheathbills begins only when penguins are feeding chicks. Earlier in the year they feed on terrestrial and intertidal invertebrates (Burger, 1981*b*). Lesser Sheathbill breeding is synchronised with penguin breeding, and the locations and sizes of territories they set up clearly foreshadow the availability of this resource later in the season (Burger, 1981*a*). Robbing by Roseate Terns occurs only when hosts are carrying fish to feed young, or in rare instances for courtship feeding (Dunn, 1973; Hulsman, 1976). Presumably Roseate Terns catch fish for themselves at other times of year.

Arctic Skuas arrive in spring at breeding sites in Shetland and Iceland after all other seabirds except terns, which return shortly after the skuas. The Arctic Skua appears to feed mainly by kleptoparasitism even at this early stage, although there have been no studies of the details of their feeding behaviour early in the breeding season. According to Grant (1971), Kittiwakes are the main source of food in spring in Iceland and puffins are only chased once their chicks have hatched. Arctic Skua hatching appears to be synchronised with puffin hatching (Arnason & Grant, 1978), and Arctic Skuas in this area of Iceland and in Shetland have such small nesting territories that, unlike their conspecifics in Alaska or elsewhere in the Arctic interior (Maher, 1974), they would be unable to obtain sufficient food within the territory to allow breeding, or perhaps even survival (Furness, 1980). They therefore depend upon kleptoparasitism from local colonial seabirds. Unlike any other kleptoparasitic species, Arctic Skuas may subsist by kleptoparasitism both during winter and during the breeding season, since they are known to rob food from terns and gulls in the southern hemisphere during the northern winter (Furness, 1983), but it is not clear whether they also need to use other feeding methods at this time or during migration.

No species of bird is an obligate kleptoparasite, and only the Arctic Skua may possibly contain some individuals which subsist throughout the year from kleptoparasitism. Most kleptoparasitic species obtain only a small proportion of their energy requirements in this way, and only for part of the annual cycle (Table 5.4). Furthermore, even those for which kleptoparasitism may be very important usually have alternative feeding methods which are employed by part of the population while others are kleptoparasitic. While during winter some Black-headed Gulls *Larus ridibundus* may obtain all their food from Lapwings *Vanellus vanellus* and Golden Plovers *Pluvialis apricaria*, others nearby may meet their energy needs by following the plough or by catching earthworms for themselves (Kallander, 1977). However, kleptoparasitism is often more common among opportunists when food is scarce (Brockman & Barnard, 1979). This accounts for the greater incidence of kleptoparasitism in juvenile gulls (Verbeek, 1977*b,c*; Greig *et al.*, 1983), and for variations in the amount of kleptoparasitism from year to year (Hays, 1970). Food shortage where other birds have large quantities of food can also explain why Lesser Black-backed Gulls stole 95% of their food and dug up only 5% on refuse sites (Verbeek, 1977*a*) since this species is usually displaced from feeding areas by Herring Gulls and so is denied direct access to the food. Bergman (1982*b*) also pointed out that Lesser Black-backed Gulls are unable to feed in flocks with Herring Gulls where dense flocks hover over the food (as at refuse tips) since they have long narrow wings which impede hovering in flocks, but pre-adapt the species for aerial chases.

Similarly, and perhaps for similar reasons, Arctic Skuas in Shetland do not attempt to splash-dive onto surface shoals of Sandeels *Ammodytes marinus* but steal the same fish from terns, Kittiwakes and auks. This has led some ornithologists to consider the Arctic Skua to be stereotyped into kleptoparasitic behaviour, but it seems more likely that it cannot compete for positions in dense seabird flocks above fish shoals and employs kleptoparasitism instead.

Duffy (1980, 1982) showed that seabirds which were only able to feed

Table 5.4. *Proportion of energy needs which can be obtained by food-robbing by various kleptoparasites*

Kleptoparasite	Percent food from robbing	Locality	Period	Reference
Arctic Skua	100	Shetland	chick-rearing	Furness, 1978
Arctic Skua	100	Iceland	chick-rearing	Arnason & Grant, 1978
Black-headed Gull	100	Midlands	winter	Barnard *et al.*, 1982
Black-headed Gull	100	S. Sweden	winter	Kallander, 1977
Lesser Black-backed Gull	95	Lancashire	all seasons	Verbeek, 1977*a*
Lesser Sheathbill	91	Marion I.	chick-rearing	Burger, 1981*a*
Greater Sheathbill	90	Signy I.	chick-rearing	Jones, 1963
Glaucous-winged Gull	35	MacKay Wash.	winter	Rockwell, 1982
Herring Gull (1 yr)	33	Durham	winter	Greig *et al.*, 1983
Herring Gull	23	Lancashire	all seasons	Verbeek, 1977*a*
Frigatebirds	< 20?	Aldabra	summer	Diamond, 1975
Frigatebirds	< 20?	Galapagos	summer	Nelson, 1976
Frigatebirds	< 20?	Christmas I.	summer	R. W. Schreiber, pers. comm.
Herring Gull (adult)	11	Durham	winter	Greig *et al.*, 1983
Great Skua	< 10	Foula	summer	Furness & Hislop, 1981
Glaucous-winged Gull	5	Mudd Bay, Wash.	winter	Rockwell, 1982
Magnificent Frigatebird	0	Barbuda	summer	Diamond, 1973

at the sea surface were more likely to exhibit kleptoparasitism than those which could dive to obtain food at depth. This is a special case of the food-availability hypothesis, since food shortage is potentially more acute for seabirds with restrictions on their foraging depths.

Choice of victim

Food visibility

Since the energy cost of an unsuccessful chase may be considerable, it is reasonable to expect kleptoparasites to be highly discriminating between potential hosts with and without food to surrender. This would suggest that species which carry food in the bill rather than swallowing it would be selected for attack, and Brockman & Barnard (1979) stated categorically that 'only individuals with visible food in the bill are attacked or pursued by a kleptoparasite'. Ten of the 74 interactions listed in Tables 5.1 and 5.2 involve hosts which swallow food for their chicks and never carry it in the bill. These hosts are Gannet, Red-footed Booby *Sula sula*, Kittiwake and Blue-eyed Shag, chased, respectively, by Great Skuas, frigatebirds, both Arctic Skuas and Great Skuas, and South Polar Skuas. In addition, Great Skuas chase Herring Gulls, Great Black-backed Gulls *Larus marinus* and Arctic Skuas at Foula (Furness, 1978), species which all regurgitate food to their young but are much less abundant at this colony than auks, so that fewer than 10 chases of each of these species were recorded in that study. In fact, the proportions of available Gannets, *Larus* gulls and Arctic Skuas chased by Great Skuas are rather high (Table 5.3), suggesting that these species are nevertheless favoured victims.

Common Guillemots *Uria aalge* carry fish to their chick inside the bill and throat, and thus hidden from view unless the fish is large enough for the tail to protrude, whereas Atlantic Puffins, Razorbills *Alca torda* and Black Guillemots *Cepphus grylle* all carry fish dangling across the bill and clearly visible to kleptoparasites. One might expect that guillemots would be chased less often or less successfully than the other auks if visibility of food were an important factor in determining kleptoparasitic behaviour, but the proportions of incoming puffins, Razorbills and guillemots chased by Arctic Skuas or Great Skuas at Foula were very similar (Table 5.3) while Black Guillemots were rarely, if ever, chased. Proportions of chases which were successful (Table 5.1) suggest that puffins are most easily robbed, but Great Skuas had a higher success against guillemots than against Razorbills (29% *versus* 18%) and not lower as would be predicted if fish visibility were critically important.

The success rate of Arctic Skuas at Heddlicliff, Foula, in robbing

Kittiwakes is consistently higher than that against puffins, guillemots or Razorbills (Table 5.1), yet Kittiwakes regurgitate to their chicks, and many of the adults close to the colony which might be chased by an indiscriminating skua would be non-breeders or off-duty adults with no food in their crops.

Taking the specific case of Great Skua chases at Foula reported by Furness (1978) we find that 93 chases were of hosts which swallowed food and 30% were successful, 113 chases were of hosts which carry food inside the bill and 29% were successful, and 261 were of hosts which carry fish across the bill and fully visible, and 35% of these chases were successful. We would expect the last class to be the largest, since most seabirds at Foula behave this way. Success rates differ little between groups, suggesting that Great Skuas find it almost as easy to discriminate between hosts with and without food whether that food is swallowed or held visible in the bill.

Perhaps it is unreasonable to expect that kleptoparasites should be unable to discriminate from a distance between hosts with or without food in their stomach, since this is likely to influence characteristics of the host's flight and behaviour, but this ability, which 'specialist' kleptoparasites clearly do have, may not be shared by 'opportunist' kleptoparasites such as gulls and terns. None of the interactions listed in Table 5.2 involves a gull or tern pirating from birds which have swallowed food. Hulsman (1976) recorded kleptoparasitism among Silver Gulls, Greater-crested *Sterna bergii*, Lesser-crested, Roseate, Black-naped and Bridled *S. anaethetus* Terns and remarked on the fact that, in marked contrast to the other species, the Bridled Tern was rarely harrassed and only once robbed of food, because, unlike the other terns, it fed its mate and chicks by regurgitation.

Although comparisons of chase success rates do not show any difference between 'specialist' and 'opportunist' kleptoparasites, this distinction may still be valuable. Skuas and frigatebirds appear to differ from gulls and terns in being able to assess whether or not birds which regurgitate to feed chicks are carrying food, and do not often chase conspecifics for food, which suggests a more highly evolved form of kleptoparasitism.

Food suitability

Although penguins robbed by sheathbills feed their chicks mainly on slightly digested Antarctic Krill *Euphausia superba*, the food is passed over in the form of large, soft balls (Burger, 1981*a*) which are intercepted by the kleptoparasites. The hosts listed in Tables 5.1 and 5.2 all lost large, discrete, food items to their kleptoparasites, with one possible exception. Kittiwakes robbed by Arctic Skuas or Great Skuas at Foula often regurgitated a well-digested mush of Sandeel or zooplankton which it was impossible for the kleptoparasite to handle. Whether the reward was complete fish or a useless liquid presumably depended on how long after food capture the Kittiwake arrived back at the colony. In spite of the high success rate of chases of Kittiwakes, only a very small proportion of returning birds are usually chased (Table 5.3), suggesting that skuas avoid wasting energy where the rewards may be in a form that they cannot utilize.

Victim suitability

Kushlan (1978) found that small species of heron or egret are more vulnerable to robbery than larger species, so that kleptoparasitism in that community tended to be size-related. Duffy (1980, 1982) suggested that this is not the case with seabirds. In Tables 5.1 and 5.2, 21 species-pairs consisted of a victim heavier than the kleptoparasite, 12 consisted of victims and kleptoparasites of equal masses, 33 consisted of victims lighter than the kleptoparasite, and 10 consisted of intraspecific interactions. While the trend is for victims to be smaller than the kleptoparasite, this is not statistically significant, and is often reversed.

Kleptoparasites might do best to chase the species which carry the largest food items but can still be robbed. Guillemots carry much larger fish than do puffins. Great Skuas' chase success rates are similar for these two species of victim (Table 5.1), so guillemots should be preferred to puffins as victims. In fact, this does not appear to be the case, although only a very small proportion of the incoming adults of either species is attacked (Table 5.3). Arctic Skuas chase guillemots less often than they chase puffins, and have a lower success rate, so that the potential advantage of selecting a species carrying larger prey is outweighed in this case by a decreased chance of successful kleptoparasitism.

Importance of kleptoparasitism to the kleptoparasite

There is a suggestion that Arctic Skua–seabird, sheathbill–penguin and Black-headed Gull–plover interactions are sufficiently highly evolved that they can persist without damage to the host but be able to sustain the parasite. None of these interactions deprives the hosts of as much as 1% of their food and none involves the attack on more than 2% of the potential victims (Table 5.3), except the case of Arctic Skuas robbing puffins in south Iceland. There, in 1973, 6% of incoming puffins were chased and 4% were robbed of food, but the Arctic Skuas had an abnormally low breeding success that year, when Arctic Terns unusually did not breed in the area (Arnason & Grant, 1978). In most years these skuas also chase Arctic Terns and this might normally have the effect of reducing their impact on puffins below that observed in 1973.

In contrast to the consistently small impact of these 'specialist' kleptoparasites on their hosts, some 'opportunist' kleptoparasites chase and rob high proportions of hosts and seem likely to cause host population instability. Hulsman (1976) found that Silver Gulls rob Lesser-crested Terns of about one-quarter of the fish they catch and attempt to bring back to chicks (Table 5.3). Herring Gulls lost one-third of their food to Lesser Black-backed Gulls (Verbeek, 1977*a*) and Atlantic Puffins at Great Island, Newfoundland, lost about one-tenth of the fish brought for chicks to Herring Gulls (Nettleship, 1972). These high rates of parasitism may not be stable over evolutionary time since they are likely to have a harmful impact on the host population. There is also reason to believe that they have not persisted at this level for long since all these gull populations have increased considerably over the last few decades so that kleptoparasitic interactions will certainly have increased. Bergman (1982*b*) found that the feeding behaviour of *Larus* gulls in the Finnish archipelago has altered as their populations changed. As numbers of Herring Gulls and Lesser Black-backed Gulls increased they changed habits, from single pairs defending a feeding range around the breeding site to colonial nesting and social feeding. With increased social feeding and food shortage, kleptoparasitism became noticeable.

Evolution of kleptoparasitism

We can trace a possible sequence of events in the evolution of kleptoparasitic feeding. A seabird breeding in isolated pairs and feeding opportunistically within a defended area around the nest increases in numbers so that social nesting and feeding become necessary. With further population increase and a resulting shortage of food, robbing other birds carrying food in the bill becomes a more profitable strategy for individuals unable to compete for food, and may increase through social learning, particularly if the species is structurally pre-adapted for agile flight. An 'evolutionary arms race' (see below) may then lead to the kleptoparasite learning to rob from a variety of species and to be able to select victims with food, even when this is not directly visible. The kleptoparasite may then adjust its breeding season, nesting patterns and migrations to fit with those of the host, and ultimately may become dependent on the host population for successful breeding.

Selection for kleptoparasitic skills will begin as soon as the habit starts to provide food for birds which are food-stressed. Once the kleptoparasites begin to steal enough food to affect the host's breeding success or survival the host should evolve defences against the kleptoparasite and an 'asymmetric evolutionary arms race' (Dawkins & Krebs, 1979) will have begun. The arms race will involve attempts by the kleptoparasite to maximise net energy gain, which will generally mean to maximise chase success rate, while the host will attempt to minimise net energy loss to the parasites, which will generally mean to minimise chase success rate and the amount of chasing. The following factors may affect the efficacy of kleptoparasitism and are therefore potentially available to be manipulated by the kleptoparasite or host.

Chase success rate increases with chase duration in many interactions (e.g. Andersson, 1976; Furness, 1978). An ability of kleptoparasites to fly faster than hosts, accelerate faster and catch up with hosts, or to be more agile in the air is likely to improve chase success, since fewer victims will escape. High speed or acceleration requires high power output and high wing-flapping speed combined with low body mass and low parasite drag (Pennycuick, 1975). These would require: streamlining of the body, strong tendon insertions, large pectoralis muscles, short wings, and large wing area (therefore a low aspect ratio). Anatomical comparisons between skuas and gulls might be used to test these possibilities. Aerial agility would be increased by reducing wing loading, and by increasing tail size. What little evidence exists suggests that wing length and aspect ratio in frigatebirds and the Arctic Skua are greater than in comparable species (Jouventin & Mougin, 1981), rather than smaller as predicted above. This suggests that these birds are adapted for slow flight, and possibly soaring. Slow flight may be important when frigatebirds are searching for flyingfish or Arctic Skuas are hunting in terrestrial environments. At present there is little to support Brockman & Barnard's assertion that anatomical adaptations to aid kleptoparasitism have taken place.

Since opportunistic kleptoparasites appear to avoid chasing species which swallow food rather than carrying it in the bill, one might expect the latter behaviour to be reduced in populations subject to attacks by kleptoparasites. I know of no cases where both methods are used by a species and thus no comparisons can be made between populations with and without kleptoparasites. Whether or not regurgitation of food to chicks evolved in response to kleptoparasitism is difficult to assess. It seems likely that other factors such as prey size, stomach volume, anatomical structure and foraging range are important.

One of the main determinants of the outcome of attacks is how quickly the host responds to the approaching kleptoparasite (Furness, 1978). Auks can fly faster than skuas in level flight, but chases are initiated by the skua stooping from above so that the auk has to accelerate fast before the skua reaches it if it is to keep its fish. It can only do so if it sees the skua in time. This is influenced by physical conditions such as visibility (Furness, 1978). Kleptoparasites increase their success rate by chasing more in conditions of poor visibility (Andersson, 1976). Chase success may be increased by kleptoparasites adopting a plumage which hosts find difficult to detect. Andersson (1976) suggests that the dark plumage of Arctic Skuas which feed by kleptoparasitism may be a form of aggressive camouflage. Certainly, in areas where the light phase predominates the species is generally a terrestrial predator rather than a kleptoparasite. Dark plumage of frigate-birds and *Catharacta* skuas may also be cryptic since most other seabirds are light in colour, but this interpretation is disputed (Nelson, 1980).

Hosts may be able to improve their detection of kleptoparasitism by forming a search image, and Arnason (1978) has suggested that the polymorphic plumage of the Arctic Skua results from apostatic selection since the unrecognised rare morph will have a higher chase success than the recognised common morph. Furness & Furness (1980) presented data showing no significant differences in chase success rates in relation to plumage colour of Arctic Skuas but Rohwer (in press) points out that this result is compatible with the theory and that the theory has not yet been tested critically.

Grant (1971) found that puffins turned away from the colony more often when the rate of attacks by skuas was high, implying that they were able to assess the risk of attack and respond accordingly. Furness (1978) found that both puffins and Kittiwakes were more likely to avoid attack when large numbers of Arctic Skuas were present. The victims can further reduce the likelihood of losing food by returning to the colony in groups, thus swamping the kleptoparasite (Grant, 1971).

Puffins are more vulnerable to kleptoparasites when nesting on flat ground than on slopes (Nettleship, 1972), when nesting at inland colonies where diving into the sea to avoid chases is impossible (Arnason & Grant, 1978; Furness, 1978), and when nesting at height above sea level

(Furness, 1978). Skuas selectively attack hosts in the more vulnerable situations which increases their success rate but reduces the relative breeding performance of hosts in poor habitat (e.g. Nettleship, 1972).

Kleptoparasites can improve their rate of food gain by learning to detect individuals with food in the stomach. At a mixed-species colony this will not only make available hosts which may provide higher food rewards (such as Gannets chased by Great Skuas), but also reduce the patrolling time required before another victim is located (Furness, 1980). Compared to the duration of chases, patrolling times are very long and may strongly influence the profitability of this feeding method.

Probability of success may be increased if several kleptoparasites chase a single victim, but reward per kleptoparasite usually decreases (Hatch, 1970, 1975; Andersson, 1976; Furness, 1978; Arnason & Grant, 1978), indicating that kleptoparasites join chases to reduce patrolling time rather than to increase yield per chase to themselves.

Kleptoparasites may be able to increase the likelihood of success by chasing birds at particular distances from the colony (Arnason & Grant, 1978) and by chasing birds flying towards, rather than away from, the colony (Maxson & Bernstein, 1982, but see also Arnason & Grant, 1978). Most attacks by skuas, frigatebirds and terns occur close to the host's colony, where the density of victims is likely to be highest so patrol times shortest, but kleptoparasites sometimes choose to fly out to host's feeding areas (Hulsman, 1976; personal observation), where they may have a higher rate of successful chasing (Hulsman, 1976).

Hosts may reduce the chances of being attacked by selecting prey species which are less conspicuous or less profitable (Fuchs, 1977) or by selecting smaller sized prey than they would in the absence of kleptoparasites (Thompson, 1983). Kleptoparasites often selectively attack hosts with large prey (Hopkins & Wiley, 1972; Hulsman, 1976; Fuchs, 1977; Arnason & Grant, 1978; Thompson, 1983) but exceptions have been reported. Roseate Terns stole fish which were similar in size to those they caught for themselves but smaller than the average carried by the hosts (Dunn, 1973) and Black-headed Gulls selected Sandwich Terns *Sterna sandivicensis* with large Sandeels but neglected birds with gadoids which, although shorter, were heavier (Fuchs, 1977). In this case the gulls appeared to be selecting for fish length or visibility rather than mass.

Few chases result in physical contact with the host, but very rarely a chase by a skua may lead to the victim being caught and killed (personal observation) and this, or physical removal of the food from the host's bill, must be the main threat which may cause hosts to release food. Maxson

& Bernstein (1982) found that physical contact, particularly if directed at the underparts, increased the chance of Blue-eyed Shags giving up food to South Polar Skuas. There is slight evidence to suggest that the 'willingness' of the host to give up food may vary. Hulsman (1976) found that chases of Lesser-crested Terns by Silver Gulls were twice as likely to be successful at the feeding area at sea than at the colony. Significantly more chases of Arctic Terns returning to the colony by Arctic Skuas at Foula were successful in 1975 and 1976 than in 1978, 1979 or 1980 (Table 5.1). Chase success rate appears to correlate both with the number of tern chicks fledged and with the breeding success of the colony (Furness, 1983) which may imply that in years when food was abundantly available close to the colony and breeding numbers and success were high, terns were more willing to lose fish than in years when food was scarce and chick survival was at risk. In the latter case terns may have been trading off greater risk of physical attack for increased chances of keeping their chicks alive.

Conclusions

A detailed study of frigatebird kleptoparasitism is needed to determine the importance of this feeding method and the way frigatebirds interact with their hosts.

We may tentatively classify kleptoparasites on a range from 'opportunists', which rob only hosts that carry visible food in the bill, to 'specialists', which show a battery of adaptations, including the ability to rob birds carrying food in the stomach, to prolong chases, to select hosts in situations where chase success is highest, to respond to changes in host availability and prey selection, to adapt their phenology to that of the hosts and, possibly, to be cryptically coloured, polymorphic and perhaps anatomically adapted for kleptoparasitism. Several of these, particularly the three last mentioned, require further study.

References

Andersson, M. (1976). Predation and kleptoparasitism by skuas in a Shetland seabird colony. *Ibis*, **118**, 208–17.

Arnason, E. (1978). Apostatic selection and kleptoparasitism in the Parasitic Jaeger. *Auk*, **95**, 377–81.

Arnason, E. & Grant, P.R. (1978). The signficance of kleptoparasitism during the breeding season in a colony of Arctic Skuas *Stercorarius parasiticus* in Iceland. *Ibis*, **120**, 38–54.

Ashmole, N.P. (1971). Sea bird ecology and the marine environment. In *Avian biology*, vol. 1, ed. D.S. Farner & J.R. King, pp. 223–86. New York: Academic Press.

Barnard, C.J. & Sibly, R.M. (1981). Producers and scroungers: a general model and its application to feeding flocks of House Sparrows *Passer domesticus* L. *Anim. Behav.*, **29**, 543–50.

Barnard, C.J. & Stephens, H. (1981). Prey size selection by Lapwings in Lapwing/gull associations. *Behaviour*, **77**, 1–22.

Barnard, C.J., Thompson, D.B.A. & Stephens, H. (1982). Time budgets, feeding efficiency and flock dynamics in mixed species flocks of Lapwings, Golden Plovers and gulls. *Behaviour*, **80**, 43–69.

Bergman, G. (1982*a*). Inter-relationships between ducks and gulls. In *Managing wetlands and their birds*, ed. D.A. Scott, pp. 241–8. Slimbridge: Int. Wildfowl Res. Bureau.

Bergman, G. (1982*b*). Population dynamics, colony formation and competition in *Larus argentatus*, *fuscus* and *marinus* in the archipelago of Finland. *Ann. Zool. Fennici*, **19**, 143–64.

Brockman, H.J. & Barnard, C.J. (1979). Kleptoparasitism in birds. *Anim. Behav.*, **27**, 487–514.

Brown, R.G.B. (1980). Seabirds as marine animals. In *Behavior of marine animals*, vol. 4, *Marine birds*, ed. J. Burger, B.L. Olla & H.E. Winn, pp. 1–39. New York: Plenum.

Burger, A.E. (1981*a*). Time budgets, energy needs and kleptoparasitism in breeding Lesser Sheathbills *Chionis minor*. *Condor*, **83**, 106–12.

Burger, A.E. (1981*b*). Food and foraging behaviour of Lesser Sheathbills at Marion Island. *Ardea*, **69**, 167–80.

Burger, J. & Gochfeld, M. (1979). Age differences in Ring-billed Gull kleptoparasitism on Starlings. *Auk*, **96**, 806–8.

Corkhill, P. (1973). Food and feeding ecology of Puffins. *Bird Study*, **20**, 207–20.

Curio, E. (1976). *The ethology of predation*. Berlin: Springer-Verlag.

Dawkins, R. & Krebs, J.R. (1979). Arms races between and within mixed species. *Proc. R. Soc. Lond., B.*, **205**, 489–511.

Diamond, A.W. (1973). Notes on the breeding biology and behaviour of the Magnificent Frigatebird. *Condor*, **75**, 200–9.

Diamond, A.W. (1975). Biology and behaviour of frigatebirds *Fregata* spp. on Aldabra Atoll. *Ibis*, **117**, 302–23.

Duffy, D.C. (1980). Patterns of piracy by Peruvian seabirds: a depth hypothesis. *Ibis*, **122**, 521–5.

Duffy, D.C. (1982). Patterns of piracy in the seabird communities of the Galapagos Islands and southern Africa. *Cormorant*, **10**, 71–80.

Dunn, E.K. (1973). Robbing behaviour of Roseate Terns. *Auk*, **90**, 641–51.

Fuchs, E. (1977). Kleptoparasitism of Sandwich Terns *Sterna sandvicensis* by Black-headed Gulls *Larus ridibundus*. *Ibis*, **119**, 183–90.

Furness, B.L. (1980). Territoriality and feeding behaviour in the Arctic Skua *Stercorarius parasiticus* (L.). Ph.D. thesis, University of Aberdeen.

Furness, B.L. (1981). Feeding strategies of Arctic Skuas *Stercorarius parasiticus* at Foula, Shetland, Scotland. In *Proceedings of the symposium on birds of the sea and shore, 1979*, ed. J. Cooper, pp. 89–98. Cape Town: African Seabird Group.

Furness, B.L. (1983). The feeding behaviour of Arctic Skuas *Stercorarius parasiticus* wintering off South Africa. *Ibis*, **125**, 245–51.

Furness, B.L. & Furness, R.W. (1980). Apostatic selection and kleptoparasitism in the Parasitic Jaeger: a comment. *Auk*, **97**, 832–6.

Furness, R.W. (1977). Effects of Great Skuas on Arctic Skuas in Shetland. *British Birds*, **70**, 96–107.

Furness, R.W. (1978). Kleptoparasitism by Great Skuas *Catharacta skua* Brunn. and Arctic Skuas *Stercorarius parasiticus* L. at a Shetland seabird colony. *Anim. Behav.*, **26**, 1167–77.

Furness, R.W. (1983). *The birds of Foula*. Ambleside: Brathay Hall Trust.

Furness, R.W. & Hislop, J.R.G. (1981). Diets and feeding ecology of Great Skuas during the breeding season in Shetland. *J. Zool., Lond.*, **195**, 1–23.

Gotmark, F., Andersson, M. & Hilden, O. (1981). Polymorphism in the Arctic Skua *Stercorarius parasiticus* in NE Norway. *Ornis Fennica*, **58**, 49–55.

Grant, P.R. (1971). Interactive behaviour of puffins *Fratercula arctica* L. and skuas *Stercorarius parasiticus* L. *Behaviour*, **40**, 263–81.

Greig, S., Coulson, J.C. & Monaghan, P. (1983). Age-related differences in foraging success in the Herring Gull *Larus argentatus*. *Anim. Behav.*, **31**, 1237–43.

Hatch, J.J. (1970). Predation and piracy by gulls at a ternery in Maine. *Auk*, **87**, 244–54.

Hatch, J.J. (1975). Piracy by Laughing Gulls *Larus atricilla*: an example of the selfish group. *Ibis*, **117**, 357–65.

Hays, H. (1970). Common Terns pirating fish on Great Gull Island. *Wilson Bull.*, **82**, 99–100.

Hopkins, C.D. & Wiley, R.H. (1972). Food parasitism and competition in two terns. *Auk*, **89**, 583–94.

Hulsman, K. (1976). The robbing behaviour of terns and gulls. *Emu*, **76**, 143–9.

Jones, N.V. (1963). The Sheathbill, *Chionis alba* (Gmelin), at Signy Island, South Orkney Islands. *Bull. Br. Antarct. Surv.*, **2**, 53–71.

Jouventin, P. & Mougin, J.-L. (1981). Les stratégies adaptives des oiseaux de mer. *Rev. Ecol.*, **35**, 217–72.

Kallander, H. (1977). Piracy by Black-headed Gulls on Lapwings. *Bird Study*, **24**, 186–94.

Kushlan, J.A. (1978). Non-rigorous foraging by robbing egrets. *Ecology*, **59**, 649–53.

Maher, W.J. (1974). Ecology of Pomarine, Parasitic and Long-tailed Jaegers in Northern Alaska. *Pacif. Coast Avifauna*, **37**, 1–148.

Maxson, S.J. & Bernstein, N.P. (1982). Kleptoparasitism by South Polar Skuas on Blue-eyed Shags in Antarctica. *Wilson Bull.*, **94**, 269–81.

Meinertzhagen, R. (1959). *Pirates and predators*. London: Oliver & Boyd.

Nelson, J.B. (1976). The breeding biology of frigatebirds – a comparative review. *Living Bird*, **14**, 113–55.

Nelson, J.B. (1980). *Seabirds, their biology and ecology*. London: Hamlyn.

Nettleship, D.N. (1972). Breeding success of the Common Puffin *Fratercula arctica* L. on different habitats at Great Island, Newfoundland. *Ecol. Monogr.*, **42**, 239–68.

Pennycuick, C.J. (1975). Mechanics of flight. In *Avian biology*, vol. 5., ed. D.S. Farner & J.R. King, pp. 1–75. New York: Academic Press.

Rockwell, E.D. (1982). Intraspecific food robbing in Glaucous-winged Gulls. *Wilson Bull.*, **94**, 282–8.

Rohwer, S. (In press). Formalizing the Avoidance-Image Hypothesis: critique of an earlier prediction. *Auk*.

Schnell, G.D., Woods, B.L. & Ploger, B.J. (1983). Brown Pelican foraging success and kleptoparasitism by Laughing Gulls. *Auk*, **100**, 636–44.

Schreiber, R.W. & Hensley, D.A. (1976). The diet of *Sula dactylatra*, *Sula sula* and *Fregata minor* on Christmas Island, Pacific Ocean. *Pacific Sci.*, **30**, 241–8.
Stonehouse, B. & Stonehouse, S. (1963). The frigatebird *Fregata aquila* of Ascension Island. *Ibis*, **103b**, 409–22.
Taylor, I.R. (1979). The kleptoparasitic behaviour of the Arctic Skua *Stercorarius parasiticus* with three species of tern. *Ibis*, **121**, 274–83.
Thompson, D.B.A. (1983). Prey assessment by plovers (Charadriidae): net rate of energy intake and vulnerability to kleptoparasites. *Anim. Behav.*, **31**, 1226–36.
Verbeek, N.A.M. (1977*a*). Interactions between Herring and Lesser Black-backed Gulls feeding on refuse. *Auk*, **94**, 726–35.
Verbeek, N.A.M. (1977*b*). Age differences in the digging frequency of Herring Gulls on a dump. *Condor*, **79**, 123–5.
Verbeek, N.A.M. (1977*c*). Comparative feeding behaviour of immature and adult Herring Gulls. *Wilson Bull.*, **89**, 415–21.

6

The food and feeding ecology of penguins

J. P. CROXALL
British Antarctic Survey, Natural Environment Research Council, High Cross, Madingley Road, Cambridge CB3 OET, UK
and

G. S. LISHMAN
British Antarctic Survey, Natural Environment Research Council, High Cross, Madingley Road, Cambridge CB3 OET, UK
and
Edward Grey Institute of Field Ornithology, Department of Zoology, University of Oxford, South Parks Road, Oxford OX1 3PS, UK

Introduction

Penguins have developed adaptations to a marine existence to a greater extent than any other group of birds. Being flightless, their wings have become wholly adapted, as stiff yet flexible paddles, powered by highly developed breast musculature, for propulsion under water; their body is nearly as dense as water, which probably facilitates diving; and their feet, which are set well back on the body act, with the tail, as rudders. Overall, their hydrodynamic design closely approaches the optimum for inanimate objects of their shape and size (Nachtigall & Bilo, 1980) but their stocky legs permit an upright stance and they are relatively agile on land.

There are about 17 species of penguins, confined to the cold- and cool-water systems of the southern hemisphere. They breed north to, or near, the equator on the Galapagos Islands (Galapagos Penguin *Spheniscus mendiculus*) and the coast of Peru (Humboldt Penguin *S. humboldti*), areas influenced by the cool Humboldt Current. The main species diversity is in the New Zealand region (with six species including three allopatric species of crested penguins *Eudyptes*) but the greatest concentrations of birds are found at sub-Antarctic islands and on the Antarctic Continent, especially in the Antarctic Peninsula area. Much information on penguin distribution and numbers has recently been collated by Wilson (1983) for the Antarctic and in various chapters in Croxall, Evans & Schreiber (1984) and is summarised in Fig. 6.1.

Particularly in the Southern Ocean, the abundance of penguins, coupled with their large biomass and high energy demands (swimming being more costly than the gliding flight typical of many of the other seabirds of the region), means that they play a major role in the avian community as consumers of marine resources. Penguins may comprise 90% of the bird

Fig. 6.1. The distribution and estimated abundance of penguins.

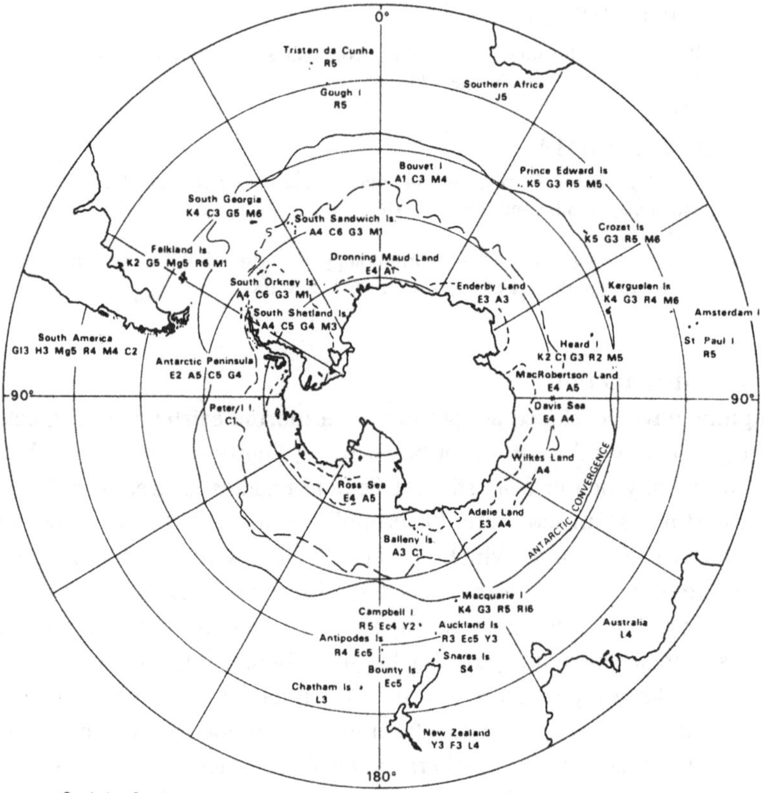

Symbol	Species
E	Emperor
K	King
A	Adelie
C	Chinstrap
G	Gentoo
R	Rockhopper
M	Macaroni
RI	Royal
S	Snares Crested
Ec	Erect-crested
F	Fiordland
Y	Yellow-eyed
L	Little blue
Mg	Magellanic
H	Humboldt
Gl	Galapagos
J	Jackass

Mean pack ice limits

— — — winter

- - - - - - - summer

Symbol	Estimated Population Size (Pairs)
1	10 - 100
2	100 - 1000
3	1000 - 10 000
4	10 000 - 100 000
5	100 000 - 1000 000
6	1000 000 +

biomass in Antarctic regions (Mougin & Prévost, 1980) and are estimated to account for 76% of the food intake by birds in the Scotia Sea (Croxall, Prince & Ricketts, 1985) and 53% at South Georgia, despite forming only 13% of breeding numbers there (Croxall, Ricketts & Prince, 1984; see also Chapter 15).

Given their potential importance in these (and many other) southern hemisphere areas, until very recently our information on their diet and feeding ecology was surprisingly sparse. Over a decade elapsed between the pioneer quantitative studies on Jackass Penguins *Spheniscus demersus* (Davies, 1955; Rand, 1960) and Adélie Penguins *Pygoscelis adeliae* (Emison, 1968) and the several studies started since the mid-1970s (Croxall & Prince, 1980*a*; Volkman, Presler & Trivelpiece, 1980; Montague, 1982; Lishman, 1985*a*; Horne, in press). Our knowledge of their feeding ecology is even more limited, as penguins are difficult to observe at sea and remote techniques with which to study their diving and at-sea behaviour are very recent developments (Kooyman, Billups & Farwell, 1983; Wilson & Bain, 1984*a,b*; Trivelpiece *et al.*, in press). Only for Jackass Penguins (Rand, 1960; Wilson, 1985) and Little Penguins *Eudyptula minor* (Montague, 1982) do we know anything about food and feeding ecology in the non-breeding season. This chapter reviews the available information on all these topics and also examines the role of diet in the ecological segregation of sympatric penguins in the breeding season.

Composition of the diet

Information from dietary studies (Appendix 6.1 and Table 6.1) indicates that a quantitative characterisation of the diet is available for eight species (nine, if the brief report on the Magellanic Penguin *Spheniscus magellanicus* samples is included), but there are only incomplete or anecdotal data for the rest. Information is particularly deficient for the New Zealand species and the two tropical *Spheniscus* species.

There are few records of penguins taking prey other than fish, squid or crustaceans. The smaller penguins breeding at sites south of the Antarctic Convergence feed mainly on crustaceans, which are almost exclusively the swarming euphausiids known collectively as krill; fish are taken frequently only by Adélie Penguins breeding on the Antarctic Continent and by Gentoo Penguins; squid are very rare. In contrast, the large *Aptenodytes* penguins in the same area take fish (and squid) and any crustaceans are probably derived from the stomach contents of these prey. For penguins breeding at sites at or north of the Antarctic Convergence, but south of the Subtropical Convergence, there is no systematic pattern to the

Table 6.1. *General characteristics of composition (% by weight) of penguin diets. (For details see Appendix 6.1; + + +: major component; + +: significant component; +: minor component or trace; A.C.: approximate position of Antarctic Convergence; S.T.C.: approximate position of Sub-tropical Convergence.)*

Species		Lat (°S)	Diet			
			Crustacean	Fish	Squid	
Emperor	*Aptenodytes forsteri*	67	2	95	3	
King	*A. patagonicus*	47	+	86	14	
Adélie	*Pygoscelis adeliae*	77	68	32	—	
		72	73–98	2–27	—	
		60–62	99	1	—	
Chinstrap	*P. antarctica*	60–62	100	—	—	
Gentoo	*P. papua*	62	85	15	—	
		54	68	32	—	
		47[a]	*c.* 30	*c.* 70	+	
Macaroni	*Eudyptes c. chrysolophus*	61	75	25	—	
		54	98	2	+	
						← A.C.
Royal	*E. c. schlegeli*	54	20	58	22	
S. Rockhopper	*E. c. chrysocome*	54	77	17	6	
		52	45	2	53	
Snares Crested	*E. robustus*	48	+ + +	?	+ + +	
Erect-crested	*E. sclateri*	45	?	+ + +	+ + +	
Fiordland	*E. pachyrhynchus*	45	+ + +	?	+ + +	
N. Rockhopper	*E. chrysocome moseleyi*	41	39+	+ + ?	+ + ?	
		37	35	10	50	
						← S.T.C.
Yellow-eyed	*Megadyptes antipodes*	45	+ ?	50	+ + + ?	
Little	*Eudyptula minor*	38	—	*c.* 90	*c.* 10	
Magellanic	*Spheniscus magellanicus*	45	10	90	+ + ?	
Jackass	*S. demersus*	33	1	94	5	
Humboldt	*S. humboldti*	13	—	+ + + ?	+ ?	
Galapagos	*S. mendiculus*	1	—	+ + + ?	+ ?	

[a] NB North of Antarctic Convergence.

importance of the three main dietary categories but squid and/or fish are usually at least as important as crustaceans (which are still largely euphausiids). Often little is known of the identity of the fish and squid prey. Further north the Little Penguin and the *Spheniscus* species all prey chiefly on abundant shoaling fish, typically anchovies *Engraulis* and sardines *Sardinops* (and *Ramnogaster*), but squid may sometimes be locally important, perhaps especially for Magellanic Penguins (see Boswall & MacIver, 1975).

Crustaceans

Euphausiid crustaceans are the main species recorded commonly in penguin stomachs. They are undoubtedly favoured prey because they are typically swarming species and are amongst the largest species in the zooplankton of surface waters. That they are especially favoured in the Antarctic presumably reflects their abundance there and the scarcity of pelagic schooling fish (Everson, 1981) and dense squid aggregations. Further north, euphausiids are smaller and less often occur in large swarms; decapods are more common and may sometimes be locally important to penguins.

Penguins eat a variety of euphausiids (Table 6.2), those taken at any site usually according well with what is known of their distribution and abundance in the area from sampling by nets. Penguins are apparently well able to cope with prey as small as about 15–20 mm (e.g. large amphipods or small euphausiids like *Thysanoessa* spp., *Euphausia lucens* and *E. crystallorophias*), presumably provided they are sufficiently abundant to compensate for the reduced individual return from capturing them. Where several euphausiid species occur together it is not always the largest that is taken most commonly (e.g. *Euphausia lucens* (11–22 mm) rather than *E. vallentinii* (17–22 mm) at the Falkland Islands, but *E. superba* (35 + mm) rather than *E. frigida* (28–38 mm) and *E. triacantha* (18–21 mm) at South Georgia). This could reflect differences in relative abundance or in tendency to form aggregations.

E. crystallorophias is the main species found in the shelf waters around the Antarctic Continent and is the staple diet of Adélie Penguins in the southern Ross Sea (Emison, 1968). Adélie Penguins caught at sea over the slope outside the shelf, however, contain mainly *E. superba* (Ainley, O'Connor & Boekelheide, 1984) and this is the major component of the diet of birds breeding at Cape Hallett in the northern Ross Sea (L. Logan, unpublished; F.C. Kinsky & P. Ensor, unpublished), where the shelf edge is only about 50 km from the colony. *E. crystallorophias* occurs north to 65° S, overlapping the range of the much larger Antarctic Krill *E. superba*

Table 6.2. *Euphausiid crustaceans eaten by penguins*

Species	Adult size (mm)[a]	Range (°S)[b]	Penguin	Site	Size (mm) of prey[c]	Reference
Euphausia crystallorophias	14–30	65–75	Adélie	C. Crozier (77° C)	9–30 (99·5%)	Emison, 1968
				C. Hallett (72° S)	12–34 (21%)	Kinsky & Ensor, unpublished
E. superba	*c.* 35–65	*c.* 54–65(–75)	Emperor	Adélie Land	40 (*c.* 90%)	Offredo & Ridoux, 1986
			Adélie	C. Crozier (77° S)	21–39 (0·5%)	Emison, 1968
				C. Hallett (72° S)	35+ (74%); < 35 (3%)	Kinsky & Ensor, unpublished
				S. Orkney Is. (60° S)	35+ (37–65%); < 35 (35–63%)	Lishman, 1985*a*
			Chinstrap	S. Shetland Is. (62° S)	35+ (98%); < 35 (2%)	Volkman *et al.*, 1980
				Elephant Is. (61° S)	35+ (83%); < 35 (17%)	Croxall & Furse, 1980
				S. Orkney Is. (60° S)	35+ (72–87%); < 35 (13–28%)	Lishman, 1985*a*
			Gentoo	S. Shetland Is. (62° S)	35+ (98%); < 35 (2%)	Volkman *et al.*, 1980
				S. Georgia (54° S)	35+ (88%); < 35 (12%)	Croxall & Prince, 1980*a*
			Macaroni	Elephant Is. (61° S)	35+ (5%); < 35 (16%)	Croxall & Furse, 1980
				S. Georgia (54° S)	35+ (22%); < 35 (78%)	Croxall & Prince, 1980*a*
E. frigida		50–65	Gentoo	S. Georgia (54° S)	28–38 (< 1%)	Croxall & Prince, 1980*a*
			Macaroni	S. Georgia (54° S)	16–22 (6%)	Croxall & Prince, 1980*a*
E. triacantha	28–32	50–65	Macaroni	S. Georgia (54° S)	18–21 (< 1%)	Croxall & Prince, 1980*a*
E. vallentinii	15–28	45–60	Royal	Macquarie I. (54° S)	?	Horne, 1985
			S. Rockhopper	Macquarie I. (54° S)	?	Horne, 1985
				Falkland Is. (52° S)	17–22 (20%)	Croxall *et al.*, 1985*b*
			Macaroni	Marion I. (47° S)	?	Williams & Laycock, 1981
E. similis	22–26	(0–)30–50	Royal	Macquarie I. (54° S)	?	Horne, 1985
E. lucens	10–18	35–50	S. Rockhopper	Falkland Is. (52° S)	1–22 (68%)	Croxall *et al.*, 1985*b*
Thysanoessa macrura	≤ 16	50–75	Adélie	C. Hallett (72° S)	— (2%)	Kinsky & Ensor, unpublished
T. sp. (cf. *macrura*)			Macaroni	Elephant Is. (61° S)	< 15 (79%)	Croxall & Furse, 1980
T. gregaria	11–16	20–56	Royal	Macquarie I. (54° S)	?	Horne, 1985
			S. Rockhopper	Macquarie I. (54° S)	?	Horne, 1985
				Falkland Is. (52° S)	12–17 (13%)	Croxall *et al.*, 1985*b*
T. vicina	≤ 16	(40–)50–75	N. Rockhopper	Gough I. (41° S)	?	Williams & Laycock, 1981
Nyctiphanes australis	≤ 14	35–50	Fiordland	S. New Zealand (45° S)	?	Warham, 1974*b*
N. capensis	≤ 13	30–40	Jackass	S. Africa (33° S)	?	Rand, 1960

[a] From Mauchline & Fisher, 1969.
[b] Southern hemisphere only, from Mauchline & Fisher, 1969.
[c] Proportion by numbers of euphausiid diet in parentheses.

but, where it occurs, the latter is the main penguin prey north to 54° S at South Georgia. Adult *E. superba* are taken in much greater quantities than juveniles, except by Macaroni Penguins at South Georgia where, however, juveniles comprise only 18% (adults 80%) by weight. Where *E. superba* does not occur (i.e. north of the Antarctic Convergence and at Macquarie Island), *E. vallentinii*, the largest and most characteristically swarm-forming of the several other euphausiids, is usually the main prey for penguins, although *Thysanoessa* spp., which are little smaller than several other *Euphausia* spp., are often important, and occasionally the dominant crustacean prey. Further north still, *Nyctiphanes* spp. may be locally important, perhaps particularly in the New Zealand region. In such regions, however, decapods are often as frequent as euphausiids in stomach contents, as e.g. *Nauticaris marionis* forming 30% by weight of Gentoo Penguin samples from Marion Island (La Cock, Hecht & Klages, 1984), juvenile *Jasus paulensis* forming 10% of the crustacean diet of Northern Rockhopper Penguins at Amsterdam Island (Duroselle & Tollu, 1977), juvenile *J. lalandii* in Jackass Penguins off South Africa (Rand, 1960), *Munida gregaria* in Magellanic Penguins in southern Chile (Venegas & Sielfeld, 1981) and megalopa larvae in Jackass and Little Penguins (Rand, 1960; Montague, 1982). The main crustacean prey of Jackass Penguins, however, was the mantis shrimp *Squilla armata* (Stomatopoda) (Rand, 1960).

Amphipods are present as traces in most penguin stomach contents. Fourteen species comprised 1–5% by numbers and 1% by volume of the diet of Adélie Penguins at Cape Crozier (Emison, 1968) with 70% belonging to the lysianassid genus *Orchomenella*, mainly *O. plebs*. Five species constituted 10% by volume of the prey of Adélie Penguins at Cape Bird (Paulin, 1975); *Hyperia macrocephala* and five gammarid species formed 3% by numbers in Emperor Penguin samples from Adélie Land (Offredo & Ridoux, 1986); and the hyperiids *Themisto gaudichaudii* (an ubiquitous Southern Ocean species) and *Primno macropa* formed 8% of the diet of Royal Penguins *Eudyptes c. schlegeli* on the west coast of Macquarie Island (Horne, 1985). Only for Southern Rockhopper Penguins at Heard Island were amphipods (*Themisto* and *Hyperia galba*) the main constituent of any sample (Ealey, 1954).

Fish

Only for Jackass Penguins (Davies, 1955, 1956; Rand, 1960; Wilson, 1985) and Little Penguins (Montague, 1982) have the fish diets of penguins been studied in detail. Rand (1960) identified at least 26 species in his Jackass Penguin samples with the pilchard *Sardinops ocellata* (32% by weight), anchovy *Engraulis capensis* (21%) and Maasbanker *Trachurus trachurus* (18%) predominating. Pilchards included many relatively large (*c* 100 g) specimens, as did the Maasbankers (mean weight of the smallest fish was 30–40 g) but the largest anchovies only weighed 7–8 g (Rand, 1960). Davies' (1956) sample was less diverse, being almost exclusively pilchard (64% by weight) and anchovy (31%), with only 3% Maasbanker. The previous year's small sample (Davies, 1955) was similar but included 26% by weight of mullet *Mugil* spp. In the last 20 years, however, there have been major changes in the abundance of fish species in South African inshore waters, involving the collapse of the pilchard fishery. Recent detailed studies have shown that, by weight, the present diet of Jackass Penguins comprises 80% anchovy, 4% Maasbanker, 3% Round Herring *Etrumeus teres* and only 1% pilchard, with about 20 other species making up the rest (Wilson, 1985).

The Little Penguin samples contain 19 identified fish species, with pilchards *Sardinops neopilchardus* (in 37% of samples) and anchovy *Engraulis australis* (34% of samples) predominating, no other species occurring in more than 6% of samples (Montague, 1982).

The Magellanic Penguin in southern Chile (45° S) took mainly the sardine *Ramnogaster arcuata* (Venegas & Sielfeld, 1981) and all reports indicate that Humboldt Penguins chiefly eat Anchoveta *Engraulis ringens* (Murphy, 1936; see also Chapter 14).

In Emperor Penguin samples (Offredo & Ridoux, 1986), about 97% of fish material was small (40–125 mm length) nototheniids (Nototheniidae), with *Trematomus borchgrevinkii* the only identified species. The rest were chaenichthyids, stomiatoids and bathydraconids (including *Gymnodraco acuticeps*).

In late winter at Marion Island the inshore demersal *Notothenia squamifrons* (originally identified as *Harpagifer georgianus* (Harpagiferidae)) was the main food of breeding Gentoo Penguins, other *Notothenia* species and lanternfishes Myctophidae (belonging to the genera *Gymnoscopelus*, *Electrona* and *Paramyctophum*) contributing only 7% of otoliths analysed (La Cock *et al.*, 1984). At South Georgia in summer the fish in Gentoo Penguin samples (Croxall & Prince, 1980*a*) were *Notothenia* spp. and ice fish *Champsocephalus gunnari* (Channichthyidae). Further south (at Argentine Islands, 65° S) *Trematomus borchgrevinkii* has also been reported (B.B. Roberts, unpublished).

The only fish identified in penguin samples from Macquarie Island were lanternfish and *Harpagifer bispinis* from Royal Penguins and *Notothenia* sp.

and *Zanclorhynchus spinifer* (Congiopodidae) from Rockhopper Penguins (Horne, 1985). Adélie Penguins took many *Pleuragramma antarcticum* (Nototheniidae) and a few *Chionodraco* (Channichthyidae) at Cape Crozier (Emison, 1968) but fish have not been reported in their diet further north; they are rare also in Chinstrap samples, with records of *Trematomus* at the South Orkney Islands (Lishman, 1985*a*), a few *Pleuragramma* at the South Shetland Islands (Volkman *et al.*, 1980) and remains of about 30 *Electrona ? carlsbergi* (Myctophidae) in one of three samples from Bouvet Island (54° S) (Cooper *et al.*, 1984).

Cephalopods

Although squid are probably eaten by most species of penguins and are significant in the diet of eight or nine species, identification of the species involved, from their keratinous beaks retained in the stomachs (see Clarke, 1962, 1980), has been attempted for only five species.

In the Jackass Penguin, for which squid are a relatively minor prey category, *Loligo reynaudi* (Loliginidae) comprised 96% of material, *Heteroteuthis* sp. (Sepiolidae) 3% and *Argonauta argo* Argonautidae) 1% (Randall, Randall & Baird, 1981). The average size of the *Loligo* taken was estimated as 116 g, that of the other two species as < 10 g. Octopus (*Octopus*: Octopodidae) and cuttlefish (*Lolliguncula*: Sepiidae) have also been recorded in Jackass Penguin samples (Rand, 1960; Wilson, 1985).

The squid diet of the Little Penguin (Montague, 1982) shows some similarities with the above, comprising 63% *Nototodarus gouldi* (Ommastrephidae), 18% *Argonauta nodosa*, 7% *Loligo* sp., 3% *Sepioteuthis australis* (Sepiolidae), 3% octopus and 7% unidentified squid. The mean size of the main prey, however, was probably < 10 g, based on mantle length measurements. Squid are much less important than fish, though the proportion of the former may increase in November–December, the breeding season for *N. gouldi* (Montague, 1982) when it may particularly tend to aggregate.

At Beauchêne Island, Falkland Islands, *Teuthowenia* sp. (Cranchiidae) was the only species found in Southern Rockhopper Penguin stomachs, with several hundred juveniles (each weighing about 0.5 g) per sample, comprising 21% by numbers and 53% by weight of the diet (Croxall *et al.*, 1985*b*).

The measurable beaks from a few Emperor Penguin stomachs from Halley Bay (75° S) comprised mainly *Psychroteuthis glacialis* (Psychroteuthidae; 45% of beaks, 23% of total estimated weight), *Alluroteuthis antarcticus* (Alluroteuthidae; 35% and 42%, respectively), *Kondakovia longimana* (Onychoteuthidae; 6% and 24%) and five other rarer species. The estimated mean weights of the main species were: *Psychroteuthis*, 140 g; *Alluroteuthis*, 320 g; *Kondakovia*, 1060 g; with an overall mean weight of 270 g (M.R. Clarke, unpublished). *Psychroteuthis* (mean weight 10–70 g) was also the main squid prey of Emperor Penguins from Adélie Land (88% of beaks, 51–68% of estimated weight), with *Kondakovia* (530–770 g; 1–3% and 14–21%, respectively), and *Gonatus antarcticus* (175–185 g; 4–16% and 18–27%) the other main constituents (Offredo, Ridoux & Clarke, 1985).

Samples from King Penguins at South Georgia (54° S) were very different, with measurable beaks consisting mainly of *Todarodes* cf. *sagittatus* (Ommastrephidae; 98% by number, 97% by weight) and six other species, of which only *Kondakovia*, *Psychroteuthis* and *Moroteuthis ingens* (Onychoteuthidae) were in common with the Emperor Penguin samples. The mean weight of the *Todarodes* was estimated as 420 g (M.R. Clarke, unpublished). Considering that King and Emperor Penguins are three and ten times the mass of, say, Jackass Penguins, the individual squid prey that they take is neither large in proportion nor significantly heavier than those taken by surface-feeding seabirds (see Chapter 7).

In addition, the squid *Loligo brasiliensis*, *L. gahi*, *Loligunculus brevis* and the octopus *Pareledone charcoti* have been reported in the diet of Magellanic Penguins (C.C. Olrog in Boswall & MacIver, 1975) and *Psychroteuthis glacialis* as the only species taken by Adélie Penguins (Offredo *et al.*, 1985).

Intraspecific variation

For five of the seven main species, studies have been made at at least two sites; for five species, data were collected over more than one season. Only three species have been studied in the non-breeding season and of the rest nearly all samples come from the chick-rearing period. This is hardly adequate to do more than indicate some of the intraspecific differences that emerge.

Locational Many differences in diet between sites seem to relate to broad-scale differences in prey availability, e.g. Adélie Penguin: presence of *Euphausia crystallorophias* rather than *E. superba* on the Antarctic continental shelf; Macaroni/Royal Penguin: absence of *E. superba* at Macquarie Island. There appear to be important regional differences nowadays in Jackass Penguin diet. Pelagic Goby *Sufflogobius bibarbatus* is the main prey (up to 71% by weight) at the more northerly breeding sites (southwest Africa, 26° S, 15° E) (Crawford & Shelton, 1981) and pilchards at the most easterly sites (Algoa Bay, 34° S, 26° E) (Randall & Randall, 1981), whereas anchovies predominate in the southwestern Cape area (33–34° S, 15–16° E) (Wilson, 1985).

In the Scotia Sea, however, the diets of Chinstrap, Adélie and Gentoo Penguins are remarkably consistent between sites and there are general similarities between Southern Rockhopper Penguin diets at the Falkland Islands and Macquarie Island.

Local differences were found with Rockhopper Penguins at Macquarie, where those on the west coast took significantly fewer euphausiids and more amphipods than those on the east (Horne, 1985), and with Macaroni Penguins at Elephant Island, where all the fish and small *E. superba* were taken by birds at Gibbs Island, while those at Clarence Island took mainly large *E. superba* and *Thysanoessa* sp. (Croxall & Furse, 1980). Such differences probably reflect local variations in prey availability or abundance.

The food of Gentoo Penguins at Marion Island, where they mainly take fish (La Cock *et al.*, 1984), is very different from their *E. superba* diet at South Georgia, but probably mainly reflects the 4–5 months difference in timing of chick rearing and the shortage of crustaceans in late winter at Marion.

Seasonal The studies of Adélie, Chinstrap, Gentoo and Macaroni Penguins showed little variation in dietary composition through the chick-rearing period, nor even a tendency to take larger krill as the season progressed (Emison, 1968; Croxall & Prince, 1980*a*; Volkman *et al.*, 1980; Lishman, 1985*a*). Little Penguins, whose egg laying starts in late September–early October, ate mainly anchovies from January to June, pilchards from July to November and a variety of other fish in December. Squid were taken chiefly in November and December (Montague, 1982). In the 1950s Jackass Penguins (with egg-laying peaks in November–December and May–July) took pilchards most frequently in winter (June–August) and spring, anchovies predominantly in summer (though very small ones were common in some autumn and winter samples) and Maasbankers mainly in summer and winter (Rand, 1960).

Recent detailed studies in the same area (Wilson, 1985) found that anchovies now comprise over 90% by weight of the diet in all months except February to May (reaching a low of 40% in April). Other fish species increase in importance during this period and Maasbankers form 30% of the April diet. This change in food, which may result from peak availability inshore of juvenile fish of many species, coincides with the period of fastest chick growth and lowest mortality (Wilson, 1985), but the reason for the switch away from anchovy remains unclear.

Annual There are few data with which to assess variations in diet between years. At Admiralty Bay (Trivelpiece *et al.*, 1983) and South Georgia (Croxall, unpublished) penguin diets have been very consistent from year to year, although the importance of small krill to Macaroni Penguins at South Georgia has varied. In two years of very different environmental circumstances and breeding success at Signy Island, South Orkney Islands, the only difference in Adélie and Chinstrap Penguin diets was that both species took more smaller krill in one year (Lishman, 1985*a*). The three sets of Jackass Penguin samples from consecutive years were generally very similar, differing mainly in the relative proportions of pilchards and anchovies and in the proportion of the total made up of less-important species like Maasbanker and mullet (Davies, 1955, 1956; Rand, 1960). Likewise, for the Little Penguin, broadly similar patterns did exist in both years, with some differences in the timing and relative abundance of the main prey (Montague, 1982).

Sexual Ainley & Emison (1972) found small, but significant, differences in the length of krill taken by male and female Adélie Penguins and suggested this reflected selection of different-sized prey. However, because krill are relatively homogeneous in length within, and heterogeneous between, swarms (Mauchline & Fisher, 1969), statistical differences between samples from different individuals, dates and locations are easy to detect and probably only reflect differences in the time and place of foraging (Croxall, Prince & Ricketts, 1985). More rigorous studies of Adélie and Chinstrap Penguins (which anyway have an 80% overlap in bill size between sexes) and of Macaroni Penguins (which have no overlap in bill size) have not revealed any sexual differences in prey size once these other factors are taken into account (Croxall & Prince, 1980*a*; Volkman *et al.*, 1980; Croxall *et al.*, 1985*c*). In Gentoo Penguins, which have a 10% overlap in bill size between sexes, the larger males ate 23–32% of fish and the females only 2–15% (Trivelpiece *et al.*, 1983) which, it is suggested, may relate to greater intraspecific competition in a species which, unlike its congeners, remains at its breeding grounds in winter when food availability may be lower and therefore potential intraspecific competition more intense.

Penguin diet in relation to prey availability and fishery operations

There has been no systematic attempt to acquire simultaneously data on the prey taken by penguins and that available in their vicinity, or to compare either in detail with the resources exploited by commercial fisheries in the same area. Some data relevant to such comparisons do, however, exist.

Croxall *et al.*, (1984*b*, 1985*c*) and Lishman (1985*a*) compared the length-frequencies of Antarctic Krill eaten by penguins (Adélie, Chinstrap, Gentoo, Macaroni) with those caught in scientific net hauls and showed that in the southern Scotia Sea there were strong similarities, allowing for the nets being unable to retain the smallest size-classes of krill. Around South Georgia, however, penguins consistently took larger krill than were found in net hauls but, lacking samples from within penguin feeding areas and from the same seasons, this could not be evaluated further. The most comparable data are those from net hauls from the enclosed Admiralty

Fig. 6.2. Comparison of length-frequency distributions of krill taken by Adélie and Chinstrap Penguins and net hauls in Admiralty Bay, South Shetland Islands.

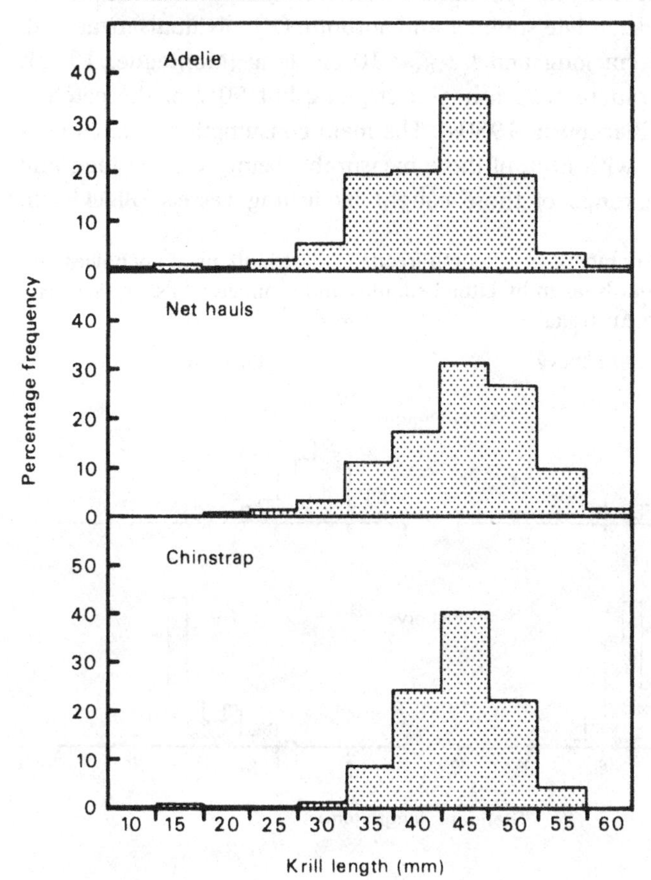

Bay, South Shetland Islands (Stepnik, 1982), where the penguins sampled by Volkman *et al.* (1980) feed (Trivelpiece *et al.*, in press). The results (Fig. 6.2) show very close similarities between the krill caught by the penguins and in the net hauls. As catches by these nets were shown to be highly comparable with those by the nets used in the commercial krill fishery (Klages & Nast, 1981), there is a strong likelihood of direct competition between penguins and krill fisheries, at least in this area.

Of the pilchards taken by Jackass Penguins, 44% were of commercial size whereas 90% of Maasbankers and 47% of anchovies were too small (< 10 cm and < *c*.5 cm long, respectively) for commercial exploitation (Rand, 1960). Of commercially fished specimens of the squid *Loligo reynaudi*, 91% are longer than the mean size taken by penguins and the difference in beak dimensions was very significant (Randall *et al.*, 1981*b*). Any direct competition with fisheries is therefore confined to the larger size-classes of pilchards and anchovies taken by penguins.

Little Penguins take adult pilchards between August and November and mainly post-larvae in late summer and autumn. Of individuals measured, 32% were < 5 cm long and 92% < 10 cm long (Montague, 1982); commercial operations take fish > 5 cm long but 90% of the catch is > 10 cm long (Blackburn, 1950*a*). The main consumption of anchovies is of post-larvae, with 68% of those measurable being < 5 cm long and outside the size range of those caught by fishing vessels (Blackburn,

Fig. 6.3. Comparison of length-frequency distributions of anchovies and pilchards taken by Little Penguins and commercial fishery vessels off South Australia.

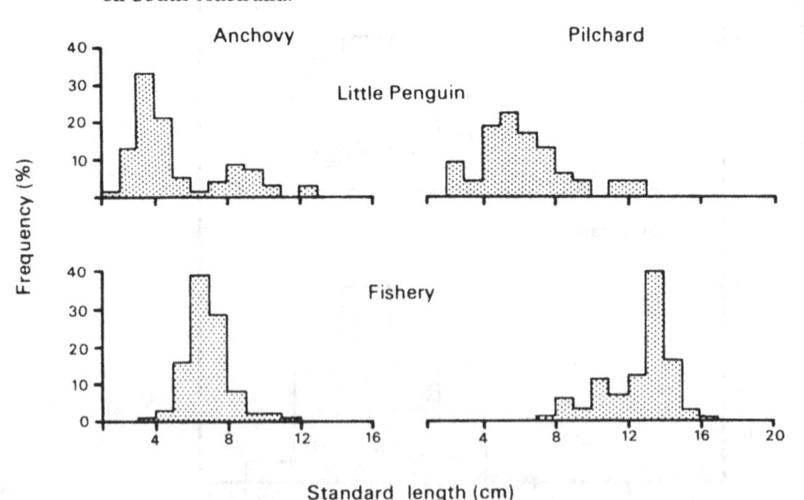

1950*b*) (Fig. 6.3). For neither species of fish is there any evidence of significant direct competition between penguin and fishery. The squid prey, *Nototodarus gouldi*, were 2.5–9.0 cm in mantle length; those taken commercially are all > 15 cm in length (Montague, 1982).

The predation by Jackass and Little Penguins on post-larval fish is also unlikely to be significant indirectly, e.g. in terms of affecting recruitment to the adult stock, given the likely great abundance and wide distribution of such juveniles.

Feeding behaviour and ecology

Feeding behaviour

There are few observations of the pursuit and capture of prey. Adélie Penguins were seen swimming in zigzag fashion at constant speed, catching prey as they moved (Falla, 1937). Species that are primarily fish-eaters often feed in groups, sometimes of up to 200 individuals (Boersma, 1976 – Galapagos Penguin), often in association with other seabirds (Boersma, 1976; Boswall & MacIver, 1975 – Magellanic Penguin) or with fish (e.g. Pacific Bonito *Sarda chiliensis*) or South American Sealions *Otaria byronia* (Murphy, 1936 – Humboldt Penguin). When water temperatures were high, Galapagos Penguins were often seen foraging inshore in pairs, scanning the bottom and the shoreline for food. With lower sea temperatures, large groups of penguins associated with huge aggregations of small fish and active pursuit of such shoals was sometimes observed (Boersma, 1976). Although penguins frequently feed in groups on discrete shoals of prey, most reports emphasise the haphazard nature of this predation (e.g. Rand, 1960) and there are no reliable indications that cooperative feeding techniques are employed (but see Boswall & MacIver, 1975).

Penguins catch fish underwater, probably often taking them from beneath (Rand, 1960), occasionally surfacing to manipulate them and then ingesting them head first (Rand, 1960; Zusi, 1975). How penguins feed in the dark, i.e. when locating prey at night or at depth, is unknown but many pelagic crustaceans, squid and fish are bioluminescent. Poulter (1969) suggested that penguins could use a form of echolocation, involving cavitation clicks produced by the bird's swimming action, to locate prey; this deserves further investigation.

Diving depth and duration

Diving depths have been studied in five species: in Emperor (Kooyman *et al.*, 1971), Little (Montague, 1982) and Gentoo (Adams & Brown, 1983) using devices giving only a single record (per foraging trip) of maximum depth attained; and in King (Kooyman *et al.*, 1982) and Chinstrap (Lishman & Croxall, 1983) with a recorder tallying the number of dives exceeding each of eight pre-set depth thresholds (Kooyman *et al.*, 1983).

The results of these studies (Fig. 6.4) show big differences between the deep-diving Emperor and King Penguins, the latter with over half its dives reaching deeper than 50 m, and the other species in which most dives (90% in Chinstrap Penguin) are shallower than this depth. This difference

Fig. 6.4. Diving-depth profiles for Emperor, King, Chinstrap, Gentoo and Little Penguins.

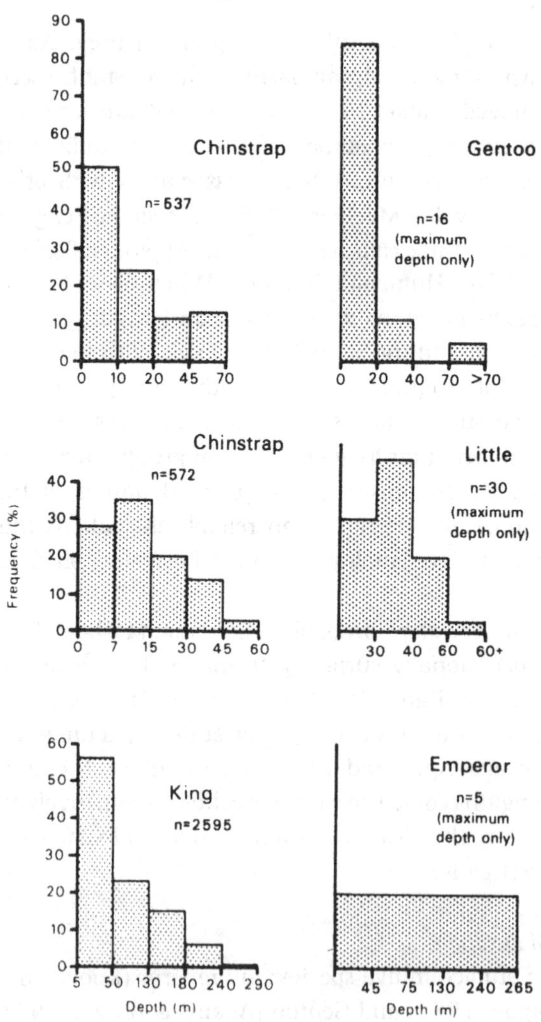

in depth attained on dives may partly be a function of size, since diving capacity is greater in larger animals (Butler & Jones, 1982), but probably also relates to diet. King and Emperor Penguins feed extensively on mesopelagic squid (and fish) which are themselves fast active carnivores. Longer, relatively deep dives may be necessary to secure such prey; even so, Kooyman *et al.* (1982) estimated that only 10% of dives need be successful in order to meet the energy requirements of an adult and its chick.

The other species mainly take smaller, less active, epipelagic prey and appear to use shorter, shallower dives to do so. Although it is difficult to compare the three smaller species because of the different recorders used, it should be noted that if Chinstrap Penguin dives were converted to maxima per trip then all but one would show a dive deeper than 45 m. This result is closer to (but more extreme than) the pattern shown by Little Penguins than that of Gentoo Penguins. This is surprising, since one might expect Little Penguins (only 20% and 25% the mass of Gentoos and Chinstraps, respectively) to make shallower dives. Gentoo Penguins, however, may feed closer to shore (perhaps within 10 km of land (Admiralty Bay, Trivelpiece *et al.*, in press; Bird Island, South Georgia, J.L. Bengtson, unpublished)) than Little Penguins (which may feed up to 30 km away from their colonies (Montague, 1982)) and thus Gentoos may normally feed in shallower water. However, recent results using electronic recorders on Gentoo Penguins at South Georgia indicate a mean dive depth of about 25 m and numerous dives exceeding 80 m (R.W. Davis, personal communication), so it may be premature to characterise species-specific diving patterns on the basis of existing data.

In their study of radio-tracked penguins, Trivelpiece *et al.*, (in press) have distinguished three types of 'diving' (underwater swimming, searching and stationary diving) by the pattern and duration of these events. They suggest that the shallow 0–7 m and 0–10 m dives of Chinstrap Penguins (Lishman & Croxall, 1983) may be equivalent to their underwater swimming and, by definition, uninvolved in feeding. Apart from the possibility that birds can swim underwater and feed simultaneously, it seems implausible that a penguin feeding exclusively on krill, which is taken at or near the sea surface by many other seabirds in the area, should not exploit this food by means of shallow dives. Penguins may, however, submerge for a variety of purposes (excluding the 'porpoising' associated with fast travelling near shore), some of which may not involve feeding.

Data on dive duration (Table 6.3) do not show any clear pattern related to the size of the bird or the prey it exploits. Thus although Emperor

Penguins have the longest dives recorded (over 18 min), their average duration is only about 2.4 min (Kooyman *et al.*, 1971). This is much the same as the mean dive duration of 2.5 min reported for Jackass Penguins (which are only about 15% of Emperor Penguin mass) by R.P. Wilson (in Adams & Brown, 1983) but very different from other data indicating that dives average only 23 s (S. Broni, unpublished). The two very different values reported for Gentoo Penguins refer to birds feeding in different habitats. The short dives were recorded for birds that could be watched feeding throughout complete dives in an inshore area (Kooyman, 1975), whereas the longer dives were those recorded for radio-tracked birds several kilometres out to sea in a bay where krill occur to 120 m depth (Trivelpiece *et al.*, in press). Similarly, the short dives of the Galapagos Penguins were from observations of birds seen from shore (Boersma, 1976) and may not necessarily be typical of the dives undertaken when pursuing fish shoals further out to sea.

Time of feeding

Apart from one record of Jackass Penguins taking bait at night (Rand, 1960), all direct observations of penguins feeding are, unsurprisingly, during daylight. Unlike surface-feeding seabirds, which may frequently forage at night to take advantage of the vertical migration of zooplankton and their associated fish and squid predators (see Chapter 7), penguins, with their pronounced diving abilities, may be much less restricted in when they feed. However, even for diving species, the optimum feeding time could still be at night, as Croxall *et al.*, (1985*a*) found with Antarctic Fur Seals *Arctocephalus gazella* at South Georgia, for which over 40 days of continuous records of diving activity were obtained. This species, which eats mainly krill, made 75% of dives at night and these were on average

Table 6.3. *Duration of natural dives by penguins*

	No. of dives	Dive duration (min) Mean (±SD)	Mode	Max.	Reference
Emperor	238	*c.* 2·4	0–1	18+	Kooyman *et al.*, 1971
Chinstrap	182	1·52±0·08	—	2·17	Trivelpiece *et al.*, in press
Gentoo	50	*c.* 0·75	0·5–1·0	1·5–2·0	Kooyman, 1975
	193	2·13±0·07	—	3·15	Trivelpiece *et al.*, in press
Jackass	?	2·5	—	?	R. P. Wilson in Adams & Brown, 1983
	138	0·38	—	—	S. Broni, personal communication
Galapagos	?	*c.* 0·5	—	1·32	Boersma, 1976

significantly shallower than daytime dives. Thus, penguins that prey predominantly on zooplankton in regions where these show vertical migration (e.g. *Eudyptes* and *Pygoscelis* species in sub-Antarctic areas) may still find it most economical to feed extensively at night. Nevertheless, in situations where their feeding trips are shorter than one day, they are most likely to feed in the daytime and in some cases this is known to be so (Trivelpiece *et al.*, in press). South of the Antarctic Circle in summer there is continuous daylight (though very perceptible differences in light intensity) and Ainley *et al.* (1984) have suggested that Adélie and Emperor Penguins feed principally during the main daylight period rather than during 'twilight'. This is based on the proportions of birds seen in the water and on ice floes, which may not necessarily be a reliable index.

For species that make relatively short trips (less than one day),the time at which individuals depart and return to feed chicks may adequately define when they were feeding. In all except two species of penguin, chicks are usually fed in the late afternoon or early evening and the adults normally stay ashore overnight. For Magellanic, Jackass and Gentoo Penguins, feeding trips usually last less than 10 h and therefore these species, all of which take significant quantities of free swimming, epipelagic fish and/or squid, must feed during the daytime. Also, Jackass Penguins caught at sea in the mornings rarely had food in their stomachs (Rand, 1960). In Galapagos and Little Penguins, adults come ashore in the late evening or after dusk (Boersma, 1976; Reilly & Balmford, 1975). Whether this is connected with facilitating feeding on fish which approach the surface waters at dusk, or whether it avoids unnecessary thermoregulatory problems for adults and chicks or diurnal predators is unknown.

Foraging range during the breeding season

Only Trivelpiece *et al.* (in press) have managed to track penguins (Gentoo and Chinstrap) throughout complete foraging trips and this was in the enclosed waters of a large bay where both species stayed within 5 km of the breeding colony. Direct observations of Galapagos Penguins indicated that when foraging singly or in pairs they stayed very close to shore, whereas groups occurred at least 0.4 km offshore (Boersma, 1976). Otherwise, assessments of potential maximum foraging ranges have been made by multiplying the time between successive feeds to the chick by the same parent by the speed of swimming and assuming that birds swim consistently in a straight line on a constant heading (Croxall & Prince, 1980*a*,*b*; Williams & Siegfried, 1980; Lishman, 1985*a*). Obviously the actual foraging range would be considerably less than this, as birds spend

Table 6.4. *Swimming speeds of penguins*

Species	Location	Distance timed over	Average speed recorded ($km\ h^{-1}$)	Reference
Emperor	Between ice-holes	27 m	7·5	Kooyman *et al.*, 1971
	Tank	8–11 m	8·1[a]	Clark & Bemis, 1975
King	Tank	8–11 m	12·1[a]	Clark & Bemis, 1975
Adélie	Between ice-holes	120 m	7·2	Kooyman, 1975
	Tank	8–11 m	7·1[a]	Clark & Bemis, 1975
Chinstrap	Return to colony	1–2 km	4·8	Trivelpiece *et al.*, in press
Gentoo	Return to colony	1–2 km	4·2	Trivelpiece *et al.*, in press
Jackass	Leaving colony	200 m	3·6	Siegfried *et al.*, 1975
	Return to colony	200 m	6·2	Siegfried *et al.*, 1975
	Tank	8–11 m	11·6[a]	Clark & Bemis, 1975
	Displacement experiment	890 km	3·4	Randall *et al.*, 1981*b*
	Foraging trip	1·3–2·3 days	5·2–7·6[b]	Nagy *et al.*, 1984
Little	At sea	*c.* 100 m	5·6	Barton, 1979
	Tank	8–11 m	6·2[a]	Clark & Bemis, 1975
Macaroni	Tank	9–11 m	8·2[a]	Clark & Bemis, 1975
Rockhopper	Tank	8–11 m	7·8[a]	Clark & Bemis, 1975

[a] Maximum speed.
[b] Excludes any periods of swimming at less than *c.* 1·5 $km\ h^{-1}$.

time resting and may deviate frequently from any particular heading. On this basis, and using a swimming speed of 7.2 km h^{-1} (and incorporating some assumptions on activity budgets), Lishman (1985*a*) estimated the mean maximum range of Adélie and Chinstrap Penguins breeding at Signy Island and making trips of duration 39–55 h and 30–42 h, respectively, to be 80–120 km and 66–92 km. For Gentoo and Macaroni Penguins at Bird Island, South Georgia, Croxall *et al.* (1984*b*) similarly estimated mean maximum ranges to be about 10 km and about 80 km, respectively. Although the swimming speed used seemed conservative for a 4 kg penguin, considering the published data then available (Table 6.4), recent estimates (Trivelpiece *et al.*, in press) suggest that for Chinstrap and Gentoo Penguins actually engaged in swimming back from their feeding grounds, an average speed of much more than 4 km h^{-1} may be unrealistic.

Table 6.5. *Intervals between successive feeds to penguin chicks by the same parent*

Feeding interval days		
⩽ 1	> 1–< 3	⩾ 3
Gentoo	Adélie	Emperor
Snares Crested	Chinstrap	King
Erect-crested	Macaroni	
Fiordland	Rockhopper	
Yellow-eyed	Royal	
Little		
Magellanic		
Galapagos		
Humboldt (?)		
Jackass		

References: Emperor: Prévost, 1961; King: Stonehouse, 1960; Adélie: Sladen, 1958; Taylor, 1962; Emison, 1968; Lishman, 1985*a*; Chinstrap: Conroy *et al.*, 1975; Croxall & Furse, 1980; Lishman, 1985*a*; Gentoo: Williams, 1980; Croxall & Prince, 1980*a*, *b*; Macaroni: Croxall & Prince, 1980*a*, *b*; Williams, 1982; S. Rockhopper: Warham, 1963; Williams, 1982; N. Rockhopper: Duroselle & Tollu, 1977; Royal: Carrick, 1972; Snares Crested: Warham, 1974*a*; Erect-crested: Richdale, 1941; Warham, 1972; Fiordland: Warham, 1974*b*; Yellow-eyed: Richdale, 1957; Little: Richdale, 1940; Kinsky, 1960; Reilly & Balmford, 1975; Galapagos: Boersma, 1976; Humboldt: no data; Magellanic: Boswall & McIver, 1975; Jackass: Rand, 1960; Nagy *et al.*, 1984.

In most cases the data are adequate only to characterise penguins as broadly inshore and offshore and this is related to the normal intervals between feeds to the chick by a single parent in Table 6.5. These two groups are not necessarily homogeneous. Whereas *Eudyptula* and all *Spheniscus* species probably feed mainly or entirely within 10km of the shore when rearing chicks (see also R.P. Wilson, in Adams & Brown (1983) for Jackass Penguins), data for the crested penguins in this category are very poor and they may well forage further afield. Conversely, except in their potential range, there may be little difference in the actual foraging ranges of Emperor and King Penguins and the other offshore species. To determine the real foraging range and to locate the actual feeding grounds of penguins remain two of the most challenging problems of penguin feeding ecology, particularly for the species that are important local consumers of marine resources.

Foraging success

Weights of stomach contents indicate that four species of Antarctic penguins (Adélie, Chinstrap, Gentoo, Macaroni) rearing chicks often transport food weighing up to about 20% (i.e. 1–1.5 kg) of their body weight (Emison, 1968; Croxall & Furse, 1980; Croxall & Prince, 1980*a*; Volkman *et al.*, 1980; Lishman, 1985*a*), although average loads are perhaps only half this. Data for Jackass and Little Penguins (Rand, 1960; Montague, 1982) are not fully comparable, as stomach samples were taken all the year round and mean values of 115 g and 80 g (5% and 7% of body weight, respectively) are probably underestimates. Maximum weights of stomach contents were 812 g (Jackass) and 379 g (Little), both about 30% of adult body weight.

There are, however, no empirical data on how long it takes penguins to catch this amount of food or on any aspect of foraging success. Only by combining information on diet, energy requirements and number and duration of dives can any indication of the possible catching rate be obtained. Sufficient data are available to do this for only three species (Table 6.6). Not surprisingly, there is a big difference in many respects between the potential performance of King Penguins, eating squid in units of about 420 g, and Chinstraps and Gentoos, taking krill in units of 0.3–1.0 g. Thus, King Penguins need only catch prey on 5% of dives to meet the combined energy requirements of adult and chick. In contrast, the krill-eaters need to catch prey at a rate of 5–7 g per dive (or perhaps double this if only half the dives are actual feeding dives), representing one krill caught every 6–30 s (or every 3–15 s if half the 'dives' are purely

Table 6.6. *Estimated food requirements and potential prey capture rates of penguins rearing chicks for one complete foraging-attendance cycle*

Species	Food required[a]				Dives		No. of food items		Average weight of food per dive (g)
	Weight (g)								
	Adult[b]	Chick[c]	Total	Number[d]	Number per trip[e]	Duration (s)	per dive	per s per dive	
Chinstrap 1	631	587	1218	3680	257	c. 130[f]	14·3	0·11	4·7
Chinstrap 2	1639	619	2258	6822	315	c. 130[f]	21·7	0·17	7·2
Chinstrap 3	378	600	978	1115	182/74	130	6·1/15·1	0·05/0·12	6·1/15·1
Gentoo	534	729	1263	1255	193/90	189	6·5/13·9	0·03/0·07	6·5/13·9
King	10163	c. 3000	13163	131·3	610	—	0·05	—	21·6

[a] Krill assumed to be sole food of Chinstraps (actually 98%) and Gentoos (*c.* 84%), and squid to be only food of Kings (actually *c.* 70%). Energy content of krill = *c.* 4·6 kJ g^{-1} (Clarke & Prince, 1980); of squid = 3·48 kJ g^{-1} (Croxall & Prince, 1982*a*). Assimilation efficiency *c.* 80% (D. P. Costa in Kooyman *et al.*, 1982).

[b] Chinstrap (weight *c.* 4 kg)

1: South Orkney Islands (Lishman & Croxall, 1983, table 1, bird 4); 18 h at sea at 8 W kg^{-1} (see Lishman & Croxall, 1983), plus 6 h ashore at 2·9 W. kg^{-1} (Lishman, 1983).

2: as above but bird 3; 48 h at sea plus 12 h ashore.

3: South Shetland Islands (Trivelpiece *et al.*, in press): 5·3 h at sea at 8 W. kg^{-1}, plus 18·7 h ashore at 2·9 W kg^{-1}.

Gentoo (weight *c.* 5·8 kg): South Shetland Islands (Trivelpiece *et al.*, in press): 6·1 h at sea at 7·5 W kg^{-1} (Davis *et al.*, 1983) plus 17·9 h ashore at 2·7 W kg^{-1} (Davis *et al.*, 1983).

King (weight *c.* 13 kg): South Georgia (Kooyman *et al.*, 1982): 3·7 days at sea at 6 W kg^{-1} (Kooyman *et al.*, 1982), plus 1·3 days ashore at 2·3 W kg^{-1}.

[c] Direct measurements of weight of meal delivered. For King see Stonehouse (1960).

[d] S. Orkney Is. krill mean length 33·6 mm (Lishman, 1985*a*); S. Shetland Is. krill mean length 48 mm (Chinstrap) and 50 mm (Gentoo), calculated from Volkman *et al.* (1980). Mean weight calculated using regression equations in Lockyer (1973). Mean squid weight 420 g (M. R. Clarke, unpublished).

[e] For Chinstrap 3 and Gentoo, first value in this and following columns is for 'searching' and 'feeding' dives combined, second value for 'feeding' dives only.

[f] Assumed.

travelling). In reality, when penguins are feeding in a krill swarm, individuals may well be caught much more rapidly than this; on the other hand, some dives may be either exploratory or entirely unsuccessful.

These calculations are based on what may be conservative values for penguin energy requirements as Nagy, Siegfried & Wilson (1984) estimated that, during a 9 h foraging trip, Jackass Penguins (weight 3.17 kg) require 1860 kJ (equivalent to 18.1 W kg^{-1}) and while brooding or resting they need 52 kJ h^{-1} (i.e. 4.6 W kg^{-1}). For a complete foraging-attendance cycle they would need to catch about 190 anchovies of 5 cm mean length.

Role of feeding ecology in ecological segregation

Breeding seabird communities with more than one species of penguin are only found south of 47° S. The most common combinations of co-occurring species are as follows (see Fig. 6.1):
Antarctic Continent: Emperor, Adélie
Sub-Antarctic Indian Ocean (and Macquarie I.): King, Gentoo, Rockhopper, Macaroni (or Royal)
New Zealand region: Rockhopper, Erect-crested, Yellow-eyed
Antarctic Peninsula and southern Scotia Sea: Adélie, Chinstrap, Gentoo (Macaroni rare and local).
Other interesting combinations are King, Gentoo and Macaroni at South Georgia (Chinstrap very local) and Gentoo, Magellanic and Rockhopper at the Falkland Islands (King and Macaroni very local).

The differences in general biology, life cycle and population dynamics between *Aptenodytes* and other penguins (see Croxall, 1984) are so substantial that their major differences in diet and feeding ecology probably play only a small role in ecological segregation. Amongst the other species above, however, most of which have broadly similar breeding timetables, differences in diet and feeding area (both vertically and horizontally) might be much more important. It is noticeable that most multi-species sites include only one 'inshore' species (usually Gentoo Penguin (which is 'replaced' by Yellow-eyed Penguin at the New Zealand sub-Antarctic Islands)), although the Falkland Islands are anomalous and the ecological interactions between the resident Gentoo Penguin and the migratory Magellanic Penguin would make a very interesting study. Except for South Georgia, most sites have two species of 'offshore' penguins, usually two *Eudyptes* species, except in the maritime Antarctic where all three *Pygoscelis* co-occur. The main interest, therefore, lies in how these pairs of offshore species interact. The only pertinent data come from the brief study of Royal and Rockhopper Penguins at Macquarie

Island (Horne, 1985) and the more detailed studies of Adélie and Chinstrap Penguins at the South Shetlands and South Orkney Islands (Volkman *et al.*, 1980; Trivelpiece, 1981; Lishman, 1985*a*,*b*).

At Macquarie, Horne (1985) found that, late in the chick-rearing period, Royal and Rockhopper Penguins prey upon similar species (though in different proportions) and may have similar foraging ranges, although the stomach contents of Royal Penguins were consistently more digested, suggesting that they feed further away from the colony. Apart from choice of breeding habitat and a 30-day difference in breeding timetable (Warham, 1963), there are few other significant differences in the breeding biology of these two species and it remains uncertain how (and to what extent) they are segregated ecologically during the breeding season.

Lishman (1985*b*) concluded that many of the biological differences between the essentially very similar Adélie and Chinstrap Penguins stem from Adélies breeding about 4 weeks earlier. The only differences in their food and feeding ecology at Signy Island are that Chinstrap chicks are fed more frequently and on a diet containing slightly more of the larger krill (Lishman, 1985*a*), which are more nutritious than juvenile krill (Clarke,

Fig. 6.5. Length-frequency distributions of krill consumed by Adélie, Chinstrap, Gentoo and Macaroni Penguins.

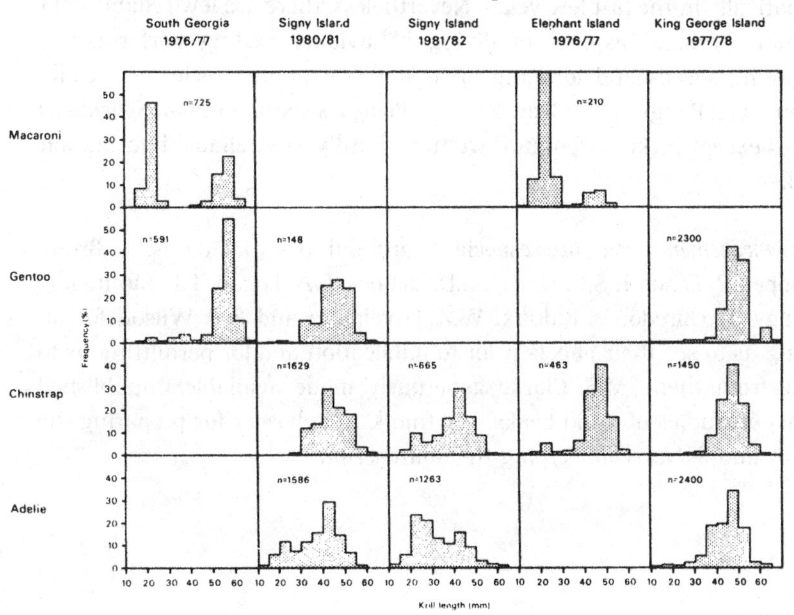

1984), the latter being more common in Adélie Penguin diets (Fig. 6.5). However, the differences in the sizes of krill taken, although persisting in successive, very different climatically, seasons at Signy Island, appear much less pronounced at King George Island (Fig. 6.5). Although Adélie Penguins have slightly smaller bills and may be better adapted to feeding on small krill (e.g. they depend on *Euphausia crystallorophias* at Cape Crozier, Ross Sea), differences of the kind found at the Antarctic Peninsula sites might also relate to the differences in breeding timetable, with perhaps more juvenile krill available earlier in the season. However, comparison of the length-frequencies of krill in the diet of Adélie and Chinstrap Penguins at Signy Island when they were both simultaneously rearing chicks, showed that Adelies still took a greater proportion of smaller krill (Lishman, 1985*a*). Unless the two species feed in different areas (which is quite possible), it may be that there is some selection for different-sized prey, which might contribute to the ecological segregation of these penguins.

It is, of course, possible that the resources exploited by *Pygoscelis* and *Eudyptes* penguins in their breeding seasons are sufficiently abundant or distributed in such a way that competition at this time rarely occurs and that it is the nature of their interactions in winter that is critical.

Considering the difficulties of studying active diving predators at sea, our knowledge of the food and feeding ecology of penguins has increased dramatically in the last few years. Nevertheless, there are few reliable data on many crucial aspects of diving behaviour, feeding performance, swimming speeds and foraging areas and, for many species, especially Yellow-eyed Penguin, the New Zealand *Eudyptes* species and all *Spheniscus* species except Jackass Penguin, we have hardly any reliable information at all.

Acknowledgements We are especially grateful to N.J. Adams, S. Broni, J. Cooper, P. Ensor, R.S.C. Horne, G.D. La Cock, L.A. Logan, T.L. Montague, K. Nagy, C. Offredo, V. Ridoux, W.Z. Trivelpiece and R.P. Wilson for allowing us to see their papers prior to publication and for permitting us to quote from them. M.R. Clarke generously made available unpublished results of studies of squid beaks. We thank A. Sylvester for preparing the figures and S. Norris for typing the manuscript.

References

Adams, N.J. & Brown, C.R. (1983). Diving depths of the Gentoo Penguin, *Pygoscelis papua*. *Condor*, **85**, 503–4.

Adams, N.J. & Klages, N.T. (In press). Diet of the King Penguin *Aptenodytes patagonicus* at sub-antarctic Marion Island. *J. Zool., Lond.*

Ainley, D.G. & Emison, W.B. (1972). Sexual size dimorphism in Adélie Penguins. *Ibis*, **114**, 267–71.

Ainley, D.G., O'Connor, E.F. & Boekelheide, R.J. (1984). *The marine ecology of birds in the Ross Sea, Antarctica*. American Ornithologists' Union, Ornithological Monographs No. 32, 97 pp.

Barton, D. (1979). Swimming speed of a Little Penguin. *Emu*, **79**, 141–2.

Blackburn, M. (1950*a*). A biological study of the anchovy *Engraulis australis* (White) in Australian waters. *Aust. J. Mar. Freshw. Res.*, **1**, 3–84.

Blackburn, M. (1950*b*). Studies on the age, growth and life history of the pilchard *Sardinops neopilchardus* (Steindachner) in Southern and Western Australia. *Aust. J. Mar. Freshw. Res.*, **2**, 179–92.

Boersma, P.D. (1976). An ecological and behavioral study of the Galapagos Penguin. *Living Bird*, **15**, 43–93.

Boswell, J. & MacIver, D. (1975). The Magellanic Penguin *Spheniscus magellanicus*. In *The biology of penguins*, ed. B. Stonehouse, pp. 271–305. London: Macmillan.

Butler, P.J.& Jones, D.R. (1982). The comparative physiology of diving in vertebrates. *Adv. Comp. Biochem. Physiol.*, **8**, 180–364.

Carrick, R. (1972). Population ecology of the Australian Black-backed Magpie, Royal Penguin and Silver Gull. In *Population ecology of migratory birds: a symposium*. United States Department of the Interior Wildlife Research Report, No. 2, pp. 41–99.

Clark, B.D.& Bemis, W. (1975). Kinematics of swimming of penguins at the Detroit Zoo. *J. Zool., Lond.*, **188**, 411–28.

Clarke, A. (1984). Lipid content and composition of Antarctic Krill *Euphausia superba* Dana. *J. Crustacean Biol.*, 4 Spec. No. 1, 283–92.

Clarke, A.& Prince, P.A. (1980). Chemical composition and calorific value of food fed to mollymauk chicks at Bird Island, South Georgia. *Ibis*, **122**, 488–94.

Clarke, M.R. (1962). The identification of cephalopod 'beaks' and the relationship between beak size and total body weight. *Bull. Br. Mus. (Nat. Hist.), Zool.*, **8**, 422–80.

Clarke, M.R. (1980). Cephalopoda in the diet of Sperm Whales of the Southern Hemisphere and their bearing on Sperm Whale biology. *Discovery Rep.*, **37**, 1–324.

Conroy, J.W.H., White, M.G., Furse, J.R.& Bruce, G. (1975). Observations on the breeding biology of the Chinstrap Penguin *Pygoscelis antarctica*, at Elephant Island, South Shetland Islands. *Bull. Br. Antarct. Surv.*, **40**, 23–32.

Cooper, J., Enticott, J.W., Hecht, T. & Klages, N. (1984). Prey from three Chinstrap Penguins *Pygoscelis antarctica* at Bouvet Island, December 1982. *S. Afr. J. Antarct. Res.*, **14**, 32–3.

Crawford, R.J.M. & Shelton, P.A. (1981). Population trends for some southern African seabirds related to fish availability. In *Birds of the sea and shore*, ed. J. Cooper, pp. 15–41. Cape Town: African Seabird Group.

Croxall, J.P. (1984). Seabirds. In *Antarctic ecology*, vol. 2, ed. R.M. Laws, pp. 533–619. London: Academic Press.

Croxall, J.P., Evans, P.G.H. & Schreiber, R.W. (eds) (1984*a*). *Status and conservation of the world's seabirds*. Cambridge: ICBP.

Croxall, J.P., Everson, I., Kooyman, G.L., Ricketts, C.& Davis, R.W. (1985*a*). Fur seal diving behaviour in relation to krill vertical distribution. *J. Anim. Ecol.*, **54**, 1–8.

Croxall, J.P. & Furse, J.R. (1980). Food of Chinstrap Penguins *Pygoscelis antarctica* and Macaroni Penguins *Eudyptes chrysolophus* at Elephant Island, South Shetland Islands. *Ibis*, **122**, 237–45.

Croxall, J.P. & Prince, P.A. (1980*a*). The food of the Gentoo Penguins *Pygoscelis papua* and Macaroni Penguin *Eudyptes chrysolophus* at South Georgia. *Ibis*, **122**, 245–53.

Croxall, J.P.& Prince, P.A. (1980*b*). Food, feeding ecology and ecological segregation of seabirds at South Georgia. *Biol. J. Linn. Soc.*, **14**, 103–31.

Croxall, J.P.& Prince, P.A. (1982*a*). Calorific content of squid (Mollusca: Cephalopoda). *Bull. Br. Antarct. Surv.*, **55**, 27–31.

Croxall, J.P. & Prince, P.A. (1982*b*). A preliminary assessment of the impact of seabirds on marine resources at South Georgia. *Com. Nat. Fr. Recherch. Antarct.*, **51**, 501–9.

Croxall, J.P., Prince, P.A., Baird, A. & Ward, P. (1985*b*). The diet of the Southern Rockhopper Penguin *Eudyptes chrysocome chrysocome* at Beauchêne Island, Falkland Islands. *J. Zool., Lond.*, **206**, 485–96.

Croxall, J.P., Prince, P.A. & Ricketts, C. (1985*c*). Relationships between prey life-cycles and the extent, nature and timing of seal and seabird predation in the Scotia Sea. In *Antarctic nutrient cycles and food webs*, ed. W.R. Siegfried, P. Condy & R.M. Laws, pp. 516–33. Berlin: Springer-Verlag.

Croxall, J.P., Ricketts, C. & Prince, P.A. (1984*b*). The impact of seabirds on marine resources, especially krill, at South Georgia. In *Seabird energetics*, ed. G.C. Whittow & H. Rahn, pp. 285–318. New York: Plenum Publishing Corporation.

Davies, D.H. (1955). The South African Pilchard *Sardinops ocellata*: bird predators, 1953–1954. *Invest. Rep. Div. Fish., Un. S. Afr.*, **18**, 1–32.

Davies, D.H. (1956). The South African Pilchard *Sardinops ocellata* and Maasbanker *Trachurus trachurus*: bird predators, 1954–1955. *Invest. Rep. Div. Fish., Un. S. Afr.*, **23**, 1–40.

Davis, R.W., Kooyman, G.L.& Croxall, J.P. (1983). Water flux and estimated metabolism of free-ranging Gentoo and Macaroni Penguins at South Georgia. *Polar Biol.*, **2**, 41–6.

Duroselle, T. & Tollu, B. (1977). The Rockhopper Penguin *Eudyptes chrysocome moseleyi* of Saint Paul and Amsterdam Islands. In *Adaptations within Antarctic ecosystems*, ed. G.A. Llano, pp. 579–604. Washington, D.C.: Smithsonian Institution.

Ealey, E.H.M. (1954). Analysis of stomach contents of some Heard Island birds. *Emu*, **54**, 204–9.

Emison, W.B. (1968). Feeding preferences of the Adélie Penguin at Cape Crozier, Ross Island. In *Antarctic bird studies*, ed. O.L. Austin, Jr, pp. 191–212. Washington: American Geophysical Union.

Everson, I. (1981). Fish. In *Biological investigations of marine Antarctic systems and stocks*, vol. II, pp. 79–97. Cambridge: SCAR/SCOR.

Falla, R.A. (1937). Birds. *Rep. B.A.N.Z. Antarct. Res. Exped.*, Series B, **2**, 1–288.

Horne, R.S.C. (1985). Diet of Royal and Rockhopper Penguins at Macquarie Island. *Emu*, **85**, 150–56.

Kinsky, F.C. (1960). The yearly cycle of the Northern Blue Penguin *Eudyptula minor novaehollandiae* in the Wellington Harbour area. *Rec. Dom. Mus. Wellington*, **3**, 145–218.

Klages, N. & Nast, N. (1981). Net selection for Antarctic Krill by the 1216 meshes krill trawl. *Arch. FischWiss.*, **31**, 169–74.

Kooyman, G.L. (1975). The physiology of diving in penguins. In *The biology of penguins*, ed. B. Stonehouse, pp. 115–37. London: Macmillan.

Kooyman, G.L., Billups, J.O. & Farwell, W.D. (1983). Two recently developed recorders for monitoring diving activity of marine birds and mammals. In *Experimental biology at sea*, ed. A.G. MacDonald & I.G. Priede, pp. 197–214. New York: Academic Press.

Kooyman, G.L., Davis, R.W., Croxall, J.P. & Costa, D.P. (1982). Diving depths and energy requirements of King Penguins. *Science*, **217**, 726–7.

Kooyman, G.L., Drabek, C.M., Elsner, R. & Campbell, W.B. (1971). Diving behavior of the Emperor Penguin *Aptenodytes forsteri*. *Auk*, **88**, 575–95.

La Cock, G.D., Hecht, T. & Klages, N. (1984). The winter diet of Gentoo Penguins at Marion Island. *Ostrich*, **55**, 188–91.

Lishman, G.S. (1983). The comparative breeding biology, feeding ecology and bioenergetics of Adélie and Chinstrap Penguins. Unpublished D.Phil. thesis, University of Oxford.

Lishman, G.S. (1985*a*). The food and feeding ecology of Adélie and Chinstrap Penguins at Signy Island, South Orkney Islands. *J. Zool. Lond.*, **205**, 245–63.

Lishman, G.S. (1985*b*). The comparative breeding biology of Adélie and Chinstrap Penguins at Signy Island, South Orkney Islands. *Ibis*, **127**, 84–99.

Lishman, G.S. & Croxall, J.P. (1983). Diving depths of the Chinstrap Penguins *Pygoscelis antarctica*. *Bull. Br. Antarct. Surv.*, **61**, 21–5.

Lockyer, C. (1973). Wet weight, volume and length correlation in the Antarctic Krill *Euphausia superba*. *Discovery Rep.*, **36**, 152–5.

Mauchline, J. & Fisher, L.R. (1969). The biology of euphausiids. *Adv. Mar. Biol.*, **7**, 1–454.

Montague, T.L. (1982). The food and feeding ecology of the Little Penguin *Eudyptula minor* at Phillip Island, Victoria, Australia. Unpublished M.Sc. thesis, Monash University.

Mougin, J.-L. & Prévost, J. (1980). Evolution annuelle des effectifs et des biomasses des oiseaux antarctiques. *Rev. Ecol., Terre Vie*, **34**, 101–33.

Murphy, R.C. (1936). *Oceanic birds of South America*. New York: Macmillan.

Nachtigall, W. & Bilo, D. (1980). Stromungsanpassung des Pinguins beim Schwimmen unter Wasser. *J. Comp. Physiol.*, **137A**, 17–26.

Nagy, K.A., Siegfried, W.R. & Wilson, R.P. (1984). Energy utilization by free-ranging Jackass Penguins *Spheniscus demersus*. *Ecology*, **65**, 1648–55.

Offredo, C. & Ridoux, V. (1986). The diet of Emperor Penguins *Aptenodytes forsteri* in Adélie Land, Antarctica. *Ibis*, **128**, 409–413.

Offredo, C., Ridoux, V. & Clarke, M.R. (1985). Cephalopods in the diet of Emperor and Adélie Penguins in Adélie Land, Antarctica. *Mar. Biol.*, **86**, 199–202.

Paulin, C.D. (1975). Feeding of the Adélie Penguin *Pygoscelis adeliae*. *Mauri Ora*, **3**, 27–30.

Poulter, T.C. (1969). Sonar of penguins and fur seals. *Proc. Calif. Acad. Sci.*, **36**, 363–80.

Prévost, J. (1961). *Ecologie du manchot empereur*. Paris: Hermann.

Rand, R.W. (1960). The biology of guano-producing seabirds. The distribution, abundance and feeding habits of the Cape Penguin, *Spheniscus demersus* off the south-western coast of the Cape Province. *Invest. Rep. Div. Fish., Un. S. Afr.*, **41**, 1–28.

Randall, R.M. & Randall, B.M. (1981). The annual cycle of the Jackass Penguin *Spheniscus demersus* at St Croix Island, South Africa. In *Proceedings of the symposium on birds of the sea and shore* 1979, ed. J. Cooper, pp. 427–50. Cape Town: African Seabird Group.

Randall, R.M., Randall, B.M. & Baird, D. (1981*a*). Speed of movement of Jackass Penguins and their possible use of ocean currents. *S. Afr. J. Sci.*, **77**, 420–1.

Randall, R.M., Randall, B.M. & Klingelhoeffer, E.W. (1981*b*). Species diversity and size ranges of cephalopods in the diet of Jackass Penguins from Algoa Bay, South Africa. *S. Afr. J. Zool.*, **16**, 163–6.

Reilly, P.N. & Balmford, P. (1975). A breeding study of the Little Penguin *Eudyptula minor* in Australia. In *The biology of penguins*, ed. B. Stonehouse, pp. 161–87. London: Macmillan.

Richdale, L.E. (1940). Random notes on the genus *Eudyptula* on the Otago Peninsula, New Zealand. *Emu*, **40**, 180–217.

Richdale, L.E. (1941). The Erect-crested Penguin *Eudyptes sclateri*. *Emu*, **41**, 25–53.

Richdale, L.E. (1957). *A population study of penguins*. Oxford: Clarendon Press.

Siegfried, W.R., Frost, P.G.H., Kinahan, J.B. & Cooper, J. (1975). Social behaviour of Jackass Penguins at sea. *Zoologica Afr.*, **10**, 87–100.

Sladen, W.J.L. (1958). The pygoscelid penguins: I. Methods of study; II. The Adélie Penguin *Pygoscelis adeliae* (Hombron & Jacquinot). *Sci. Rep. Falkl. Isl. Depend. Surv.*, No. 17, 97 pp.

Stepnik, R. (1982). All-year populational studies of Euphausiacea (Crustacea) in the Admiralty Bay (King George Island, South Shetland Islands, Antarctica). *Pol. Polar Res.*, **3**, 49–68.

Stonehouse, B. (1953). The Emperor Penguin *Aptenodytes forsteri*. I. Breeding behaviour and development. *Sci. Rep. Falkl. Isl. Depend. Surv.*, No. 6, 1–33.

Stonehouse, B. (1960). The King Penguin *Aptenodytes patagonicus* of South Georgia. I. Breeding behaviour and development. *Sci. Rep. Falkl. Isl. Depend. Surv.*, No. 23, 1–81.

Taylor, R.H. (1962). The Adélie Penguin *Pygoscelis adeliae* at Cape Royds. *Ibis*, **104**, 176–204.

Trivelpiece, W.Z. (1981). Ecological studies of pygoscelid penguins and Antarctic skuas. Unpublished Ph.D. thesis, State University of New York.

Trivelpiece, W.Z., Bengston, J.L., Trivelpiece, S.G. & Volkman, N.J. (In press). Foraging behavior of Gentoo and Chinstrap Penguins. *Auk*.

Trivelpiece, W.Z., Trivelpiece, S.G., Volkman, N.J. & Ware, S.H. (1983). Breeding and feeding ecologies of pygoscelid penguins. *Antarct. J. U.S.*, **18**, 209–10.

Venegas, C. & Sielfeld, W. (1981). Utilizacion de aves como indicadores de presencia y potencialidad de recursos marinos eventualmente manejables. *Jornadas Cienc. Mar*, **12–14**, 83.

Volkman, N.J., Presler, P. & Trivelpiece, W.Z. (1980). Diets of pygoscelid penguins at King George Island, Antarctica. *Condor*, **82**, 373–8.

Warham, J. (1963). The Rockhopper Penguin *Eudyptes chrysocome* at Macquarie Island. *Auk*, **80**, 229–56.

Warham, J. (1972). Aspects of the biology of the Erect-crested Penguin *Eudyptes sclateri*. *Ardea*, **60**, 145–84.

Warham, J. (1974*a*). Breeding biology and behaviour of the Snares Crested penguin. *J. Roy. Soc. N.Z.*, **4**, 63–108.
Warham, J. (1974*b*). The Fiordland Crested Penguin *Eudyptes pachyrhynchus*. *Ibis*, **116**, 1–27.
Westerkov, K. (1960). Birds of Campbell Island. *N.Z. Dep. Intern. Aff., Wildl. Publ.*, No. 61, 1–83.
Williams, A.J. (1980). Aspects of the breeding biology of the Gentoo Penguin *Pygoscelis papua*. *Gerfaut*, **70**, 283–95.
Williams, A.J. (1982). Chick feeding rates of Macaroni and Rockhopper Penguins at Marion Island. *Ostrich*, **53**, 129–34.
Williams, A.J. & Laycock, P.A. (1981). Euphausiids in the diet of some Subantarctic *Eudyptes* penguins. *S. Afr. J. Antarct. Res.*, **10/11**, 27–8.
Williams, A.J. & Siegfried, W.R. (1980). Foraging ranges of krill-eating penguins. *Polar Rec.*, **125**, 159–62.
Wilson, G.J. (ed.) (1983). Distribution and abundance of Antarctic and Subantarctic penguins: a synthesis of current knowledge. *BIOMASS Sci. Ser.*, No. 4, 46 pp.
Wilson, R.P. (1985). Seasonality in diet and breeding success of the Jackass Penguin *Spheniscus demersus*. *J. Orn.*, **126**, 53–62.
Wilson, R.P. & Bain, C.A.R. (1984*a*). An inexpensive depth gauge for penguins. *J. Wildl. Mgmt*, **48**, 1077–84.
Wilson, R.P. & Bain, C.A.R. (1984*b*). An inexpensive speed meter for penguins at sea. *J. Wildl. Mgmt*, **48**, 1360–4.
Zusi, R.L. (1975). An interpretation of skull structure in penguins. In *The biology of penguins*, ed. B. Stonehouse, pp. 59–84. London: Macmillan.

Appendix 6.1. *Synopsis of quantitative dietary data for penguins*

Species and site	Latitude °S	Sample		% Krill				% Other crustacea		% Fish		% Squid		Main prey	References
				Adult/large		Juv/small									
		Size	Period	By wt.	By no.	By wt.	By no.	By wt.	By no.	By wt.	By no.	By wt.	By no.		
Emperor															
Adélie Land	67	29	Chick	2	31	—	—	< 0·1	3	95	65	3	1	*Tr. borchgrevinkii*	1
King															
Marion I.	47	120	All year	—	—	—	—	—	—	86·5	74·2	13·5	19·1	Myctophidae	31
S. Georgia	54	10	Chick	—	—	—	—	—	—	30	10[a]	70	90[a]	Squid	2, 3
Gentoo															
Marion I.	47	64	Creche	—	—	—	—	*c.* 30	75[b]	*c.* 70	72[b]	?	13[b]	*Notothenia squamifrons*	4
S. Georgia	54	43	Creche	66	88[c]	2	12[c]	—	—	32	?[d]	—	—	*E. superba*	5
S. Shetland Is.	62	46	Pre-hatch	84·5[e]	98·2	?	0·2	0·1	0·1	15·4	1·5	—	—	*E. superba*	6
Adélie															
S. Orkney Is.	60	15	Pre-hatch	75·7	64·6	22·6	34·8	0·3	0·6	1·4	< 0·1	—	—	*E. superba*	7
	60	13	To fledge	72·4	36·9	26·8	62·8	0·4	0·3	0·4	< 0·1	—	—	*E. superba*	7
S. Shetland Is.	62	48	To creche	99·1[e]	94·7	?	5·0	0·5	0·2	0·4	< 0·1	—	—	*E. superba*	6
Cape Hallett	72	76	To creche	*c.*65[f]	77	*c.* 7[f]	23	*c.* 1[f]	3	27	9	—	—	*E. superba*	8
		73	Creche	56	100[b]	1	?	1	62[b]	2	48[b]	—	—	*E. superba*	9
Beaufort I.	76	5	To fledge	—	—	?	63·7	?	4·8	?	31·5	—	—	*E. crystallorophias*	10
Cape Bird	77	15	Chick	—	—	46[g]	75·0	10[g]	3·0	44[g]	22·0	—	—	*E. crystallorophias*	11
Cape Crozier	77	16	Pre-hatch	—	—	?	91·5	?	1·2	?	7·3	—	—	*E. crystallorophias*	10
	77	15	To fledge	—	—	?	94·6	?	1·4	?	4·0	—	—	*E. crystallorophias*	10
Ross Sea	77[h]	3	?	—	—	99·9	99·6	< 0·1	0·3	0·1	0·1	—	—	*E. crystallorophias*	12
	77[j]	2	?	97·7	99·3	—	—	< 0·1	0·6	2·3	0·1	—	—	*E. superba*	12
Chinstrap															
S. Orkney Is.	60	21	Pre-hatch	92·2	86·8	4·8	13·0	< 0·1	0·1	3·0	< 0·1	—	—	*E. superba*	7
	60	14	To fledge	89·8	72·0	10·0	28·0	< 0·1	< 0·1	0·2	< 0·1	—	—	*E. superba*	7
Elephant Is.	61	46	Pre-hatch	94·0	83·0[c]	2·0	17·0[c]	< 0·1	< 0·1	4·0	?[d]	—	—	*E. superba*	13
S. Shetland Is.	62	29	To creche	99·6	98·0	?	2·0	0·1	< 0·1	0·3	< 0·1	—	—	*E. superba*	6
Macaroni															
S. Georgia	54	40	Creche	80·0	21·0	18·0	77·0	< 0·1	2·0	2·0	< 0·1	—	—	*E. superba*	5
Elephant Is.	61	37	Guard/ creche	29·0	5·0[c]	46·0	95·0[c]	—	—	25·0	?[d]	—	—	*E. superba*, *Thysanoessa* sp.	13
Royal															
Macquarie I.	54	29	Creche	3	?	?	?	8	?	54	?	36	?	Myctophidae	14
	54	21	Creche	29	?	?	?	0·5	?	62	?	9	?	Myctophidae	14

N. Rockhopper															
Amsterdam I.	37	100	Guard/ creche	?	25·0	?	?	?	10	?	10	?	50	Squid	15
Gough I.	41	5	Chick	?	—	39[g]	—	—	—	?	—	?	—	*Thy. vicina*	16
S. Rockhopper															
Falkland Is.	52	29	Early guard	—	—	45	79[c]	—	—	2	?	53	21[c]	Squid, *E. lucens*	17
Macquarie I.	54	19	Creche	—	—	77	?	0·3	?	17	?	6	?	*E. vallentinii*	14
Erect-crested															
S. New Zealand	45	?	?	—	—	—	—	—	—	—	—	—	—	Fish &/or squid	18
Snares Crested															
Snares Is.	48	?	?	—	—	—	—	—	—	—	—	—	—	Squid &/or crustaceans	19
Fiordland															
S. New Zealand	45	?	?	—	—	—	—	—	—	—	—	—	—	Squid &/or krill	20
Yellow-eyed															
S. New Zealand	45	?	?	—	—	—	—	—	—	?	?	50	?	Fish and squid	21, 22
Little (Blue)															
S. Australia	38	700	All year	—	—	< 0·1[g]	< 0·1[b]	< 0·1[g]	< 1·0[b]	90–97[g]	73[b]	3–10[g]	25[b]	*Engraulis australis* *S. neopilchardus*	23
Magellanic															
S. Chile	45	32	?	—	—	—	—	9·5	?	90·5	?	?	?	*Ramnogaster arcuata*	24
Humboldt															
Peru	13	?	?	—	—	—	—	—	—	—	—	—	—	*En. ringens*	25
Galapagos															
Galapagos	1	?	?	—	—	—	—	—	—	—	—	—	—	Fish	25, 26
Jackass															
S. Africa	33	16	All year	—	—	—	—	c. 2	c. 8[b]	98	92[b]	—	—	*S. ocellata*	27
	33	112	All year	—	—	—	—	< 1	1[b]	98	97[b]	2	2[b]	*S. ocellata*	28
	33	247	All year	—	—	< 0·1	4	1	13	94	68	5	11	*S. ocellata* *En. capensis*	29
	33	566	All year	—	—	—	—	—	—	c. 98	c. 90	c. 2	c. 8	*En. capensis*	30

[a] Qualitative. [b] Frequency of occurrence. [c] Does not account for the unknown number of fish. [d] Too digested to count. [e] Includes weight of juvenile krill. [f] Estimated subdivision of total krill plus other crustacea = 73% by weight. [g] By volume. [h] Shelf. [j] Slope.

References: 1, Offredo & Ridoux, 1986; 2, Croxall & Prince, 1980*b*; 3, Croxall & Prince, 1982*b*; 4, La Cock *et al.*, 1984; 5, Croxall & Prince, 1980*a*; 6, Volkman *et al.*, 1980; 7, Lishman, 1985*a*; 8, F. C. Kinsky & P. Ensor, unpublished; 9, L. Logan, unpublished; 10, Emison, 1968; 11, Paulin, 1975; 12, Ainley *et al.*, 1984; 13, Croxall & Furse, 1980; 14, Horne, 1985; 15, Duroselle & Tollu, 1977; 16, Williams & Laycock, 1981; 17, Croxall *et al.*, 1985*b*; 18, Richdale, 1941; 19, Warham, 1974*a*; 20, Warham, 1974*b*; 21, Westerkov, 1960; 22, J. T. Darby, personal communication; 23, Montague, 1982; 24, Venegas & Sielfeld, 1981; 25, Murphy, 1936; 26, Boersma, 1976; 27, Davies, 1955; 28, Davies, 1956; 29, Rand, 1969; 30, Wilson, 1985; 31, Adams & Klages, in press.

7

Diet and feeding ecology of Procellariiformes

P. A. PRINCE and R. A. MORGAN
British Antarctic Survey, Natural Environment Research Council, High Cross, Madingley Road, Cambridge CB3 0ET, UK

Introduction

The order Procellariiformes comprises four families: Diomedeidae (albatrosses; 14 species in 2 genera), Procellariidae (petrels and shearwaters; *c.* 65 species in 12 genera), Hydrobatidae (storm petrels; 20 species in 8 genera) and Pelecanoididae (diving petrels; 4 species). The range of body mass, from 25 g storm petrels to 12 kg albatrosses is the largest of any order of birds. The body mass and bill dimension characteristics of most species in the order are illustrated in Fig. 7.1, which emphasises the essential structural homogeneity of the group – except for diving petrels whose body form shows strong convergence with the auks Alcidae.

A diagnostic characteristic of the order is the possession of external tubular nostrils, associated with extensive development of the olfactory part of the brain. All species have grooved, hooked bills and their webbed feet are set well back on the body. Only albatrosses and giant petrels show any real competence in walking on land. Albatrosses, petrels and shearwaters are particularly adapted for long-distance and wide-ranging economic flight, having proportionately large wings with high aspect ratios, and rely extensively on gliding flight. Diving petrels are quite different, having stocky bodies propelled by 'whirring' flight which readily continues underwater. Diving petrels are also the only members of the order that do not accumulate stomach oils in the proventriculus (Warham, 1977*a*). Stomach oils result from the breakdown of ingested food; following Ashmole & Ashmole (1967) they have usually been regarded as an adaptation for combining the provision of a concentrated high-energy food to the chick with a reduction in the costs of transport to the adult (by eliminating a large volume of water). Many studies (reviewed in Warham (1977*a*) and Jacob (1982)) have confirmed that the decanted oil

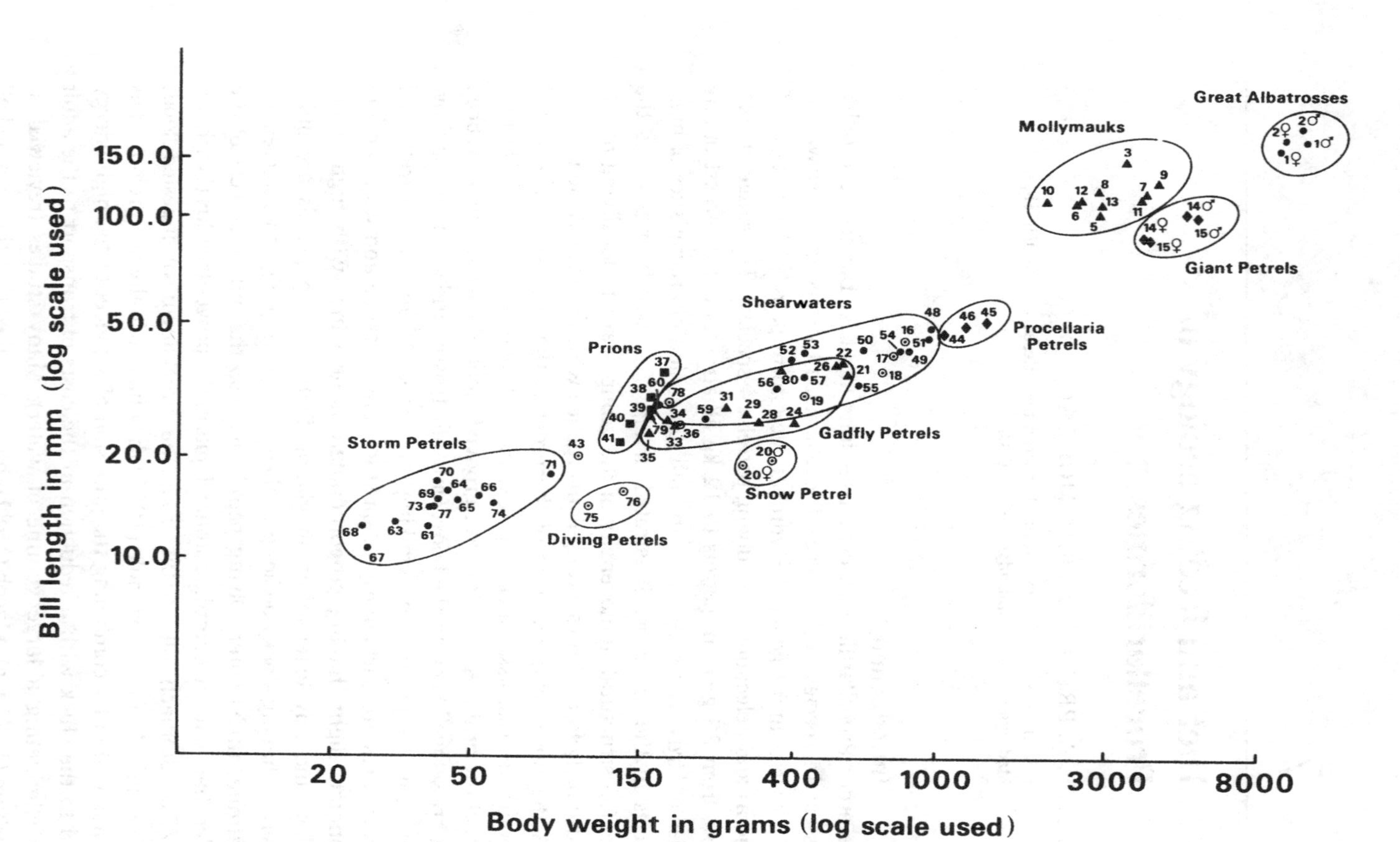

Fig. 7.1. Relations between body weight and bill length in Procellariiformes. Species numbers as in Table 7.1 with the addition of *Oceanodroma monorhis* (77), *Bulweria fallax* (78), *Pterodroma pycrofti* (79), *P. rostrata* (80).

is rich in lipids, but of the liquid passed to albatross chicks (50% by weight of feeds) up to 70–90% was water (Clarke & Prince, 1980), so the whole topic still needs critical investigation.

Nearly two-thirds of all Procellariiformes breed in the southern hemisphere. Some species are comparatively sedentary, while others perform circumpolar movements or migrate between hemispheres. Generally, species occurring at high latitudes move towards more temperate regions during winter, and species breeding in the tropics are normally fairly sedentary. The wide distribution of Procellariiformes and the abundance of many species suggests that they play a potentially important role as top consumers in pelagic marine systems. Until very recently, knowledge of their diet was limited to information derived from birds taken for museum specimens, and reports on feeding ecology were largely anecdotal. Even after the pioneering work by Ashmole & Ashmole (1967) and Imber (1973), quantitative studies of procellariiform diet were slow to develop, and only in this decade has really significant progress been made in investigating their diet and feeding ecology. This chapter reviews available information on their diet and the methods by which they obtain food. Most of this information is summarised in Table 7.1. For some species this is still based entirely on anecdotal information, for others on data for selected components of the diet; only for a few species have more comprehensive analyses been undertaken. Most of these are treated in detail later in the chapter.

Diet

Albatrosses Diomedeidae

Squid are important food for albatrosses and recent improvements in the identification of squid beaks (which are retained in the birds' stomachs and periodically regurgitated) have provided detailed information on this element of their diet (see below). However, quantitative studies of the entire diet are limited to four albatross species at South Georgia (Subantarctic Atlantic Ocean) and two species at the Hawaiian Islands (tropical Pacific Ocean).

At South Georgia, three species of similar size breed in the summer and were sampled while rearing chicks. Even when comparing the proportion of the main prey classes in the diet (Table 7.2) there are some significant differences. Thus, squid forms almost half the food fed to chicks of Grey-headed Albatrosses *Diomedea chrysostoma* and Light-mantled Sooty Albatrosses *Phoebetria palpebrata* but only 20% of food fed to Black-browed Albatross *Diomedea melanophrys* chicks. By weight fish is most important

Table 7.1. *Feeding methods and diet of procellariiform seabirds. (Symbols for feeding methods and type of prey: 1 minor importance, 2 moderate importance, 3 major importance, (1) importance unknown, ? feeding method probably used.)*

Species	Feeding methods: Diving		Plunging		Surface feeding					Flight feeding				Diet										References
	Pursuit-dive	Surface-dive	Surface-plunge	Shallow-plunge	Pursuit-plunge	Surface-seize	Surface-filter	Hydroplane	Foot paddling	Aerial pursuit	Dipping	Pattering	Piracy	Squid	Fish	Crustacea: Euphausiids	Amphipods	Copepods	Unidentified	Other: Molluscs	Other	Carrion/Refuse	
Diomedeidae – Albatrosses																							
1 *Diomedea exulans*				1	1	3								3	2	1			1			1	24. 32. 78. 132.
2 *D. epomophora*				1		3								3	3				1		1		1. 119. 125.
3 *D. irrorata*						3							1	3	3	3			2	1			5. 63.
4 *D. albatrus*						3								3	3				3			1	66.
5 *D. nigripes*						3								3	3	1	1	1	2	1	1	1	28. 65. 93.
6 *D. immutabilis*						3								3	2	1	1		2	1	2		65.
7 *D. melanophrys*		1	1		1	3								3	3	3	1	1	1		1	1	103. 112. 125. 130
8 *D. bulleri*		1				3								2	2	1							41. 125. 141.
9 *D. cauta*		1			1	3					1			3	1							1	9. 50. 103. 125.
10 *D. chlororhynchos*		1	1	1	1	3					1			3	1								9. 50. 52. 120. 137.
11 *D. chrysostoma*		1		1		3								3	3	2	1		1		1	1	32. 112. 130.
12 *Phoebetria fusca*						3								3								1	12. 50. 76. 137
13 *P. palpebrata*		1	1	1		3								3	2	3			1			1	1. 12. 48. 76. 128.
Procellariidae – Petrels etc.																							
14 *Macronectes halli*		1			1	3								2	2	3	2	1	2		2	3	3. 67. 82. 133.
15 *M. giganteus*		1				3								2	2	3	2	1	1		2	3	27. 67. 82.
16 *Fulmarus glacialoides*		1				3								3	2	3	2		3			3	13. 46. 125. 137.
17 *F. glacialis*					1	3								3	3	1	3	1	3	1	3	3	15. 16. 30. 45. 121.
18 *Thalassoica antarctica*		1	2	1	1	3					1	1		3	3	3	1		3	1	2	1	3. 13. 40. 49. 94. 134.
19 *Daption capense*		2			1	3			2			2		2	2	3	1	1			1	1	10. 13. 32. 39. 48. 50. 99.
20 *Pagodroma nivea*		2				3					3	1		3	3	3			2	1	1	2	3. 13. 33. 40. 48. 49. 91. 145.
21 *Pterodroma macroptera*						3					?			3	2			1	2		1		50. 52. 71. 108. 123.
22 *P. lessonii*						3								3	2								3. 34. 50. 71. 137.
23 *P. cahow*						2					2	1		3	1				1				5. 102. 142.
24 *P. phaeopygia*						2					2	1		3	2				1		1		5. 62. 88. 90.
25 *P. externa*						1					1	1		1	1								5. 75
26 *P. incerta*						3								3	2		1		2				48. 50. 140
27 *P. alba*						1					1	1		3	2	1			1		1		5. 6
28 *P. inexpectata*			2		1	3								2	3				1				3. 4. 75.
29 *P. brevirostris*						3								3	1				2				13. 34. 50. 108. 123.
30 *P. neglecta*						2					2	1		1					1				5. 106.
31 *P. mollis*						3								3	1	2							30. 34. 50. 52. 123. 137.
32 *P. cooki*						3								3	2	1	1		2			2	71. 75. 84.
33 *P. leucoptera*						?					?	?		1	1								5. 75.
34 *P. hypoleuca*						(1)					(1)			3	3	1	1	1	3		3		65.
35 *P. nigripennis*						2					2	1		(1)					(1)				5. 75.

Table 7.1. (*cont.*)

Species	Feeding methods: Diving		Plunging		Surface feeding					Flight feeding				Diet: Squid	Fish	Crustacea				Other			References
	Pursuit-dive	Surface-dive	Surface-plunge	Shallow-plunge	Pursuit-plunge	Surface-seize	Surface-filter	Hydroplane	Foot paddling	Aerial pursuit	Dipping	Pattering	Piracy			Euphausiids	Amphipods	Copepods	Unidentified	Molluscs	Other	Carrion/Refuse	
36 *Halobaena caerulea*		1				3					2			2	2	3	1	1	1				13. 32. 50. 111.
37 *Pachyptila vittata*		1				2	3	3			1			3	2	3	3	3	3	1	1		29. 40. 57. 73. 117. 118. 125. 137.
38 *P. salvini*		1		1	1	2	2	2						3	1	3				1	1		13. 57. 108. 125. 137.
39 *P. desolata*		1				3	1	1			2			1	1	3	1	2		1			3. 32. 36. 38. 54. 73. 111. 129.
40 *P. belcheri*		1				3					3	3		1	1		3						54. 127.
41 *P. turtur*		2				3					3	2		1	1	3	3			1	1		55. 73. 98. 118. 125. 131. 137.
42 *P. crassirostris*						3								2	1	3				1	1	2	38. 73. 137.
43 *Bulweria bulwerii*						3					1			2	3		1	1	1		1		19. 30. 65. 146.
44 *Procellaria cinerea*					2	3								3	2				1				40. 50. 137. 140.
45 *P. aequinoctialis*		3			2	3								3	2	2	1	1	1				32. 34. 72. 100.
46 *P. westlandica*		3			2	3								3	2								72.
47 *P. parkinsoni*		2			2	2								3	2				1		1		5. 72.
48 *Calonectris diomedea*			3	2		3								3	3				3			3	30.
49 *Puffinus creatopus*					3	3								3	3				1				4. 7. 28. 76.
50 *P. carneipes*		1	2		3									3	3	3							4. 18. 84.
51 *P. gravis*		3		1	3									3	2	2				1			17. 18. 48. 84. 137.
52 *P. pacificus*					3	3					3			3	3			1			1		47. 65.
53 *P. bulleri*		1				3	1				2			3	3	3	2		1				5. 39. 58. 81.
54 *P. griseus*		3		2	3	1		1						3	3	3			2		1		3. 4. 7. 17. 84. 86. 122. 124.
55 *P. tenuirostris*		2			3	1		1						3	3	3	3	3		1			4. 7. 18. 28. 84. 86. 95. 97. 98. 122.
56 *P. nativitatis*	1				3	1								3	3				1				5. 6. 65.
57 *P. puffinus*	3				3	3								2	3				2			2	30.
58 *P. huttoni*	2				3	2								1	3	3							5. 53. a
59 *P. assimilis*		3			3	2								3	1								18. 30. 39. 50. 74. 137.
60 *P. lherminieri*	2				2	3								3	3				3				5. 35. 61.
Hydrobatidae – Storm Petrels																							
61 *Oceanites oceanicus*		1		1		2					3	3		2	1	3	1					2	1. 11. 32. 39. 48. 50. 97. 125.
62 *O. gracilis*												(1)			(1)							(1)	60.
63 *Garrodia nereis*				2							3	3			1	1	1				3		32. 73. 137.
64 *Pelagodroma marina*											1	3			3	3	2	1			1		30. 50. 73. 125. 137.
65 *Fregetta grallaria*											3	3											1. 50. 137.
66 *F. tropica*				2							3	3		2	1								13. 52. 50. 108. 137.
67 *Hydrobates pelagicus*												3		3	3				3		3	3	30.
68 *Oceanodroma tethys*						2					2	2		2	3				1			1	5. 60.
69 *O. castro*						2					2	2		3	3				3			3	30. 60.
70 *O. leucorhoa*						3					3	3		3	3	3	2	2	3	1	3	3	4. 5. 30. 89. 139.
71 *O. tristrami*												3		3	3	1	1		1				65.
72 *O. melania*						2								(1)	(1)				(1)			(1)	92. 107
73 *O. homochroa*														(1)	(1)				(1)				2. 107.
74 *O. furcata*						3					3			3	3	2	2	2	3		1	3	4. 121.
Pelecanoididae – Diving petrels																							
75 *Pelecanoides georgicus*	3	2		2	3	2									1	2	2	3					32. 50. 77. 109.
76 *P. urinatrix*	3	2		2	3	2									1	2	2	3					32. 50. 109.

[a] West & Imber (1985: *Notornis* 32: 333–6).

to *D. melanophrys* (38%) and least to *P. palpebrata* (11%) with *D. chrysostoma* intermediate (24%) but this species also takes substantial quantities of Southern Lampreys *Geotria australis* (10%). In view of the abundance of krill in this region it is not surprising that it is represented in all three species' diet, being of particular importance to *D. melanophrys* (40% by weight). These broad-scale differences have been confirmed by samples taken in subsequent seasons (Prince, 1980*b*; Thomas, 1982, Croxall & Prince, unpublished).

It is difficult to establish the extent to which these dietary differences represent some degree of specialisation on different prey or are simply attributable to foraging in different areas. The prevalence of lampreys in the *D. chrysostoma* diet (absent from *P.palpebrata* samples; present in < 5% of *D. melanophrys* samples) is particularly intriguing as the size and developmental stage of these lampreys show that they are sub-adults, presumably free swimming and in the process of returning to their natal streams in Patagonia (Potter, Prince & Croxall, 1979). This and the restricted distribution of *D. chrysostoma* colonies at South Georgia (all clustered at the northwest end of the island, in contrast to *D. melanophrys* and *P. palpebrata* which are widely dispersed) suggest that this species may forage predominantly in a westerly direction to areas less frequented by the other two species. Antarctic Krill *Euphausia superba* seems particularly important to *D. melanophris* as severely reduced breeding success results

Table 7.2. *Composition of food samples from albatrosses on South Georgia (percentage by weight; data from Prince, 1980b; Thomas, 1982; Croxall & Prince, unpublished) and on the Hawaiian Islands (percentage by volume; data from Harrison* et al., 1983)

Species	Sample size	Squid	Fish	Lamprey	Crustacea	Other
South Georgia						
D. exulans	79	39·5	41·5	—	0·2	18·8[a]
D. melanophris	138	20·6	38·0	1·5	39·9	0
D. chrysostoma	132	49·0	24·1	10·4	16·5	0
P. palpebrata	46	45·5	10·9	—	39·5	4·1[a]
Hawaiian Islands						
D. nigripes	172	32·0	50·0	—	5·0	13·0[b]
D. immutabilis	183	65·0	9·0	—	—	17·0[c]

[a] Carrion.
[b] 12% unidentified, 1% other taxa.
[c] 11% unidentified, 4% coelenterates, 2% other taxa.

in seasons when it is absent or rarely forms surface swarms in the South Georgia area (Prince, 1985). In most of these seasons the squid-feeding *D. chrysostoma* was much less affected and it seems that if *D. melanophrys* switches to a squid diet it can no longer maintain the provisioning rate necessary to sustain the intrinsic growth rate of its chicks, which is significantly higher than that of *D. chrysostoma* chicks (Prince & Ricketts, 1981). In addition to the advantage provided by the higher energy content of krill, *D. melanophrys* feeds extensively over shelf and shelf-slope areas; e.g. around the Crozet Islands (Jouventin *et al.*, 1982) and thus in areas closer to the breeding colony than *D. chrysostoma*. It is not clear why the latter species catches so few krill but it is possible that, where birds congregate to feed on swarms, *D. melanophrys* is more aggressive (as it is in the breeding colonies) and is able to restrict its congeners' access to patchy resources at sea. *P. palpebrata* probably forages more widely and in a more southerly direction towards the pack-ice edge than the other species (Thomas, 1982) and the absence of ommastrephid squid and the abundance of cranchids and especially of *Psychroteuthis* in its diet probably reflects this difference.

Wandering Albatrosses *Diomedea exulans* rear their chicks during the winter and the data on their diet cannot, therefore, be compared directly with the other species. In the samples obtained throughout two winters, krill are absent, fish and squid predominate, and the proportion of carrion is the highest for any albatross (Croxall & Prince, unpublished).

In the Hawaiian Islands, Black-footed Albatrosses *D. nigripes* and Laysan Albatrosses *D. immutabilis* breed at the same sites and at similar times of the year. Harrison, Hida & Seki (1983) showed that *D. nigripes* feeds fish, especially flyingfish larvae (44% by volume), to its chicks but that fish are much less important (only 9% by volume) in the diet of *D. immutabilis* (Table 7.2). Conversely, squid formed 65% by volume of the latter's diet, twice as much as for *D. nigripes*. Of the squid so far identified, Ommastrephidae of similar sizes were the main prey of both species. Harrison *et al.* (1983) concluded that both species fed offshore, and probably in similar areas, but that *D. immutabilis* was better adapted for feeding at night, having a four-times greater concentration of rhodopsin in its retina, facilitating detection of prey under low light intensities. This might account for the preponderance of squid in its diet, as these are likely to be available at the surface mainly at night (Clarke, 1966; Imber, 1973).

Recent studies of squid beaks allow comparisons of the species taken by five species of albatrosses in three different regions of the Southern Ocean (Table 7.3). At South Georgia the squid mainly taken by *D. chrysostoma*

Table 7.3. *Composition by percentage numbers and weight of the cephalopod diet of albatrosses. Data compiled from references cited, with revised identifications from Clarke (1986)*

Predator species (Study site):	Wandering Albatross (New Zealand)		Wandering Albatross (South Georgia)		Grey-headed Albatross (South Georgia)		Black-browed Albatross (South Georgia)		Light-mantled Sooty Albatross (South Georgia)		Light-mantled Sooty Albatross (Marion Island)		Sooty Albatross (Marion Island)	
Sample size (beaks):	281		527		322		100		1162		411		2789	
References	1		2		3		3		4		5		5	
	No.	Wt.	No.	Wt.	No.	Wt.	No.	Wt.	No.	Wt.	No.	Wt.	No.	Wt.
Squid														
Architeuthidae														
Architeuthis sp.	0·4	2·6	—	—	—	—	—	—	—	—	—	—	—	—
Ommastrephidae														
Todarodes ? sagittatus	—	—	0·4	0·1	88·2	91·4	68·0	75·9	6·0	4·7	—	—	—	—
Nototodarus sp.	—	—	—	—	—	—	—	—	—	—	—	—	0·1	0·2
Unidentified sp.	0·4	1·2	—	—	—	—	—	—	—	—	—	—	—	—
Onychoteuthidae														
Onychoteuthis banksi	0·4	< 0·1	—	—	—	—	—	—	—	—	—	—	—	—
Moroteuthis ingens	3·5	21·8	—	—	—	—	—	—	—	—	—	—	0·3	2·4
M. robsoni	1·1	3·3	1·9	2·0	—	—	—	—	—	—	—	—	0·2	0·7
M. knipovitchi	1·1	1·2	0·6	0·1	—	—	—	—	1·2	3·0	6·3	8·8	8·6	19·2
Kondakovia longimana	0·7	11·0	40·0	81·1	0·6	3·9	1·0	1·2	1·6	13·5	12·7	64·3	5·6	47·4
Psychroteuthidae														
Psychroteuthis glacialis	—	—	—	—	—	—	—	—	19·9	47·5	4·1	5·6	0·3	0·4
P. sp. B.	—	—	—	—	—	—	—	—	—	—	0·2	—	0·1	< 0·1
P. sp.	0·4	0·5	—	—	—	—	—	—	—	—	—	—	—	—
Gonatidae														
Gonatus antarcticus	1·1	0·5	3·6	0·8	—	—	1·0	1·2	7·2	5·4	1·9	0·8	1·3	0·6
G. sp. A.	—	—	2·1	0·2	—	—	—	—	1·2	0·7	—	—	—	—
G. sp. C.	0·4	0·1	–	—	—	—	—	—	—	—	—	—	—	—
G. sp. D.	1·4	0·5	—	—	—	—	—	—	—	—	—	—	—	—
G. phoebetriae	—	—	—	—	—	—	—	—	—	—	—	—	0·1	—
G. sp. (cf. *separata*)	2·1	0·5	—	—	—	—	—	—	—	—	—	—	—	—
? Berryteuthis anonychus	5·3	1·0	—	—	—	—	—	—	—	—	—	—	—	—
Enoploteuthidae														
? Enoploteuthis sp.	1·1	0·1	—	—	—	—	1·0	1·2	—	—	—	—	—	—
Ancistrocheirus lesueuri	—	—	1·5	1·4	—	—	1·0	6·6	—	—	—	—	0·1	< 0·1
Lycoteuthidae														
Oregoniateuthis sp.	—	—	—	—	—	—	—	—	0·1	—	—	—	—	—
Octopoteuthidae														
Octopoteuthis ? longiptera	2·5	2·1	—	—	—	—	—	—	—	—	—	—	—	—
O. rugosa	—	—	1·5	0·3	—	—	—	—	—	—	—	—	—	—
O. sp.	—	—	—	—	—	—	—	—	—	—	—	—	0·1	0·2
Taningia ? danae	0·7	2·5	1·3	5·5	—	—	—	—	—	—	0·2	0·3	—	—
Lepidoteuthidae														
Lepidoteuthis grimaldi	—	—	0·9	0·8	—	—	—	—	0·3	1·4	—	—	—	—
Histioteuthidae														
Histioteuthis ? eltaninae	29·5	11·2	8·7	0·4	0·6	0·2	—	—	0·8	0·2	24·8	4·0	25·6	5·3

H. ? atlantica	11·7	5·1	5·9	0·7	—	—	—	—	—	—	—	—	0·2	0·1
H. ? macrohista	4·6	1·1	—	—	—	—	—	—	—	—	—	—	0·3	0·2
H. ? dofleini	0·7	0·2	—	—	—	—	—	—	—	—	—	—	0·1	—
H. ? miranda	6·0	4·9	—	—	—	—	—	—	—	—	0·2	0·1	0·3	—
H. ? meleagroteuthis	—	—	—	—	—	—	—	—	—	—	0·2	< 0·1	2·5	1·7
H. sp. G.	0·4	< 0·1	—	—	—	—	—	—	—	—	—	—	—	—
H. sp.	—	—	—	—	—	—	—	—	—	—	—	—	0·1	< 0·1
Neoteuthidae														
Alluroteuthis antarcticus	—	—	—	—	—	—	—	—	1·5	2·1	0·2	< 0·1	0·1	—
Cycloteuthidae														
Cycloteuthis ? akimushkini	1·4	5·2	—	—	—	—	—	—	—	—	—	—	—	—
C. serventi	—	—	—	—	—	—	—	—	—	—	—	—	0·1	0·7
Discoteuthidae														
Discoteuthis sp.	—	—	—	—	—	—	—	—	—	—	—	—	0·1	0·9
Mastigoteuthidae														
Mastigoteuthis sp. A.	—	—	2·8	0·7	0·6	0·4	—	—	—	—	—	—	—	—
M. sp. B.	1·4	0·8	—	—	—	—	—	—	—	—	—	—	—	—
M. sp. C.	—	—	—	—	—	—	—	—	—	—	0·2	0·1	0·1	—
M. sp. D.	0·7	0·1	—	—	—	—	—	—	—	—	—	—	—	—
Chiroteuthidae														
Chiroteuthis veranyi	0·2	0·2	1·0	< 0·1	0·3	0·2	2·0	1·2	—	—	—	—	0·1	< 0·1
C. macrosoma	—	—	—	—	—	—	—	—	—	—	1·7	1·2	0·8	0·9
C. picteti	—	—	—	—	—	—	—	—	—	—	1·5	0·2	0·8	0·3
C. sp. A.	7·1	7·5	—	—	—	—	—	—	—	—	—	—	—	—
C. sp. D.	0·7	0·8	—	—	—	—	—	—	—	—	—	—	—	—
C. sp. E.	—	—	—	—	—	—	—	—	—	—	—	—	1·8	0·7
C. sp.	—	—	—	—	—	—	—	—	0·3	0·1	—	—	—	—
Cranchiidae														
Bathothauma lyromma	—	—	—	—	—	—	—	—	—	—	—	—	0·1	0·3
Taonius pavo	10·3	12·6	16·8	3·4	—	—	—	—	—	—	0·5	0·2	1·1	0·6
T. belone	—	—	—	—	—	—	—	—	—	—	—	—	0·1	< 0·1
T. cymoctypus	—	—	—	—	—	—	—	—	—	—	0·5	0·4	0·8	1·2
Teuthowenia pellucida	0·7	0·2	—	—	—	—	—	—	—	—	—	—	0·1	< 0·1
T. antarctica	—	—	—	—	—	—	—	—	—	—	25·5	6·6	42·5	14·4
Galiteuthis armata	—	—	—	—	—	—	—	—	—	—	—	—	0·1	< 0·1
G. glacialis/Teuthowenia	—	—	8·0	0·9	7·5	3·9	25·0	12·0	59·8	21·5	18·5	7·3	5·6	2·6
Mesonchyoteuthis hamiltoni	—	—	0·2	0·4	—	—	—	—	—	—	0·5	—	0·1	0·3
Unidentified	0·4	0·1	—	—	—	—	—	—	—	—	—	—	—	—
Octopus														
Octopodidae	—	—	—	—	—	—	1·0	0·6	—	—	—	—	—	—
Argonautidae														
Argonauta nodosa	1·1	0·4	—	—	—	—	—	—	—	—	—	—	—	—
A. ? argo	0·4	0·3	—	—	—	—	—	—	—	—	—	—	—	—
Alloposidae														
Alloposus mollis	—	—	1·7	1·0	—	—	—	—	—	—	—	—	0·1	—
Other	—	—	1·2	0·5	2·2	—	—	—	—	—	—	—	0·1	—

1. Imber & Russ, 1975. 2. Clarke, Croxall & Prince, 1981. 3. Clarke & Prince, 1981. 4. Thomas, 1982; M. R. Clarke, unpublished data. 5. Berruti & Harcus, 1978; Imber & Berruti, 1981.

and *D. melanophrys* is the medium-sized ommastrephid *Todarodes sagittatus*. *Phoebetria palpebrata*, breeding at the same time, feeds predominently on *Psychroteuthis glacialis*, *Galiteuthis glacialis/Teuthowenia* sp. and *Kondakovia longimana*; all these, except *Galiteuthis*, are unimportant to mollymauks. *K. longimana*, however, is the main squid prey of *D. exulans*. The squid beaks from this species at South Georgia were regurgitated by chicks in November and therefore represent squid taken throughout the preceding winter (Clarke, Croxall & Prince, 1981).

There is a considerable diversity of squid in these Wandering Albatross samples, including a number of warmer-water species that do not occur south of the Antarctic Convergence. All this indicates that Wandering Albatrosses are particularly wide-ranging (as suggested by the very long interval between successive feeds to the chick) and that they may be feeding in areas to the north of South Georgia. This is, of course, not unexpected at a time of year when the pack-ice edge is often within 100 km of the south of the island. Squid beaks from a sample of Wandering Albatrosses from the New Zealand region (Imber & Russ, 1975) showed some broad similarities with the South Georgian material but the significance of the differences can only be evaluated when more samples are available.

The most detailed study of the two *Phoebetria* albatrosses comes from Marion Island (Berruti & Harcus, 1978; Imber & Berruti, 1981). Despite the diversity of squid recorded, and circumstantial evidence indicating that *P. palpebrata* mainly forages south and Sooty Albatrosses *P. fusca* north of Marion Island, there were few major differences in the composition of their squid prey, with *Kondakovia* the most important. Apart from the rarity of *Psychroteuthis* in Marion Island samples, the squid taken by *Phoebetria palpebrata* there and at South Georgia are broadly similar, some of the apparent differences probably being due to differences in taxonomic interpretation and nomenclature.

Some of the beaks found in albatross stomachs belong to squid of a size which it is very unlikely that albatrosses could take as live prey (Imber & Russ, 1975; Clarke *et al.*, 1981). With *D. chrysostoma* and *D. melanophrys*, however, where food samples came direct from adults about to feed chicks, the squid were usually intact. The average weight of the main prey species (*Todarodes sagittatus*) was 410 g, approximately 11% of the albatross' body weight. As Clarke *et al.* (1981) argue, by analogy one would expect *D. exulans* to be able to capture squid up to about 1000 g. This would imply that most *Moroteuthis ingens*, *Kondakovia longimana* and *Taonius pavo* in *D. exulans*' diet were probably scavenged and the same may be true of *Kondakovia* specimens taken by *P. palpebrata*.

Giant petrels Macronectes

The two species of giant petrels are confined to the southern hemisphere where they are circumpolar in distribution. The Southern Giant Petrel *Macronectes giganteus* is widespread while the Northern Giant Petrel *M. halli* is much less numerous, breeding only at a few island groups, always in association with *M. giganteus*. There have been two main studies, comparing the food of both species at a single site during the breeding season, by Johnstone (1977) at Macquarie Island, south of New Zealand, and by Hunter (1983) at South Georgia. Less extensive studies elsewhere, chiefly of *M. giganteus*, were reviewed by Hunter (1985). While both species are justly regarded as the main avian scavengers in the Southern Ocean, such habits are easy to observe and it is clear that, even during the breeding season, often quite substantial quantities of prey are caught alive; in winter such prey are likely to be of even greater importance. As Hunter (1985) suggests, it is best to regard giant petrels as opportunists, able to tackle a wide variety of prey but possessing some specific adaptations for dealing with large carrion, particularly seal carcasses.

At South Georgia more than 50% of the weight of food regurgitated by chicks of both species was penguin carrion, probably taken mainly from birds killed by Antarctic Fur Seals *Arctocephalus gazella* (Bonner & Hunter, 1982). At Macquarie Island penguin carrion also predominated, occurring in 96% and 87% of samples from *M. giganteus* and *M. halli* chicks, respectively. At South Georgia, fur seal carrion was an important food source for *M. halli* in December, at the peak of the seal breeding season, when the birds were rearing young chicks. Later on krill, *Euphausia superba* is frequently taken, and in some seasons may form up to 22% by weight of the diet. During the birds' breeding season, fish and squid occur infrequently in the diet but are probably more important in the winter when carrion is largely unavailable.

One of the interesting findings of Hunter's (1983) study was that intersexual differences in diet were of greater significance than interspecific ones. In both species males took more penguin and seal carrion than females, which took more live marine prey. This undoubtedly reflects the substantial difference in size between the sexes (the greatest in the Procellariiformes), whereby the larger males are dominant over females for access to seal carcasses, and possibly also whenever there is competition to secure penguin carrion.

Fulmars Fulmarus

Although increased availability of food (particularly fish offal) was suggested as a main reason for the rapid spread of Northern Fulmars

Fulmarus glacialis southwards through the British Isles to northern France (Fisher, 1952, 1966; Harris, 1976; Cramp & Simmons, 1977; Furness, 1981), until very recently there was no quantitative information on their diet.

Furness & Todd (1984) analysed food samples collected from chicks on the Scottish offshore islands of St Kilda and Foula, approximately 450 km apart. On Foula, small fish occurred in 72% of samples compared with only 8% on St Kilda, where zooplankton occurred in 71% of samples (11% of samples on Foula). Fish offal was not very common in samples from either Foula (14%) or St Kilda (1%). At both sites the commonest species in the zooplankton was the euphausiid crustacean *Meganyctiphanes norvegica*. The mysid *Gnathophausia zoea* and the isopod *Idotea metallica* were frequent in both sets of samples but the pelagic decapod *Parapasiphaea sulcatifrons* was only found in regurgitations from birds on St Kilda. On Foula all the fish taken were Sandeels *Ammodytes marinus* while on St Kilda the fish taken were either Herring *Clupea harengus* or Sprats *Sprattus sprattus*. Chick feeding frequency and diurnal attendance was greater at Foula than St Kilda, implying that food abundance was greater in the waters around Shetland. Furness & Todd (1984) concluded that the feeding behaviour and food in Shetland was atypical in that the diet is modified to diurnal catching of live Sandeels which have greatly increased due to overfishing of their natural predators, Whitefish, as well as Herring and Mackerel *Scomber scombrus*. On St Kilda the high occurrence of zooplankton in the diet, coupled with daytime nest attendance, suggested that this population forages nocturnally, perhaps in a manner typical of ancestral populations.

In contrast, 43 Northern Fulmar stomachs from the Gulf of Alaska contained mainly squid (63% by volume), some fish (27%, mainly unidentified) and a few crustaceans (Sanger, 1983). Bradstreet & Cross (1982) examined 31 Fulmar stomachs collected in the Canadian Arctic and by frequency of occurrence the most common items were copepods (67%) and cephalopods (27%). Because squid material comprised solely beaks it was unlikely to have been taken by birds feeding at the ice edge; when squid were excluded from dry weight analysis, Arctic Cod *Boreogadus saida* accounted for 97% of the diet.

Gadfly petrels Pterodroma

This is the largest genus of the family Procellariidae. Most species breed in the southern hemisphere on remote islands, and for many species even the basic details of their breeding biology are poorly known (Imber, 1985).

Table 7.4. *Composition (percentage by weight) of diet of* Pterodroma *petrels*

Species	Locality	Sample size	Squid	Fish	Crustacea	Other	Unidentified	Reference
P. macroptera	New Zealand	85	58	28	12	2	—	Imber, 1973
P. macroptera	Marion Island	20	90	4	6	—	—	Schramm, 1982
P. brevirostris	Marion Island	22	70	6	24	—	—	Schramm, 1982
P. mollis	Marion Island	9	89	1	10	—	—	Schramm, 1982
P. alba	Christmas Island	95	78	14	2	6	—	Ashmole & Ashmole, 1967
P. hypoleuca[a]	Hawaiian Islands	144	21	47	7	—	24	Harrison *et al.*, 1983
P. incerta	Gough Island	13	70	17	13	—	—	Williams & Imber, 1982

[a] Percentage by volume.

Quantitative dietary data are available for only a few species (Table 7.4), but squid seem to be of particular importance in their diet.

In New Zealand the Grey-faced Petrel *Pterodroma macroptera gouldi* preyed extensively on the small *Spirula spirula* (mean estimated weight 10 g) and members of the family Histioteuthidae and Cranchiidae. As with other squid-feeding Procellariiformes, the large estimated size of some of the squid suggested that they had been scavenged. However, of 25 identified species, most were under 200 g in weight and were believed to have been taken live at night (Imber, 1973). Ommastrephidae were the only squid family identified by Ashmole & Ashmole (1967) in Phoenix Petrel *P. alba* regurgitations, although other squid families were taken. Apart from squid, these two *Pterodroma* species preyed upon small fish. Members of the family Myctophidae and Gonostomatidae were frequently taken by *P. macroptera*. A variety of crustaceans and other invertebrates occurred in the diets of both species.

The Bonin Petrel *P. hypoleuca* is the only *Pterodroma* reported mainly to take fish; of 12 families identified, lanternfishes Myctophidae and hatchetfishes Sternoptychidae were the most important (Harrison *et al.*, 1983). However, although fish formed nearly half the *P. hypoleuca* diet by volume, squid, especially small Ommastrephidae (mean mantle length 46 mm), were the most abundant prey class.

These studies suggest that *Pterodroma* species feed extensively on small squid (and fish), mainly caught at night as indicated by Imber (1983), perhaps supplemented by any opportunities to scavenge on moribund squid. *P. alba* and *P. hypoleuca* also exploited a more unusual prey, taking small (1 cm long) marine sea-striders Gerridae which, although unimportant in biomass, demonstrated the versatility of gadfly petrel foraging abilities (Cheng & Harrison, 1983).

Prions Pachyptila

Imber (1981) has recently reviewed the diet of this group and details are summarised in Table 7.5, together with new data from Fairy Prions *Pachyptila turtur* at South Georgia. The diet of the Blue Petrel *Halobaena caerulea*, often regarded as intermediate between prions and gadfly petrels, is treated for convenience in this section. Of the six prion species usually recognized (Kinsky *et al.*, 1970; Serventy, Serventy & Warham, 1971; Harper, 1980), Thin-billed Prions *P. belcheri*, Fairy Prions and Fulmar Prions *P. crassirostris* have the least-modified bills, lack palatal lamellae and hence have no specialised filtering apparatus. In Antarctic *P. desolata*, Salvin's *P. salvini* and Broad-billed *P. vittata* Prions, the width of the bill is progressively larger and all species have well-developed comb-like lamellae inside the upper mandible (illustrated in Prince, 1980*a*) which are used to filter small particles of marine food from ingested sea water.

P. turtur feeds extensively on *Nyctiphanes australis* in the New Zealand region where this euphausiid is abundant and also common in the diet of other seabirds (Harper, 1976). *Nyctiphanes* does not occur in sub-Antarctic waters and at South Georgia another euphausiid, *Euphausia superba*, formed 79% (by weight) of the meals fed to *P. turtur* chicks. Three species of amphipods occurred in the samples, *Themisto gaudichaudii* being the most important (14% by weight). Copepods occurred in six samples but totalled only 4% by weight and a few barnacle larvae and fish made up the rest (P.A. Prince & P.G. Copestake, unpublished).

At South Georgia, Blue Petrel diet consisted of 91% crustaceans, 8% fish and 1% squid, by weight. However, as Prince (1980*a*) pointed out, weight analysis alone may significantly underestimate the importance of fish, which occurred in 83% of samples and probably formed the bulk of unidentified material. (The values in Table 7.5 have been modified to allow for this.) By comparison, Antarctic Prions fed almost exclusively on crustaceans (96% by weight). Fish occurred in only 13% of samples and crustaceans formed the bulk of unidentified material. Of the crustaceans consumed by *H. caerulea*, 86% by weight were Euphausiaceae, the rest being mysids, decapods, amphipods and copepods. There is little doubt that the specialised feeding methods used by Antarctic Prions facilitate the capture of so many small copepods. These adaptations may be of even greater importance in seasons when *Euphausia superba* is unavailable to surface-feeding birds (Croxall & Prince, 1980) and perhaps also in the winter. Overall, however, Blue Petrels tended to take larger crustaceans than Antarctic Prions. The differences, which were statistically significant for *Thysanoessa*, *Antarctomysis* and three species of amphipod (*Hyperoche*, *Hyperiella* and *Hyperia*), were suggested to relate to differences in feeding technique. A variety of evidence indicated that Blue Petrels foraged over a much wider area than Antarctic Prions.

The other two prions with well-developed lamellae, *Pachyptila salvini* and *P. vittata*, both probably feed principally by filtering out copepods, which formed 70% by weight of the diet of *P. vittata* at the Chatham Islands (Imber, 1981).

Fulmar Prions at the Chatham Islands take substantial quantities of adults of the pelagic barnacle *Lepas australis*, pteropods, some squid and fish (Imber, 1981). At Heard Island this species takes a variety of amphipods (mainly *Themisto gaudichaudii* and *Hyperiella antarctica*), and pteropods; barnacles were not recorded (Ealey, 1954).

Table 7.5. *Composition (% by weight) of diet of prions* Pachyptila *and Blue Petrel* Halobaena

Species:	Blue Petrel	Fairy Prion	Antarctic Prion	Fulmar Prion	Fulmar Prion	Broad-billed Prion	Fairy Prion	Thin-billed Prion	Thin-billed Prion
	H. caerulea	*P. turtur*	*P. desolata*	*P. crassirostris*	*P. crassirostris*	*P. vittata*	*P. turtur*	*P. belcheri*	*P. belcheri*
Site:	South Georgia	South Georgia	South Georgia	Heard I.	Chatham Is.	Chatham Is.	New Zealand	Subantarctic Pacific Ocean	Falkland Is.
Sample size:	156	40	90	38	16	57	151	11	5
Reference:	Prince, 1980*a*[a]	Prince & Copestake, unpublished	Prince, 1980*a*	Ealey, 1954	Imber, 1981	Imber, 1981	Harper, 1976 (& personal communication)	Harper, 1972 (& personal communication)	Strange, 1980
Euphausiacea									
Euphausia superba	34·1	78·9	56·9	—	—	—	—	—	+
Thysanoessa macrura	1·7	—	0·3	—	—	—	—	—	—
Nyctiphanes australis	—	—	—	—	—	5·8	94·2	—	—
Mysidacea									
Antarctomysis maxima	1·5	—	0·7	—	—	—	—	—	—
Boreomysis rostrata	—	—	—	+	—	—	—	—	—
Amphipoda									
Platyscelus ovoides	—	—	—	—	—	10·9	—	—	—
Eupronoe minuta	—	—	—	—	—	4·1	—	—	—
Cyllopus magellanicus	—	—	—	—	—	3·8	—	—	—
C. macropis	—	—	—	—	—	0·4	—	—	—
C. lucasii	0·2	> 0·1	0·9	—	—	—	—	—	—
Vibilia antarctica	0·5	—	0·5	—	—	—	—	—	—
V. armata	—	—	—	+	—	+	—	—	—
Themisto gaudichaudii	0·2	13·7	4·5	+ +	—	0·5	—	—	+
T. australis	—	—	—	—	—	+	—	99·1	—
Hyperia spp.	—	—	—	+	—	—	—	—	—
H. spinigera	—	—	—	—	—	+	—	—	—
H. macrocephala	> 0·1	—	0·1	—	—	—	—	—	—
Hyperiella antarctica	0·2	—	1·7	+ +	—	—	—	—	—
Hyperoche medusarum	0·8	—	0·4	—	—	—	—	—	—
Tryphosella barbatipes	—	—	—	+	—	—	—	—	—
Eurythenes gryllus	0·1	—	—	—	—	—	—	—	—
Lestrigonus sp.	—	—	—	—	—	+	—	—	—

Copepoda									
Clausocalanus sp.	—	—	—	—	—	2·0	—	—	—
Rhincalanus gigas	0·6	2·8	31·5	—	—	—	—	—	—
Calanoides acutus	—	0·8	—	—	—	—	—	—	—
Calanus tonsus	—	—	—	—	—	68·0	—	—	—
Decapoda	—	—	—	—	—	+	2·0[b]	—	—
Acanthephyra sp.	1·8	—	—	—	—	—	—	—	—
Brachyura									
Nectocarcinus antarcticus	—	—	—	—	—	1·3	—	—	—
Pteropoda	—	—	—	+ +	+	—	—	—	—
Cirripedia									
Lepas sp. (cf. *australis*)	—	> 0·1	—	—	+ +	+	—	—	—
Cephalopoda	0·3	1·2	0·6	—	—	0·6	—	0·3	+
Histioteuthis atlantica	—	—	—	—	+	—	—	—	—
Mollusca	—	—	—	—	—	0·2	—	—	—
Tunicata									
Pyrosoma sp.	—	—	—	—	+	—	—	—	—
Stomatopoda									
Squilla armata	—	—	—	—	—	+	—	—	—
Teleostei	58·3	2·4	1·8	+	+	2·3	4·0	0·6	+

+ + : major constituent: + : minor constituent.

[a] Original data modified to account for unidentifiable material, assumed to be largely fish.

[b] Nauplii of *Leptograpsus variegatus*.

Petrels Procellaria

Quantitative dietary information is available for three of the four *Procellaria* species: Black Petrel *P. parkinsoni* and Westland Black Petrel *P. westlandica* in New Zealand (Imber, 1976) and White-chinned Petrel *P. aequinoctialis* at South Georgia (Croxall & Prince, 1980; Prince, unpublished). In New Zealand waters the three species form a latitudinal series in respect of foraging ranges, that of *P. aequinoctialis* being in the sub-Antarctic zone (44–55° S) and of *P. parkinsoni* in the sub-tropical zone (30–40° S), with *P. westlandica* in the temperate waters in between. Squid are the main element in the diet of all *Procellaria* so far studied, but fish and crustaceans are usually also present. Around northern New Zealand, *P. parkinsoni* took mainly Ommastrephidae (33% by weight), Cranchiidae (17%) and Histioteuthidae (14%). Farther south, *P. westlandica* took mainly Histioteuthidae but also Ommastrephidae and Enoploteuthidae; at even higher latitudes Histioteuthidae predominated in the diet of *P. aequinoctialis*.

At South Georgia *P. aequinoctialis* takes squid (47% by weight), fish (24%) and crustaceans (29%), chiefly *E. superba* (Croxall & Prince, 1980). Squid beaks from South Georgia birds are mainly *Teuthowenia* (Cranchiidae), *Todarodes* (Ommastrephidae) and *Kondakovia* (Onychoteuthidae), the last, of an estimated 8 kg weight, certainly scavenged (Prince, unpublished).

Petrels Bulweria

Of the two species of the genus, Jouanin's Petrel *Bulweria fallax* is restricted to the northwest Indian Ocean and little known. Bulwer's Petrel *B. bulweri* occurs in the Atlantic, Pacific and Indian Oceans and its diet has been studied in detail at the Hawaiian Islands (Harrison *et al.*, 1983). Of the identifiable material, fish formed 71% by volume, squid 22%, crustaceans 4% and sea-striders 3%. Seven fish families were identified, the most important being lanternfishes Myctophidae and hatchetfishes Sternoptychidae. Only ommastrephid squid were identified. Some seasonal variation was noted, squid being more important in spring and fish in summer. Hatchetfishes were the commonest prey in a small series of samples from Bulwer's Petrel at Madeira (Zonfrillo, in press). Other fish reported included Myctophidae, Gonostomatidae and flyingfish Exocoetidae. Squid were also common, with *Mastigoteuthis* and *Leachia* being identified.

Shearwaters Puffinus

This genus includes some of the commonest and most widely distributed of procellariiform seabirds. Quantitative dietary data from the breeding grounds are only available, however, for two tropical species, the Wedge tailed Shearwater *Puffinus pacificus* and Christmas Shearwater *P. nativitatis*, and for one cool temperate species, the Short-tailed Shearwater *P. tenuirostris* in South Australia. This last species, together with the Great Shearwater *P. gravis* (breeding at the Tristan da Cunha group) and Sooty Shearwater *P. griseus* (widespread in the Southern Ocean but especially in New Zealand and southern South America) undertake major trans-equatorial migrations and such are their numbers that they make substantial contributions to the northern hemisphere seabird biomass in the northern summer (Wiens & Scott, 1975; see also Chapters 11 and 12). Consequently, their diet has recently been studied at a number of localities in both the Pacific and Atlantic Oceans. With better data from their breeding grounds these would have been amongst the few seabird species whose year-round diet could realistically be described.

P. tenuirostris has recently been studied in Australia during its breeding season (September–April). Fish occurred in 73% of samples (46% by weight), crustaceans in 56% (45% by weight) and squid in 13% (9% by weight) (Montague, Cullen & Fitzherbert, in press). During the early part of the breeding cycle (September/October) fish were uncommon and crustaceans, especially *Nyctiphanes australis* (51% by occurrence), predominated. During the chick-rearing period (January onwards) fish increased in abundance with schooling post-larval anchovies *Engraulis australis* being most important followed by leatherjackets Aluteridae. Squid continued to be taken in small numbers, *Nototodarus gouldi* being commonest. At the same time of the year, large flocks of *P. tenuirostris* are feeding hundreds of kilometres to the south, in Antarctic waters, on Antarctic krill, *Euphausia superba*. Flocks of over a thousand birds have been observed during February and the few birds collected had regressed gonads, suggesting that they were either immatures, non-breeders or failed breeding birds (Kerry, Horne & Dorward, 1983). It is possible that some form of segregation may exist for food resources between birds actively involved in breeding, and thus confined to waters near the colony and those not of breeding status, which are able to travel further afield (Montague *et al.*, in press).

P. tenuirostris and *P. griseus* leave their breeding grounds in April and May and migrate north. In the North Pacific Ocean, their main wintering area, Ogi, Kubodera & Nakamura (1980) and Ogi (1984) have extensively studied their diets, using birds collected from salmon gill nets in which they had become entangled and drowned. The three main areas sampled were the Okhotsk Sea, the North Pacific Ocean and the Bering Sea. In the

Okhotsk Sea birds were sampled from over the continental shelf. Euphausiids (*Thysanoessa raschii*) made up 83% (by weight) of *P. tenuirostris* diet, with decapod larvae (in 12% of samples) the next most abundant item. Interestingly, Thick-billed Murres *Uria lomvia* in the area were feeding on fish. Farther east in the North Pacific Ocean, fish comprised 63% (by weight), squid 19% and euphausiids (*T. longipes*) only 9% of *P. tenuirostris* diet. The fish were mainly *Pleurogrammus* sp. and myctophids. Thick-billed Murres in this region preyed extensively on squid. Farther north in the Bering Sea euphausiids dominated *Puffinus tenuirostris* diet (73% by weight), with squid (14%) and amphipods (7.5%) less important. The differences in diet in each of these areas indicate the ability of *P. tenuirostris* to adapt to locally abundant food resources. This contrasts strongly with *P. griseus* whose feeding flocks are usually associated with Pacific Saury *Cololabis saira* (83% by weight of the diet overall), itself closely linked to a specific range of sea temperatures. In the north, *P. griseus* tends to occur in dense feeding flocks where water temperature ranges from 9 to 13° C. Its penetration into sub-Arctic waters corresponds with warming surface waters during early May. The only marked change in diet noted by Ogi (1984) was with increased latitude where Pacific Sauries decreased and were replaced by squid.

Other studies on these two shearwaters in the Arctic region of the Alaskan continental shelf show *P. griseus* feeding predominantly on fish, mainly Capelin *Mallotus villosus*, while *P. tenuirostris* fed on the euphausiid *Thysanoessa inermis* (see Chapter 10).

In the east Pacific Ocean off the California coast in Monterey Bay, *P. griseus* feed primarily on juvenile rockfish *Sebastes* spp. from May through July. In August and September the Northern Anchovy *Engraulis mordax* is the main food, although Market Squid *Loligo opalescens* and euphausiids are also commonly taken (Chu, 1984). Chu suggested that the switch in diet during August to *E. mordax* enables *P. griseus* to increase its own fat deposits, owing to the high lipid content of the anchovy, and thus facilitates the long migration back to the breeding colonies.

The number of *P. griseus* migrating into the North Atlantic is small compared with the Pacific Ocean but *P. gravis* numbers are large. During April and May these two species move towards New England and Newfoundland where, according to Jangaard (1974), they are attracted to spawning Capelin and immature squid *Illex illecebrosus*. Off Nova Scotia and Newfoundland Brown *et al.* (1981) investigated the diet and feeding behaviour of these two shearwaters. While there was broad overlap in their diets, *P. gravis* generally took more squid while *P. griseus* took more crustaceans. In seasons when *Meganyctiphanes* did not swarm near the surface, *P. gravis* fed almost entirely on squid; in seasons when *Meganyctiphanes* and sandlances *Ammodytes* were available, *P. gravis* mainly took fish and *P. griseus* took euphausiids. *P. griseus* dived more frequently and, judged by submersion times etc., was more mobile under the water than *P. gravis*, in line with Kuroda's (1954) deductions from anatomical studies. The difference in foraging behaviour and consequent energetic costs were possible reasons for difference in diet. Their respective movements around the North Atlantic, 'a series of journeys from one concentration to another' (Brown *et al.*, 1981), is not unlike the Pacific situation already described. The general tendency for *P. gravis* to reach high latitudes compared to the more southerly distribution of *P. griseus* may also minimise interspecific competition.

Tropical *Puffinus* shearwaters have only been studied at the Hawaiian Islands and Christmas Island (Pacific Ocean). At Hawaii, *P. pacificus* feed extensively on fish (66% by volume), comprising species from 21 fish families with Mullidae and Carangidae the commonest. Ommastrephid squid form 28% of their diet (Harrison *et al.*, 1983). Ommastrephids were also taken by *P. nativitatis* and formed almost half their diet (48% by volume), the rest being almost entirely fish from 17 families, with Mullidae and flyingfish Exocoetidae being the most frequent (Harrison *et al.*, 1983). On Christmas Island, *P. nativitatis* mainly takes ommastrephids (71% by volume) and small fish, especially Exocoetidae, although Scombridae and Gonostomatidae were also important (Ashmole & Ashmole, 1967).

Storm petrels Hydrobatidae

Quantitative dietary data are available for only four species of storm petrels. Imber (1981) studied the food of Grey-backed Storm Petrels *Garrodia nereis* and White-faced Storm Petrels *Pelagodroma marina* breeding on the Chatham Islands, New Zealand. The former fed extensively on the larvae of the pelagic barnacle *Lepas australis* (85.5% by weight), the rest of the diet including the euphausiid *Nyctiphanes australis* (7.3%), four species of amphipod and the occasional isopod and fish. *P. marina*, breeding at the same time, fed mainly on *Nyctiphanes* (35% by weight), fish (30%) and the crustacean *Nematoscelis megalops* (10%). The only identifiable fish were *Lampichthys procerus* and *Maurolicus muelleri*. Barnacle larvae were hardly taken by *P. marina* and Imber (1981) believes that this dietary specialisation of *G. nereis* (*c.* 30 g) helps avoid competition with the larger (*c.* 40 g) and more numerous *P. marina*. Furthermore, Imber (1981) suggested that the *G. nereis* breeding range is linked to the distribution of these small pelagic barnacles which are distributed between the Subtropical and Antarctic Convergences. All five food samples collected from *G. nereis* at other localities (Marion Island and southern New Zealand) consisted almost entirely of barnacle larvae.

Sooty Storm Petrels *Oceanodroma tristrami* on Hawaii took 29% by volume squid, 23% fish, 12% coelenterates (*Velella velella*) and 5% crustaceans (Harrison *et al.*, 1983). Linton (1978) collected food samples from Leach's Storm Petrel *Oceanodroma leucorhoa* at Nova Scotia and Newfoundland. At both sites fish formed the most important component of the diet (69% and 66% by volume, respectively). The euphausiid *Meganyctiphanes norvegica* was the second most important component in Nova Scotia, while amphipods (*Hyperia galba*) took this rank in Newfoundland. Of fish taken, Myctophidae were the most abundant (52%) but Cod *Gadus morhua* were also common, especially in Nova Scotia (40%). The size range of Myctophidae taken varied from 10 to 69 mm with 44% (by volume) being between 50 and 59 mm in length. Linton (1978) thought it likely that some of these prey were caught on the edge of the Scotian shelf, some 180km from the colony.

Diving petrels Pelecanoididae

The four species of diving petrels are confined to the southern hemisphere and little is known about the Peruvian *Pelecanoides garnoti* and Magellanic *P. magellani* Diving Petrels. The only quantitative dietary information available is from Payne & Prince's (1979) study at South Georgia of the two circumpolar species, South Georgia Diving Petrel *P. georgicus* and Common Diving Petrel *P. urinatrix exsul*. The last species breeds nearly one month earlier than *P. georgicus* and chick stomach contents showed it to feed extensively on copepods (68% by volume), amphipods (17%) and euphausiids (15%). The later-breeding *P. georgicus* fed mainly on euphausiids (76% by volume) and copepods (20%), with amphipods rarely present (4%). Without knowledge of prey availability, it is not known whether this difference in food is simply a result of differing food availability at the time of chick rearing or of genuine prey selection.

Feeding methods

Table 7.1 summarised the various feeding methods procellariiforms use to obtain their food. Definitions follow those given by Ashmole (1971) with modifications based on Cramp & Simmons (1983) and Harper, Croxall & Cooper (1985). Because procellariiforms are highly pelagic and many species may typically feed at night, there have been no quantitative studies on feeding methods and nearly all information comes from anecdotal observations. In addition, many records relate to birds feeding on offal and waste, and techniques used may not necessarily reflect those employed to catch natural prey.

Albatrosses Diomedeidae

Most observations of albatrosses feeding are of birds sitting on the water and seizing squid, often thought to be moribund. Circumstantial data indicate that albatrosses feed extensively at night when vertical migration brings prey to the water surface and Wandering Albatrosses have been observed to 'surface-seize' squid at night (P.C. Harper, personal communication). The smaller albatrosses are also capable of shallow plunging and have been seen to dive from the air and plunge 1 m or so beneath the surface water to catch sinking fish thrown overboard. White-capped Albatrosses *Diomedea cauta* have been seen to catch, while in gliding flight, Jack Mackerel *Trachurus declivis* as they 'ripple' the surface of the water and similar observations are reported for Yellow-nosed Albatrosses *D. chlororhynchos* with schools of the pilchard *Sardinops neopilchardus* (Barton, 1978). Waved Albatrosses *D. irrorata* sometimes kleptoparasitise boobies *Sula* spp. (Harris, 1973; Duffy, 1980) by gliding up to them as they take off from the water after a dive. The close presence of albatrosses was sufficient to induce them to regurgitate. Boobies in flight were rarely attacked.

Petrels and shearwaters Procellariidae

Giant Petrels *Macronectes giganteus* and *M. halli* are unique amongst Procellariidae in being able to walk strongly (although awkwardly) on land and thus exploit terrestrial prey, especially carrion. (Albatrosses, however, which are also capable of walking and standing upright, do not take prey on land.) At sea, Giant Petrels 'surface-seize' their food but they have been observed 'surface-diving' and 'pursuit-plunging'.

Surface-seizing is a virtually universal technique amongst procellariiforms. Species for which other techniques seem important include Snow Petrels *Pagodroma nivea*, Cape Pigeons *Daption capense*, and Antarctic Petrels *Thalassoica antarctica* which feed while in flight by 'dipping'. Such a feeding method seems particularly characteristic of gadfly petrels *Pterodroma*, 'surface-plunging' or 'diving' being unrecorded (apart from Mottled Petrels *P. inexpectata* which have been observed surface-plunging). Amongst the more specialised members of the family are the larger prion species (*Pachyptila vittata*, *P. salvini* and *P. desolata*) which have deep broad bills and lamellae fringing the inside of the upper mandible through which they filter out small prey organisms by expelling water. This they carry out while sitting on the water or, when wind conditions are suitable, by 'hydroplaning' into the wind. The smaller prions (e.g. *P. belcheri* and *P. turtur*) feed by 'pattering' and 'dipping' while all members of the group 'surface-dive' and 'surface-seize' prey such as squid. The shearwaters, especially the Short-tailed Shearwater, use 'pursuit-diving', sometimes down to depths of 20 m (Skira, 1979).

The shearwaters show a range of morphological variation from the lightly built Cory's *Calonectris diomedea* and Streaked *C. leucomelas* Shearwaters, best adapted to buoyant flight, through the largely surface-feeding Flesh-footed *Puffinus carneipes* and Pink-footed *P. creatopus* Shearwaters, to the more sturdily built Wedge-tailed, Great, Sooty, Short-tailed and Manx *P. puffinus* Shearwaters, well adapted for swimming underwater, as are the smaller species like Little *P. assimilis* and Audubon's *P. lherminieri* Shearwaters (Kuroda, 1954; Brown, Bourne & Wahl, 1978).

Storm petrels Hydrobatidae

These small petrels are characteristically seen 'dipping' or 'pattering' above the surface water, using their long legs and webbed feet as a brake to their forward momentum, enabling them to lean forward and seize small prey items. They also engage in shallow plunging but, unlike other members of the order, rarely sit on the water.

Diving petrels Pelecanoididae

Diving petrels typically undertake active 'pursuit-dives' in search of food, as well as sitting on the surface and seizing prey. Their dives are distinctive in being essentially a continuation of their rapid flight low over the wave tops. This foraging behaviour is unique in the Procellariiformes but very similar to that of the smaller alcids of the northern hemisphere.

Detection of prey

Procellariiforms have numerous adaptations conferring the ability to cover long distances and/or large areas in search of prey (see Chapter 3). However, except for diving petrels and perhaps some *Puffinus* shearwaters, they have restricted abilities to exploit the water column to any great depth, especially compared with penguins and alcids, and even with specialised plunge-diving pelecaniforms. Consequently, procellariiforms chiefly depend on finding food at or near the sea surface. Many marine organisms migrate towards the surface at night and the nature of the prey taken by many procellariiforms (especially *Pterodroma* species) has been an important part of the circumstantial evidence, suggesting that they feed extensively at night (Imber, 1973, 1976). This is strongly supported by activity budget data for albatrosses (see later) and numerous direct observations of various *Pterodroma* species catching squid at night (P.C. Harper, personal communication), which raises the question of how such seabirds detect their prey.

Sight

Imber (1973, 1976) noted that a high proportion of prey were bioluminescent and suggested that such items might be selected because of their visibility. However, for squid at least, Clarke *et al.* (1981) suggest that this may not be the case. They note that the proportions of bioluminescent and non-bioluminescent prey taken by albatrosses were consistent with the relative abundance of these two categories of squid, that several important prey species were non-bioluminescent and, furthermore, that squid use their luminescence to break up or hide their silhouettes from below against the downwelling light. It is unlikely that light emitted could be observed from above because squid light organs are specially screened to avoid this (Young, 1977). Whatever the importance of bioluminescence in prey detection, we need to know much more about the visual capacities of seabirds, especially in view of intriguing species-specific differences reported for albatrosses. Harrison *et al.* (1983) found that the levels of rhodopsin (measured in optical density units) in the retinas of Laysan and Black-footed Albatrosses *Diomedea immutabilis* and *D. nigripes* were 16.3 and 3.9 D g^{-1} and that this fitted well with circumstantial evidence suggesting that *D. immutabilis* fed chiefly at night. For comparison, values for Barn Owls *Tyto alba* and Great Horned Owls *Bubo virginianus* are 19.5 and 46.6 D g^{-1}, respectively. Further research is clearly indicated, especially involving genera like *Pterodroma* which have very large eyes, particularly likely to be associated with nocturnal vision.

Harper (1979) conducted some simple field experiments to test the ability of various procellariiforms to perceive colour. He found that birds were preferentially attracted to red and orange, with lower levels of response to pink, yellow, blue, green, and white, in that order. Only giant petrels showed little obvious colour discrimination. Harper suggested that the preference for bright colours (especially reds) reflects the colour of crustacean prey typically taken by many species; certainly bright, inedible objects not infrequently turn up in food samples from a variety of seabirds.

Smell

The importance of smell to procellariiforms as a means of locating their food has regained prominence following some recent important field studies (Hutchinson & Wenzel, 1980; Hutchinson *et al.*, 1984). Despite evidence from Wood Jones (1937) and Grubb (1972), and field observations by many other seabird workers during the early part of the century (Murphy 1936), the significance of procellariiforms locating their food by odour had become overlooked. Wenzel's (1980) excellent review indicated that while all birds possess olfactory mucosa typical of vertebrates, it is more pronounced in seabirds. When ranked by relative olfactory mass (the ratio of olfactory bulb diameter to the largest diameter of the cerebral hemisphere (Cobb, 1960)), 10 of the first 12 bird species are procellariiforms, the others being kiwis *Apteryx* (second) and Turkey Vulture *Cathartes aura* (tenth). The only anomalous procellariiform was the South Georgia Diving Petrel which was ranked fifty-fourth (Bang, 1971).

Field experiments off the Californian coast clarified the nature of odours to which procellariiforms are attracted. Hutchinson & Wenzel (1980) presented stimuli in translucent plastic enclosures floating on the ocean, in surface slicks and as saturated wicks on free-floating rafts. Test stimuli included substances with different visual characteristics and with odorous properties related to natural food of seabirds (e.g. cod liver), as well as those unrelated to food such as petroleum and mineral oil. When food-related substances were used, procellariiforms were rapidly attracted from downwind whereas other seabirds were unaffected and approached upwind as often as downwind. Recently, Hutchinson *et al.* (1984) showed that Northern Fulmars and Sooty Shearwaters arrived more rapidly and consistently from downwind when cod liver oil was used, compared with the whole oil and squid and krill homogenates. However species-specific differences were noted: for example Northern Fulmars made few approaches to squid or krill homogenates whereas Sooty Shearwaters were attracted to both. Such use of olfactory guidance to foods implies foraging in areas with enough wind to disperse odorous molecules but indicates that odour quality may be as important as intensity (Hutchinson *et al.*, 1984).

Activity patterns at sea

Obtaining quantitative information on the activity patterns of foraging Procellariiformes has only become feasible very recently, with the development of recorders measuring the amount of time birds spend on the sea during daylight and at night and therefore the proportion of time spent in flight (Prince & Francis, 1984). These devices have so far only been used

Table 7.6. *Activity budgets of albatrosses on foraging trips from Bird Island, South Georgia. Data from Prince & Francis (1984) and unpublished*

Species	Date	Sample size	Mean (±s.d.) hours darkness per day	Mean (±s.d.) hours on sea: Day	Night	Total	% day on water	% night on water	% time on water	Mean (±s.d.) hours in flight
D. chrysostoma	Feb.–Mar. 1982	13	9·0±1·4	2·3±1·4	4·3±1·4	6·6±2·3	15	50	28	17·4±2·3
D. chrysostoma	Feb.–Mar. 1984	13	9·9±0·4	0·9±0·5	7·6±0·7	8·4±0·8	6	76	36	15·6±0·8
D. melanophris	Feb.–Mar. 1983	2	7·5±0·03	4·6±1·4	6·4±1·0	10·9±0·4	28	85	49	13·1±0·4
D. exulans	May–July 1983	9	16·1±0·6	1·4±0·4	8·8±1·0	10·2±1·1	17	56	43	13·9±1·2
D. exulans	Dec. 1983	3	6·7±0·3	6·0±0·7	2·8±1·2	8·8±1·1	35	42	36	15·2±1·1

on albatrosses at South Georgia (Table 7.6) but a modified all-electronic instrument will be deployable on a variety of much smaller species. There are a number of interesting features in the albatross data. First, with the exception of Wandering Albatrosses during incubation, all birds spend at least three times as much time on the sea at night as they do during the day. This provides further support for the importance of nocturnal foraging in procellariiforms. During incubation, foraging trips to sea are several times as long as during chick rearing and adults have only to catch enough food to meet their own requirements. It would not be surprising, therefore, if Wandering Albatrosses were able to spend more time on largely 'non-feeding' (daytime on water) activities at this time.

Second, the limited data for *Diomedea melanophrys* suggest that they spend much more time on the sea (and less time flying) than *D. chrysostoma*. This is in line with other indications that *D. melanophrys* forage closer to South Georgia, perhaps even mainly confined to the vicinity of the continental shelf, at least during the chick-rearing period. Thirdly, Wandering Albatrosses rearing chicks spend proportionately the least amount of time flying and the most time on the sea at night of any of these albatrosses at a similar stage of their breeding cycle. Wandering Albatrosses, of course, are rearing chicks during the winter, and, if it takes the average bird eight night-time hours to acquire enough food, it may be an important feature of the breeding timetable that many more hours of darkness are available for feeding in the winter.

Finally, the differences between the 1982 and 1984 Grey-headed Albatross data probably reflect very real differences in the prevailing environmental conditions. Thus, in 1984, Grey-headed Albatross breeding success (chicks fledged from eggs laid) was only 28%, the second lowest ever recorded (and only the third time in 10 years that breeding success has been less than 45%), compared with 53% in 1982. We interpret the combination of near doubling in night-time hours spent on the sea and halving of daytime hours spent on the sea, together with the reduction in chick provisioning rate from 85% of days in 1982 to 54% in 1984, as indicating the much greater difficulty in catching prey in the latter season.

Routine use of these recorders clearly has considerable potential for illuminating the at-sea activities of seabirds generally, in addition to the value of such activity budgets in interpreting the energy costs of foraging trips (D.P. Costa & P.A. Prince, unpublished).

In all these interpretations we do not know whether the duration of time spent on the sea at night directly reflects the duration of feeding activity, but it seems the most reasonable hypothesis, especially for birds engaged in rearing chicks. Non-breeding birds should be able to spend much more time on the sea and there are some anecdotal data which suggest that this might be the case. Fig. 7.2 shows a series of observations from *RRS John Biscoe* in the early afternoon (1400–1546 GMT) of 14 February 1982 near the South Shetland Islands (64° 32'–64° 57'S, 64° 51'W). *D. chrysostoma* and *D. melanophrys* were present on the sea throughout, but their numbers were only recorded during the 10-min sampling periods shown. The two species were present in about equal numbers and some 30% of birds were in immature plumage. The albatrosses were not feeding and were presumably awaiting the nocturnal vertical migration of suitable prey. Although only 30% of the albatrosses in the flocks were identifiable as immatures it is possible that many of the remaining birds were either failed breeders or non-breeders. Because Grey-headed Albatrosses breed biennially (Prince, 1985) about 40% of adult birds in any population will be non-breeding. In addition, by February, about half the breeding pairs will have failed (Prince, 1980*b*, 1985). Thus, no more than 15% of the other

Fig. 7.2. Sightings of albatrosses on the sea surface on 14 February 1982 in relation to krill density and distribution. Krill data, starting here at 1400 GMT and ending at 1545 GMT, were obtained using a Simrad EKS-120 Scientific Echosounder and a QM Mk II Analogue Integrator. The numbers of albatrosses recorded during 10-min observation periods are shown at the top of the diagram.

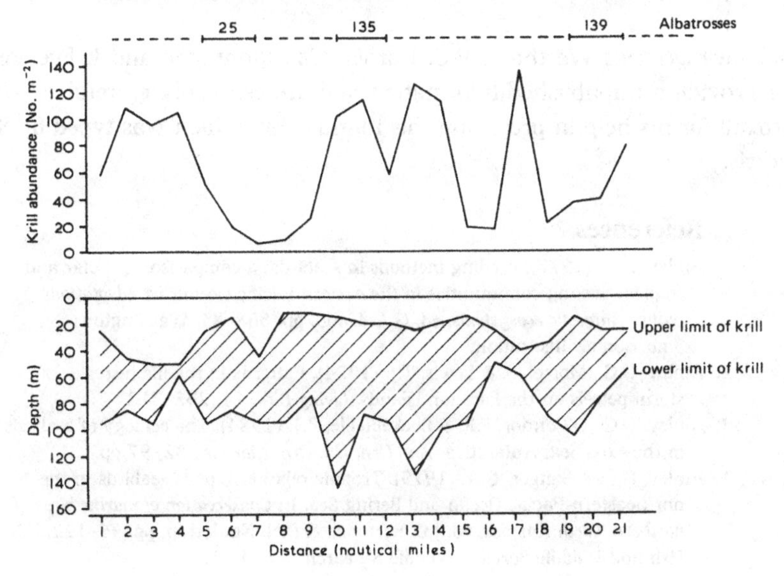

birds in these flocks were likely still to be rearing chicks. Daytime aggregations of non-breeders, however, could provide valuable information on the potential location of prey for breeding birds covering the area.

Conclusions

Although our knowledge of seabird food and feeding ecology has advanced in recent years, information for the majority of species still remains anecdotal. Even for species where quantitative data are available, a more realistic picture will only emerge as information from other sites and seasons becomes available. In particular, data on species' diets in winter is the most serious gap in our knowledge. Paradoxically, for the few transequatorial migrants for which good data on winter diet are available, we have little quantitative information on their summer foods.

Dietary studies are likely to develop steadily, particularly as improved techniques for identifying squid beaks and otoliths become available, and as more work is done on digestion rates. Studies of feeding ecology, however, will remain much more difficult, which is unfortunate, as the development of feeding skills is likely to be an important feature in the long deferment of sexual and social maturity in Procellariiformes. Additional data may come through the use of night-observation equipment but major developments are more likely to come from the development of integrated systems of satellite tags, prey-catching sensors and activity pattern recorders which can be deployed for long periods on birds of known status whose subsequent performance can be monitored accurately.

Acknowledgements We thank P.C. Harper, T.L. Montague and I. Everson for providing unpublished information and are especially grateful to J.P. Croxall for his help in preparing this manuscript, which was typed by S. Norris.

References

1. Ainley, D.G. (1977). Feeding methods in seabirds: a comparison of polar and tropical nesting communities in the eastern Pacific Ocean. In *Adaptations within Antarctic ecosystems*, ed. G.A. Llano, pp. 668–85, Washington: Smithsonian Institution.
2. Ainley, D.G., Morrel, S. & Lewis, T.J. (1974). Patterns in the life histories of storm petrels on the Farallon Islands. *Living Bird*, **14**, 295–312.
3. Ainley, D.G., O'Connor, E.F. & Boekelheide, R.J. (1984). The ecology of seabirds in the Ross Sea, Antarctica. *Am. Orn. Un. Orn. Monogr.*, **32**, 97 pp.
4. Ainley, D.G. & Sanger, G.A. (1979). Trophic relationships of seabirds in the northeastern Pacific Ocean and Bering Sea. In *Conservation of marine birds of northern North America*, ed. J.C. Bartonek & D.N. Nettleship, pp. 95–122. U.S. Fish and Wildlife Service, Wildlife Research Report 11.

5. Ashmole, N.P. (1971). Seabird ecology and the marine environment. In *Avian biology*, vol. 1, ed. D.S. Farner & J.R. King, pp. 224–71. New York: Academic Press.
6. Ashmole, N.P. & Ashmole, M.S. (1967). Comparative feeding ecology of seabirds of a tropical oceanic island. *Peabody Mus. Nat. Hist., Bull.*, **24**, 1–131.
7. Baltz, D.M. & Morejohn, G.V. (1977). Food habits and niche overlap of seabirds wintering on Monterey Bay, California. *Auk*, **94**, 526–43.
8. Bang, B.G. (1971). Functional anatomy of the olfactory system in 23 orders of birds. *Acta Anat.*, **58**, Suppl., 1–76.
9. Barton, D. (1978). Albatrosses in the western Tasman Sea. *Emu*, **79**, 31–5.
10. Beck, J.R. (1969). Food, moult and age of first breeding in the Cape Pigeon, *Daption capensis* Linnaeus. *Bull. Br. Antarct. Surv.*, **21**, 33–44.
11. Beck, J.R. & Brown, D.W. (1972). The biology of Wilson's Storm Petrel *Oceanites oceanicus* (Kuhl), at Signy Island, South Orkney Islands. *Sci. Rep. Br. Antarct. Surv.*, **69**, 1–54.
12. Berruti, A. & Harcus, T. (1978). Cephalopod prey of the Sooty Albatrosses, *Phoebetria fusca* and *P. palpebrata* at Marion Island. *S. Afr. J. Antarct. Res.*, **8**, 99–103.
13. Bierman, W.H. & Voous, K.H. (1950). Birds observed and collected during the whaling expedition of the Willem Barendz in the Antarctic 1946–1947. *Ardea*, **37** (extra no.), 1–123.
14. Bonner, W.N. & Hunter, S. (1982). Predatory interactions between Antarctic Fur Seals, Macaroni Penguins and giant petrels. *Bull. Br. Antarct. Surv.*, **56**, 75–9.
15. Bradstreet, M.S.W. (1976). *Summer feeding ecology of seabirds in eastern Lancaster Sound, 1976*. Report for Norlands Petroleum, Calgary, LGL Ltd. Toronto: Environmental Research Associates.
16. Bradstreet, M.S.W. & Cross, W.E. (1982). Trophic relationships at high Arctic ice edges. *Arctic*, **35**, 1–12.
17. Brown, R.G.B., Barker, G.B., Gaskin, D.E. & Sandeman, M.R. (1981). The foods of Great and Sooty Shearwaters, *Puffinus gravis* and *P. griseus* in eastern Canadian waters. *Ibis*, **123**, 19–30.
18. Brown, R.G.B., Bourne, W.R.P. & Wahl, T.R. (1978). Diving by shearwaters. *Condor*, **80**, 123–5.
19. Cheng, L. & Harrison, C.S. (1983). Seabird predation on the sea-skater *Halobates sericeus* (Heteroptera: Gerridae). *Mar. Biol.*, **72**, 303–9.
20. Chu, E.W. (1984). Sooty Shearwaters off California: diet and energy gain. In *Marine birds: their feeding ecology and commercial fisheries relationships*, ed. D.N. Nettleship, G.A. Sanger & P.F. Springer, pp. 64–71. Ottawa: Can. Wildl. Serv. Spec. Publ.
21. Clarke, A. & Prince, P.A. (1980). Chemical composition and calorific value of food fed to mollymauk chicks *Diomedea melanophrys* and *D. chrysostoma* at Bird Island, South Georgia. *Ibis*, **122**, 488–94.
22. Clarke, M.R. (1966). A review of the systematics and ecology of oceanic squids. *Adv. Mar.Biol.*, **4**, 91–300.
23. Clarke, M.R. (1986) *A handbook for the identification of cephalopod beaks*. Oxford: Clarendon Press.
24. Clarke, M.R., Croxall, J.P. & Prince, P.A. (1981). Cephalopod remains in regurgitations of the Wandering Albatross at South Georgia. *Bull. Br. Antarct. Surv.*, **54**, 9–22.
25. Clarke, M.R. & Prince, P.A. (1981). Cephalopod remains in regurgitations of the

Black-browed and Grey-headed Albatrosses at South Georgia. *Bull. Br. Antarct. Surv.*, **54**, 1–7.

26. Cobb, S. (1960). Observations on the comparative anatomy of the avian brain. *Perspect. Biol. Med.*, **3**, 383.
27. Conroy, J.W.H. (1972). Ecological aspects of the biology of the Giant Petrel, *Macronectes giganteus* (Gmelin), in the maritime Antarctic. *Sci. Rep. Br. Antarct. Surv.*, **75**, 1–74.
28. Cottam, C. & Knappen, P. (1939). Food of some uncommon north American birds. *Auk*, **56**, 138–69.
29. Cox, J.B. (1980). Some remarks on the breeding distribution and taxonomy of the prions. *Rec. Aust. Mus.*, **18**, 91–121.
30. Cramp, S. & Simmons, K.E.L. (eds) (1977). *The birds of the western Palearctic*, vol. 1. Oxford: Oxford University Press.
31. Cramp, S. & Simmons, K.E.L. (eds) (1983). *The birds of the western Palearctic*, vol. 3. Oxford: Oxford University Press.
32. Croxall, J.P. & Prince, P.A. (1980). Food, feeding ecology and ecological segregation of seabirds at South Georgia. *Biol. J. Linn. Soc.*, **14**, 103–31.
33. Croxall, J.P. & Prince, P.A. (1982). A preliminary assessment of the impact of seabirds on marine resources at South Georgia. *Com. Nat. Fr. Recherch. Antarct.*, **51**, 501–9.
34. Despin, B., Mougin, J.-L. & Segonzac, M. (1972). Oiseaux et mammifères de l'Ile de l'Est, archipel Crozet. *Com. Nat. Fr. Recherch. Antarct.*, **31**, 1–106.
35. Diamond, A.W. (1971). The ecology of the seabirds of Aldabra. *Phil. Trans. R. Soc. Lond. B.*, **260**, 561–71.
36. Downes, M.C., Ealey, F.H.M., Gwynn, A.N. and Young, P.S. (1959). The birds of Heard Island. *A.N.A.R.E. Rep.*, **13**, 1–135.
37. Duffy, D.C. (1980). Patterns of piracy by Peruvian seabirds: a depth hypothesis. *Ibis*, **122**, 521–5.
38. Ealey, E.H.M. (1954). Analysis of stomach contents of some Heard Island birds. *Emu*, **54**, 204–10.
39. Falla, R.A. (1934). The distribution and breeding habits of petrels in Northern New Zealand. *Rec. Auck. Inst. Mus.*, **1**, 245–60.
40. Falla, R.A. (1937). Birds. *B.A.N.Z.A.R.E. Rep.*, **BII**, 1–304.
41. Fenwick, G.D. (1978). Plankton swarms and their predators at the Snares Islands. *N.Z. J. Mar. Freshw. Res.*, **12**, 223–4.
42. Fisher, J. (1952). *The Fulmar*. London: Collins.
43. Fisher, J. (1966). The Fulmar population of Britain and Ireland, 1959. *Bird Study*, **13**, 5–76.
44. Furness, R.W. (1981). Seabird populations of Foula. *Scott. Birds*, **11**, 237–53.
45. Furness, R.W. & Todd, C.M. (1984). Diets and feeding of Fulmars *Fulmarus glacialis* during the breeding season: a comparison between St. Kilda and Shetland colonies. *Ibis*, **126**, 379–87.
46. Furse, J.R. (1976). The Antarctic Fulmar *Fulmarus glacialoides*, at home near Elephant Island. *Sea Swallow*, **27**, 7–10.
47. Gould, P.J. (1967). Nocturnal feeding of *Sterna fuscata* and *Puffinus pacificus*. *Condor*, **69**, 529.
48. Griffiths, A.M. (1982). Observations of pelagic seabirds feeding in the African sector of the Southern Ocean. *Cormorant*, **10**, 9–14.
49. Griffiths, A.M. (1983). Factors affecting the distribution of the Snow Petrel *Pagodroma nivea* and the Antarctic Petrel *Thalassoica antarctica*. *Ardea*, **71**, 145–50.

50. Griffiths, A.M., Siegfried, W.R. & Abrams, R.W. (1982). Ecological structure of a pelagic seabird community in the Southern Ocean. *Polar Biol.*, **1**, 39–46.
51. Grubb, T.C., Jr (1972). Smell and foraging in shearwaters and petrels. *Nature*, **237**, 404–5.
52. Hagen, Y. (1952). Birds of Tristan da Cunha. *Res. Norw. Sci. Exp. Tristan da Cunha 1937–1938*, **20**, 1–256.
53. Halse, S.A. (1981). Migration by Hutton's Shearwater. *Emu*, **81**, 42–4.
54. Harper, P.C. (1972). Field identification and distribution of the Thin-billed Prion *Pachyptila belcheri* and the Antarctic Prion *Pachyptila desolata*. *Notornis*, **19**, 140–75.
55. Harper, P.C. (1976). Breeding biology of the Fairy Prion *Pachyptila turtur* at the Poor Knights Islands, New Zealand. *N.Z. J. Zool.*, **3**, 351–71.
56. Harper, P.C. (1979). Colour vision in the Procellariiformes. *Mauri Ora*, **7**, 151–55.
57. Harper, P.C. (1980). The field identification and distribution of the prions with particular reference to the identification of storm-cast material. *Notornis*, **27**, 235–86.
58. Harper, P.C. (1983). Biology of the Buller's Shearwater *Puffinus bulleri* at the Poor Knights Islands, New Zealand. *Notornis*, **30**, 299–318.
59. Harper, P.C., Croxall, J.P. & Cooper, J. (1985) A guide to foraging methods used by marine birds in Antarctic and subantarctic seas. *BIOMASS Handbook*, No. 24, 22 pp.
60. Harris, M.P. (1969*a*). The biology of storm petrels in the Galapagos Islands. *Proc. Calif. Acad. Sci.*, **37**, 95–166.
61. Harris, M.P. (1969*b*). Food as a factor controlling the breeding of *Puffinus l'herminieri*. *Ibis*, **111**, 139–56.
62. Harris, M.P. (1970). The biology of an endangered species, the Dark-rumped Petrel *Pterodroma phaeopygia* in the Galapagos Islands. *Condor*, **72**, 76–84.
63. Harris, M.P. (1973). The biology of the Waved Albatross *Diomedea irrorata* of Hood Island, Galapagos. *Ibis*, **115**, 483–510.
64. Harris, M.P. (1976). The seabirds of Shetland in 1974. *Scott. Birds*, **9**, 37–68.
65. Harrison, C.S., Hida, T.S. & Seki, M.P. (1983). Hawaiian seabird feeding ecology. *Wildl. Monogr.*, **85**, 1–71.
66. Hasegawa, H. & DeGange, A.L. (1982). The Short-tailed Albatross, *Diomedea albatrus*, its status, distribution and natural history. *Am. Birds*, **36**, 806–14.
67. Hunter, S. (1983). The food and feeding ecology of the Giant Petrels *Macronectes halli* and *M. giganteus* at South Georgia. *J. Zool., Lond.*, **200**, 521–38.
68. Hunter, S. (1985). The role of Giant Petrels in the Southern Ocean ecosystem. In *Antarctic nutrient cycles and food webs*, ed. W.R. Siegfried, P.R. Condy & R.M. Laws, pp534–42. Berlin: Springer-Verlag.
69. Hutchinson, L.V. & Wenzel, B.M. (1980). Olfactory guidance in foraging by procellariiforms. *Condor*, **82**, 314–19.
70. Hutchinson, L.V., Wenzel, B.M., Stager, K.E. & Tedford, B.L. (1984). Further evidence for olfactory foraging in Sooty Shearwaters and Northern Fulmars. In *Marine birds: their feeding ecology and commercial fisheries relationships*, ed. D.N. Nettleship, G.A. Sanger & P.F. Springer, pp. 72–7. Ottawa: Can. Wildl. Serv. Spec. Publ.
71. Imber, M.J. (1973). The food of Grey-faced Petrels *Pterodroma macroptera gouldi* with special reference to diurnal vertical migration of their prey. *J. Anim. Ecol.*, **42**, 645–62.

72. Imber, M.J. (1976). Comparison of the prey of the Black *Procellaria* Petrels of New Zealand. *N.Z. J. Mar. Freshw. Res.*, **10**, 119–30.
73. Imber, M.J. (1981). Diets of storm petrels *Pelagodroma* and *Garrodia* and of prions *Pachyptila* (Procellariiformes): ecological separation and bill morphology. In *Proceedings of the symposium on birds of the sea and shore, 1979*, ed. J. Cooper, pp. 63–88. Cape Town: African Seabird Group.
74. Imber, M.J. (1983). The lesser petrels of Antipodes Islands, with notes from Prince Edward and Gough Islands. *Notornis*, **30**, 283–98.
75. Imber, M.J. (1985). Origins, phylogeny and taxonomy of the gadfly petrels *Pterodroma* spp. *Ibis*, **127**, 197–229.
76. Imber, M.J. & Berruti, A. (1981). Procellariform seabirds as squid predators. In *Proceedings of the symposium on birds of the sea and shore, 1979*, ed. J. Cooper, pp. 43–61. Cape Town: African Seabird Group.
77. Imber, M.J. & Nilsson, R.J. (1980). South Georgia Diving Petrels breeding on Codfish island. *Notornis*, **27**, 11–20.
78. Imber, M.J. & Russ, R. (1975). Some foods of the Wandering Albatross *Diomedea exulans*. *Notornis*, **22**, 27–36.
79. Jacob, J. (1982). Stomach oils. In *Avian biology*, vol. 6, ed. D.S. Farner, J.R. King & K.C. Parkes, pp. 325–40. New York: Academic Press.
80. Jangaard, P.M. (1974). The Capelin *Mallotus villosus*: biology, distribution, exploitation, utilization and composition. *Bull. Fish. Res. Bd. Can.*, **186**, 1–70.
81. Jenkins, J.A.F. (1974). Local distribution and feeding habits of Buller's Shearwaters. *Notornis*, **21**, 109–20.
82. Johnstone, G.W. (1977). Comparative feeding ecology of the Giant Petrels *Macronectes giganteus* (Gmelin) and *M. halli* (Mathews). In *Adaptations within Antarctic ecosystems*, ed. G.A. Llano, pp. 647–68. Washington: Smithsonian Institution.
83. Jouventin, P., Mougin, J.-L., Stahl, J.-C., Bartle, J.A. & Weimerskirch, H. (1982). Données preliminaires sur la distribution pelagiques des oiseaux des T.A.A.F. *Com. Nat. Fr. Recherch. Antarct.*, **51**, 427–36.
84. Kerry, K.R., Horne, R.S.C. & Dorward, D. (1983). Records of the Short-tailed Shearwater *Puffinus tenuirostris* in Antarctic waters. *Emu*, **83**, 35–7.
85. Kinsky, F.C., Bell, B.D., Braithwaite, D.H., Falla, R.A., Sibson, R.B. & Turbott, E.G. (1970). *Annotated checklist of the birds of New Zealand including the birds of the Ross Dependency*. Wellington: A.H. & A.W. Reed.
86. Krasnow, L.D., Sanger, G.A. & Wiswar, D.W. (1979). Nearshore feeding ecology of marine birds in the Kodiak area, 1978. *Annual Report NOAA – OCSEAP. U.S. Fish Wildl. Ser. Biol. Ser. Prog.*
87. Kuroda, N. (1954). *On the classification and phylogeny of the order Tubinares, particularly the shearwaters* Puffinus. Tokyo: Herald Co.
88. Larson, J.W. (1967). *The Dark-rumped Petrel in Haleakala Crater, Maui, Hawaii.* Report to the National Parks Service. U.S. Dept. Interior.
89. Linton, A. (1978). The food and feeding of Leach's Storm Petrel *Oceanodroma leucorhoa* at Pearl Island, Nova Scotia and Middle Lawn Island, Newfoundland. Unpublished M.Sc. thesis, Dalhousie University, Halifax, Nova Scotia.
90. Loomis, L.M. (1918). A review of the albatrosses, petrels and diving petrels. *Proc. Calif. Acad. Sci.*, **2**, 1–187.
91. Maher, W.J. (1962). Breeding biology of the Snow Petrel near Cape Hallett, Antarctica. *Condor*, **64**, 488–99.
92. Miller, L. (1936). Some maritime birds observed off San Diego, California. *Condor*, **38**, 9–16.

93. Miller, L. (1940). Some tagging experiments with Black-footed Albatrosses. *Condor*, **44**, 3–9.
94. Montague, T.L. (1984). The food of Antarctic Petrels *Thalassoica antarctica*. *Emu*, **84**, 244–5.
95. Montague, T.L., Cullen, J.M. & Fitzherbert, K. (In press) The diet of the Short-tailed Shearwater *Puffinus tenuirostris* during its breeding season. *Emu*.
96. Morejohn, G.V., Harvey, J.T. & Krasnow, L.T. (1978). The importance of *Loligo opalescens* in the food web of marine vertebrates in Monterey Bay California. *Calif. Dep. Fish Game, Fish. Bull.*, **169**, 67–98.
97. Morgan, W.L. (1982). Feeding methods of the Short-tailed Shearwater *Puffinus tenuirostris*. *Emu*, **82**, 226–7.
98. Morgan, W.L. & Ritz, D.A. (1982). Comparison of the feeding apparatus in the Mutton Bird *Puffinus tenuirostris* and the Fairy Prion *Pachyptila turtur* in relation to the capture of the krill *Nyctiphanes australis*. *J. Exp. Mar. Biol. Ecol.*, **59**, 61–75.
99. Mougin, J.-L. (1968). Étude écologique de quatre espèces de pétrels Antarctiques. *Oiseau*, **38**, 2–52.
100. Mougin, J.-L. (1970). Le pétrel à menton blanc *Procellaria aequinoctialis* de l'Ile de la Possession (Archipel Crozet). *Oiseau*, **40**, 62–96.
101. Murphy, R.C. (1936). *Oceanic birds of South America*, vols I and II. New York: Macmillan.
102. Murphy, R.C. & Mowbray, L.S. (1951). New light on the Cahow. *Auk*, **68**, 266–80.
103. Nicholls, G.H. (1979). Underwater swimming by albatrosses. *Cormorant*, **6**, 38.
104. Ogi, H. (1984). Feeding ecology of the Sooty Shearwater in the western subarctic North Pacific Ocean. In *Marine birds: their feeding ecology and commercial fisheries relationships*, ed. D.N. Nettleship, G.A. Sanger & P.F. Springer, pp. 78–85. Ottawa: Can. Wildl. Serv. Spec. Publ.
105. Ogi, H., Kubodera, T. & Nakamura, K. (1980). The pelagic feeding ecology of the Short-tailed Shearwater *Puffinus tenuirostris* in the Subarctic Pacific Region. *J. Yamashina Inst. Ornithol.*, **12**, 157–82.
106. Oliver, W.R.B. (1930). *New Zealand birds*. Wellington: A.H. & A.W. Reed.
107. Palmer, R.S. (ed.) (1962). *Handbook of North American Birds*, vol. 1. New Haven & London: Yale University Press.
108. Paulian, P. (1953). Pinnipèdes, cétacés, oiseaux des Iles Kerguelen et Amsterdam. *Mem. Inst. Sci. Madagasc.*, **A8**, 111–234.
109. Payne, M.R. & Prince, P.A. (1979). Identification and breeding biology of the diving petrels, *Pelecanoides georgicus* and *P. urinatrix exsul* at South Georgia. *N.Z. J. Zool.*, **6**, 299–318.
110. Potter, I.C., Prince, P.A. & Croxall, J.P. (1979). Data on the adult marine and migratory phases in the life cycle of the Southern Hemisphere Lamprey *Geotria australis* Gray. *Environ. Biol. Fish.*, **4**, 65–9.
111. Prince, P.A. (1980*a*). The food and feeding ecology of Blue Petrel *Halobaena caerulea* and Dove Prion *Pachyptila desolata*. *J. Zool., Lond.*, **190**, 59–76.
112. Prince, P.A. (1980*b*). The food and feeding ecology of Grey-headed Albatross *Diomedea chrysostoma* and Black-browed Albatross *Diomedea melanophris*. *Ibis*, **122**, 476–88.
113. Prince, P.A. (1985). Population and energetic aspects of the relationship between Black-browed and Grey-headed Albatrosses and the Southern Ocean marine environment. In *Antarctic nutrient cycles and food webs*, ed. W.R. Siegfried, P.R. Condy & R.M. Laws, pp. 473–77. Berlin: Springer-Verlag.

114. Prince, P.A. & Copestake, P.G. (unpublished) Food and feeding ecology of the Fairy Prion *Pachyptila turtur* at Bird Island, South Georgia.
115. Prince, P.A. & Francis, M.D. (1984). Activity budgets of foraging Grey-headed Albatrosses. *Condor*, **86**, 297–300.
116. Prince, P.A. & Ricketts, C. (1981). Relationships between food supply and growth in albatrosses: an interspecies chick fostering experiment. *Ornis Scand.*, **12**, 207–10.
117. Richdale, L.E. (1944). The Parara or Broad-billed Prion, *Pachyptila vittata*. *Emu*, **43**, 191–217.
118. Richdale, L.E. (1965). Breeding behaviour of the Narrow-billed Prion and the Broad-billed Prion on Whero Island, New Zealand. *Trans. Zool. Soc. Lond.*, **31**, 87–155.
119. Robertson, C.J.R. (ed.) (In press) *Handbook of the birds of New Zealand.*
120. Rowan, M.K. (1951). The Yellow-nosed Albatross *Diomedea chlororhynchos* at its breeding grounds in the Tristan da Cunha group. *Ostrich*, **22**, 139–55.
121. Sanger, G.A. (1983). Diets and food web relationships of seabirds in the Gulf of Alaska and adjacent marine regions. In *Environmental assessment of the Alaskan Continental Shelf*. Boulder, Colorado: NOAA.
122. Sanger, G.A. & Baird, P.A. (1977). Trophic relationships of marine birds in the Gulf of Alaska and the southern Bering Sea. *Ann. Rep. NOAA–OCSEAP. U.S. Fish Wildl. Ser.*
123. Schramm, M. (1982). The breeding biologies of the petrels *Pterodroma macroptera*, *P. brevirostris* and *P. mollis* at Marion Island. *Emu*, **83**, 75–81.
124. Sealy, S.G. (1973). Interspecific feeding assemblages of marine birds off British Columbia. *Auk*, **90**, 796–802.
125. Serventy, D.L., Serventy, V. & Warham, J. (1971). *The handbook of Australian seabirds*. Sydney: A.H. & A.W. Reed.
126. Skira, I. (1979). Underwater feeding by Short-tailed Shearwaters. *Emu*, **79**, 43.
127. Strange, I.J. (1980). The Thin-billed Prion, *Pachyptila belcheri* at New Island, Falkland Islands. *Gerfaut*, **70**, 411–45.
128. Thomas, G. (1982). The food and feeding ecology of the Light-mantled Sooty Albatross at South Georgia. *Emu*, **82**, 92–100.
129. Tickell, W.L.N. (1962). The Dove Prion *Pachyptila desolata* (Gmelin). *Sci. Rep. Falkl. Is. Depend. Surv.*, **33**, 1–55.
130. Tickell, W.L.N. (1964). Feeding preferences of the albatrosseses *Diomedea melanophris* and *D. chrysostoma* at South Georgia. In *Biologie Antarctique*, ed. R. Carrick, M.W. Holdgate & J. Prevost, pp. 383–7. Paris: Hermann.
131. Vernon, D.P. (1978). Food of prions with particular reference to Fairy Prion, *Pachyptila turtur*. *Sunbird*, **9**, 7–9.
132. Voisin, J.F. (1981). A pursuit plunging Wandering Albatross *Diomedea exulans*. *Cormorant*, **9**, 136.
133. Voisin, J.F. & Shaughnessy, P.D. (1980). Diving by giant petrels *Macronectes*. *Cormorant*, **8**, 25–6.
134. Voous, K.H. (1949). The morphological, anatomical and distributional relationship of Arctic and Antarctic Fulmars (Aves, Procellariidae). *Ardea*, **37**, 113–22.
135. Warham, J. (1977*a*). The incidence, functions and ecological significance of petrel stomach oils. *Proc. N.Z. Ecol. Soc.*, **24**, 84–93.
136. Warham, J. (1977*b*). Wing loadings, wing shapes and flight capabilities of Procellariiformes. *N.Z. J. Zool.*, **4**, 73–83.
137. Watson, G.E. (1975). *Birds of the Antarctic and Subantarctic*. Washington, D.C.: American Geophysical Union.

138. Wenzel, B.M. (1980). Chemoreception in seabirds. In *Behaviour of marine animals*, vol. 4., ed. J. Burger, B.L. Olla & H.E. Winn, pp. 41-67. New York: Plenum Press.
139. Wiens, J.A. & Scott, J.M. (1975). Model estimation of energy flow in Oregon coastal seabird populations. *Condor*, **77**, 439–52.
140. Williams, A.J. & Imber, M.J. (1982). Ornithological observations at Gough Island in 1979, 1980 and 1981. *S. Afr. J. Antarct. Sci.*, **12**, 40–6.
141. Wilson, G.J. (1973). Birds of the Solander islands. *Notornis*, **20**, 318–23.
142. Wingate, D.B. (1972). First successful hand-rearing of an abandoned Bermuda Petrel chick. *Ibis*, **114**, 97–101.
143. Wood Jones, F. (1937). The olfactory organ of the Tubinares. *Emu*, **36**, 281–6.
144. Young, R.E. (1977). Ventral bioluminescent countershading in midwater cephalopods. *Symp. Zool. Soc. Lond.*, No. 38, 161–90.
145. Zink, R.L. (1981). Observations of seabirds during a cruise from Ross Island to Anvers Island, Antarctic. *Wilson Bull.*, **93**, 1–20.
146. Zonfrillo, B. (In press) Diet of Bulwer's Petrel *Bulweria bulwerii* in the Madeiran archipelago. *Ibis*.

8

Pelecaniform feeding ecology

RALPH W. SCHREIBER
Los Angeles County Museum of Natural History, 900 Exposition Boulevard, Los Angeles, California 90007, USA
and
ROGER B. CLAPP
Museum Section, US Fish & Wildlife Service, National Museum of Natural History, Washington DC 20560, USA

Introduction

The Pelecaniformes is primarily a pantropical group composed of four closely related families (Phalacrocoracidae, 27 spp.; Anhingidae, 2 spp.; Sulidae, 9 spp.; and Pelecanidae, 7 spp.) and two families whose affinities are unclear (Fregatidae, 5 spp.; Phaethontidae, 3 spp.). The most accurate and up-to-date general descriptions and distributions for all species except the anhingas are presented by Harrison (1983). All pelecaniforms have totipalmate feet (all four toes united by a web) and desmognathous palates but most other anatomical characteristics are highly variable (Sibley & Ahlquist, 1972), which may indicate polyphyly.

Most aspects of the biology of this order were succinctly summarized by Ashmole (1975). More recently, feeding methods and the published literature on one or more species of the order have been summarized (Nelson, 1978, 1984; Clapp *et al.*, 1982; various chapters in Haley, 1984). In Fig. 8.1 we illustrate the head and bill of representative species of each family to show size and shape variation in the primary feeding apparatus. Details of morphology as it relates to feeding behavior and locomation of most species are lacking with the exception of the cormorants and anhingas (Owre, 1967; Burger, 1978). Ashmole's (1971) review emphasized the interrelationships of seabirds and their environment, and provided a useful classification and illustration of feeding methods and techniques. Virtually all methods of catching fish and squid are practiced by one or more pelecaniform, either primarily or occasionally (Table 8.1).

Our aim here is to summarize briefly some of this diversity and thus to indicate the amount of additional new research needed, even though the

Fig. 8.1 Heads and bills of representatives of the Pelecaniformes families. All drawn to scale.

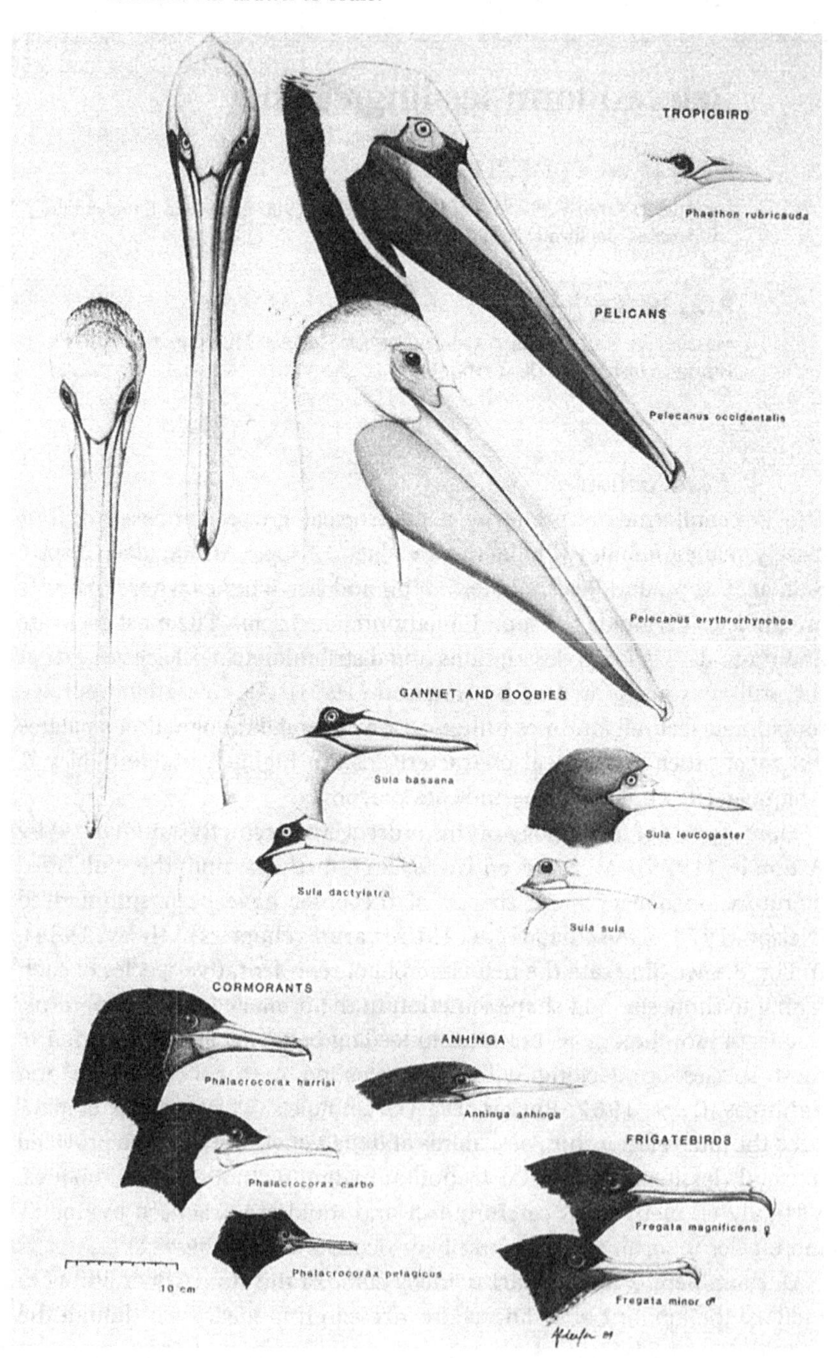

Table 8.1. *Summary of feeding biology of the Pelecaniformes (modified from Ashmole, 1971, and references in text). The primary purpose of this table is to show that while generalities are known, little progress has been made since Ashmole's classification and many questions remain to be studied*

	Feeding method						Gregariousness		Time			Habitat						Diet			
	Foot propelled underwater	Deep plunging	Surface plunging	Settled on water	While flying at or near water surface	Kleptoparasitism	Flock	Individual	Day	Crepuscular	Night	Bottom	Midwater	Surface	Near shore, within 20 km of land	Pelagic, out of sight of land	Fresh-water	Fish	Squid	Other invertebrates	Other items (turtles, bird eggs, nestling birds, scavenge)
Phaethontidae	—	+	?	—	—	—	?	+	+	+	?	—	—	+	—	+	—	+	+	?	—
Fregatidae	—	—	—	—	+	+	+	+	+	+	?	—	—	+	+	+	+	+	+	?	+
Phalacrocoracidae	+	—	—	—	—	—	+	+	+	+	?	+	+	+	+	—	+	+	—	+	—
Anhingidae	+	—	—	—	—	—	—	+	+	+	?	+	+	—	—	—	+	+	—	—	—
Sulidae	—	+	+	?	+	—	+	+	+	+	?	?	+	+	+	+	—	+	+	—	—
Pelecanidae	—	—	+	+	—	—	+	+	+	+	+	—	—	+	+	—	+	+	—	+	+

literature on feeding is extensive. In our discussion we explore aspects of pelecaniform feeding ecology in the hope that future studies will shed sufficient light on feeding mechanisms and ecology to allow valid evolutionary conclusions. Recent compendia provide useful general summaries of the diet of many species (e.g. Palmer, 1962; Cramp & Simmons, 1977; Brown, Urban & Newman, 1982).

The anhingas, several cormorants, and most pelicans are primarily inland/fresh-water forms and are here considered only in passing.

Diet and feeding behavior

Tropicbirds Phaethontidae

The three species of tropicbirds are widely distributed in tropical seas: Red-tailed and White-tailed Tropicbirds *Phaethon rubricauda* and *P. lepturus* broadly overlap in breeding range in the Pacific and Indian Oceans and White-tailed Tropicbirds are also found in the Atlantic. Red-billed Tropicbirds *P. aethereus* tend to be coastal in the eastern Pacific and Caribbean and overlap broadly with *P. lepturus* in the Atlantic but occur farther north than that species in the Indian Ocean. Tropicbirds are probably the most pelagic of the Pelecaniformes and are absent from nesting locations during the non-breeding season. Flyingfish (Exocoetidae) and squid (Ommastrephidae), the primary items of diet in all areas studied, are obtained by plunging from a considerable height apparently to some depth. Little is known of actual feeding behavior and distribution at sea (Gould, King & Sanger, 1974). Details of diet in a few locations are known but much remains to be studied. Tropicbirds take quite remarkably large fish (up to 15–18% of their body weight) for their size (Harris, 1969; Clapp *et al.*, 1982; Harrison, Hida & Seki, 1983; Diamond, 1984; Schreiber, unpublished data). Diamond (1975*b*) found considerable differences in diet and nest sites between Red-tailed and White-tailed Tropicbirds on Aldabra Atoll.

Frigatebirds Fregatidae

Frigatebirds have a pantropical distribution, and are found in areas of consistent, moderate winds which are necessary for their sustained flight (Harrington, Schreiber & Woolfenden, 1972; Pennycuick, 1983). One species (*Fregata aquila*) is confined to Ascension Island, Atlantic Ocean; another (*F. andrewsi*) is confined to Christmas Island, Indian Ocean; Magnificent Frigatebirds *F. magnificens* occur along the coasts of the New World; and Greater and Lesser Frigatebirds, *F. minor* and *F. ariel*, are broadly sympatric in the equatorial Pacific and Indian Oceans, with small populations in the southwest Atlantic (Harrison, 1983). At least some populations undergo long migrations (Sibley & Clapp, 1967), but while frigatebirds are primarily diurnal and remain 'close' to breeding colonies, they can be found far at sea, usually as individuals (Pitman, personal communication). Frigatebirds are extremely light in weight for their wing area, the distinctive bent wings and forked tails allowing ready manoeuverability. These characters and the hooked beak are apparently adaptations for grasping food from close to the water's surface while remaining in the air (Nelson, 1976); frigatebirds virtually never enter or alight on the water. Frigatebirds primarily eat flyingfish and squid. Other items, including nestlings of Sooty Terns *Sterna fuscata*, Sooty Storm Petrels *Oceanodroma tristrami* and turtles *Chelonia mydas* (Cornelius, unpublished) are eaten, and they sometimes eat the eggs and young of their own species (Schreiber & Ashmole, 1970; Schreiber, unpublished data). Eating their own eggs and young may be the result of human disturbance (Diamond, 1973, 1975*a*; Schreiber & Hensley, 1976; Harrison *et al.*, 1983) but this is as yet unproven. Considerable variation in diet exists from locality to locality. Frigatebirds frequently steal food from other species but this habit probably provides important amounts of food only when other food is scarce (Schreiber & Hensley, 1976; Schreiber & Schreiber, unpublished data). In the only published comparative study of frigatebirds' diet, Diamond (1975*a*) showed that Greater and Lesser Frigatebirds on Aldabra ate the same species of fish and squid in the same proportions. Size of prey *per se* was not important in determining what the birds captured and the overlap in feeding was greater than in any other study of coexisting congeneric seabirds.

Cormorants Phalacrocoracidae

The 27 species of cormorants are primarily all black or black above with white below; many are polymorphic. Several species breed both inland and along coasts. Some are highly migratory, primarily the more northerly distributed species. Many species are broadly sympatric with one or more other species. All nest colonially and huge colonies exist for a few species. Although many species are widely distributed others have quite restricted ranges. All marine forms tend to remain coastal and feed in shallow (< 5–8m) water and only a few species venture far from nesting colonies or mainland regions.

Owre's (1967) classic study compared adaptations in skeleton and muscle systems for locomotion and especially feeding in the American Anhinga *Anhinga anhinga* and Double-crested Cormorant *Phalacrocorax auritus*. The anhinga feeds by stealthily approaching fish and spearing them with the tip of the mandibles. The cormorant pursues fish by active, rapid swimming, and grasps the prey with the bill. The anhinga flies and soars easily but the cormorant flies laboriously. Regarding the cormorant, Owre comments that modifications for an aquatic existence are close to the maximum potential for such modifications concomitant with 'proficient flight'. The labored flight of cormorants may partially explain why they occur primarily on or near continental areas although their feeding adaptations are undoubtedly also involved. Further studies relating comparative morphology to ecology/behavior, such as Owre (1967) and Burger (1978), including measurements of wing loading, are required for other cormorants and would be especially instructive for members of the other four pelecaniform families.

The diet of many species of cormorants has been extensively studied, primarily because these birds were viewed as competitors with man for food (Mendall, 1936; Winkler, 1983). Few deleterious effects of cormorant predation on fish eaten by man have been detected. This is particularly true of cormorants in marine waters (McNally, 1957; Robertson, 1974). On the other hand, commercial fisheries combined with changes in oceanic conditions have adversely affected several pelecaniform populations. Overfishing of Anchoveta *Engraulis ringens* in the Humboldt Current system of South America has drastically affected the local Guanay Cormorants *P. bougainvillei* (and boobies and pelicans) in Peru (Ashmole, 1975; Furness, 1982; Barber & Chavez, 1983) and overfishing of Pilchard *Sardinops ocellata* and Cape Anchovy *E. capensis* in South African waters has contributed to reductions in populations of cormorants and gannets (and other species) (Crawford & Shelton, 1981).

Bartholomew (1942) and Hubbs, Kely & Limbaugh (1970), respectively, provided useful examples of the feeding behavior/ecology of Double-crested and Brandt's *P. penicillatus* Cormorants on the southwest coast of the USA. Both species usually feed in flocks, flying in large groups from communal roosts. Depending on the density and availability of prey, they will feed individually, and in various portions of the water column, changing feeding locations frequently. In southern California, 15 species of fishes were recorded in 12 stomach samples of Brandt's Cormorants. Palmer (1962) and Clapp *et al.* (1982) summarized feeding of Double-crested and Olivaceous *P. olivaceus* Cormorants on the east and south coasts of the USA. Little obvious ecological segregation was found in feeding between inland Reed Cormorants *P. africanus* and the African Darter *Anhinga melanogaster* in Zimbabwe (Birkhead, 1978), but Miller (1979) found differences in feeding habitat between the Little Pied *P. melanoleucos* and Little Black *P. sulcirostris* Cormorants in inland Australia. Knopf & Kennedy (1981) also found significant segregation between Double-crested Cormorants and American White Pelicans *Pelecanus erythrorhynchos* in Nevada, the pelicans taking significantly larger Tui Chub *Gila bicolor* and other species of fish from fresh-water lakes, while the cormorants only ate chub. Pioneering studies by Lack (1945) on ecological segregation, showed that Great Cormorants *Phalacrocorax carbo* and Common Shags *P. aristotelis* in Britain differed in feeding locations, since cormorants fed chiefly in shallow waters of estuaries, harbors, rivers and reservoirs, and rarely at sea. Shags fed mainly at sea, and rarely on fresh water, eating primarily Sandeels *Ammodytes* sp., while cormorants feeding in the same area took no Sandeels and were catholic in their diet. Williams & Burger (1978) found obvious differences in feeding ecology between Cape *P. capensis* and Bank *P. neglectus* Cormorants in South Africa. Cape Cormorants primarily ate pelagic shoaling fish and Bank Cormorants fed mainly in the littoral zone, especially among kelp beds. Only in inshore waters over a sandy bottom was there any overlap in feeding ecology between these two species. Stonehouse (1967) noted differences in feeding habits and habitats among Little Black, Little Pied, Greater Pied *P. varius* and Spotted *P. punctatus* Shags in New Zealand, primarily in diving and resting periods, with small species feeding in shallow water and larger species feeding in deeper, offshore waters. However, Robertson's (1974) preliminary data on Double-crested and Pelagic *P. pelagicus* Cormorants indicate that competition may exist in British Columbia: both species feed in the littoral-benthic zone and to a considerable extent on the bottom, although Pelagic Cormorants may prefer rocky bottoms. Both species are food generalists but Robertson notes that individual birds may well be specialists. He further indicates that indices and measures of overlap may be misleading, and since a 4-week difference in timing of nesting exists between species in this study region, the level of competition may well be lower than he computed. Similar studies are needed on other species. It would be particularly instructive to study the diet of the uncommon species found in isolated locations, as well as studying potential competition and niche overlap in sympatric species, especially over a large geographic range.

The studies of Dunn (1975) and Ricklefs (1984) provide excellent general summaries of examples of how feeding studies can be related to energy requirements of piscivorous birds and Williams, Siegfried & Cooper's (1982) summary of egg composition data provide useful insights. This type of integration of field studies with physiological analyses should prove extremely useful for further understanding of the biology of the Pelecaniformes and especially in respect of their interactions with the other components of marine ecosystems where they may be major consumers (e.g. Furness & Cooper, 1982).

Looking for methods to study food of Cape Cormorants, Duffy & Laurenson (1983) examined the birds' habit of regurgitating pellets. They found that pellets (containing otoliths and eye lenses) provided no reliable data on daily food intake. However, Ainley, Anderson & Kelly (1981) used pellet contents to show differential use of feeding habitats between Pelagic, Brandt's and Double-crested Cormorants in the northeastern Pacific Ocean. Pellets are an important topic for further investigation because otoliths, squid beaks and crustacean carapaces can all be used to identify prey and, in many cases, to estimate the size (and even age) of these prey.

Gannets Sulidae

This family comprises three species of gannets (found in cold waters of the North Atlantic, along the coast of South Africa, and around southern Australia and New Zealand) and six species of boobies. The Masked *Sula dactylatra*, Red-footed *S. sula*, and Brown *S. leucogaster* Boobies are essentially sympatric, pantropical, and pelagic in warm waters. The Peruvian Booby *S. variegata* is restricted to the coasts of Peru, Ecuador, and Chile influenced by the Humboldt Current. Blue-footed Boobies *S. nebouxi* range from the Sea of Cortez in Mexico to Ecuador and the Galapagos. Abbott's Booby *S. abbotti* breeds only on Christmas Island, Indian Ocean.

Nelson (1978) extensively reviewed the biology of the Sulidae, pointing out that feeding ecology is still poorly known. Clapp *et al.* (1982) summarized feeding methods and the variations on the usual sulid deep plunge-diving, vertically from a considerable height, singly or in large flocks. Gannets are highly pelagic, typically feed in flocks, and are primarily vertical divers, frequently from the same height, and may use their wings as well as feet to gain additional depth in the water. They also scavenge from fishing vessels, as do Brown Boobies, at least in Mexican waters. Masked Boobies are highly pelagic, primarily vertical plunge-divers and mainly feed singly, although groups are recorded. Red-footed Boobies are probably the most pelagic of the boobies and primarily plunge-dive although they also catch flyingfish from the air and while sitting on the water. Brown Boobies are basically nearshore feeders although they do travel long distances out to sea where they plunge vertically or in a slanting dive. It has been suggested that they may capture a high percentage of their fish by aerial persuit and take flyingfish in flight, but data are few. Blue-footed Boobies primarily plunge-dive in flocks but they also make surface plunges and while most dives are accomplished by gravity alone, they also commonly power-dive. They are known to seize flyingfish from the air, and occasionally they feed cooperatively. Peruvian Boobies mainly power-dive, often in large flocks nearshore. Little is known about the feeding methods of Abbott's Booby.

Gannets chiefly eat fish, with numerous species recorded. They also eat some squid and, at least in South African waters, their diet reflects the relative abundance of the various species taken by commercial fisheries (Nelson, 1978). The gannets apparently feed primarily on shoaling fishes within 30–40 km of the coast. However they also wander much further offshore and away from breeding colonies (Clapp *et al.*, 1982)

Flyingfish and squid form the major diet items of the Masked, Red-footed, and Brown Boobies but many other items are eaten with considerable geographic and seasonal variability throughout their range (Diamond, 1974; Schreiber & Hensley, 1976; Nelson, 1978; Clapp *et al.*, 1982; Harrison *et al.*, 1983). Nelson (1978) has speculated extensively on the feeding range, diet and population of these species. Ecological segregation is apparent in regurgitated samples from sympatric species. Thus Masked Boobies eat larger fish and Red-footed Boobies eat more squid, based on data from one year on Christmas Island, Pacific Ocean. No explanation of booby evolutionary biology incorporates sufficient data on actual behavior and distribution at sea to be entirely satisfactory. For instance, Brown Boobies are uncommon on some islands in the Pacific where Red-footed and Masked Boobies are common (Schreiber & Ashmole, 1970). Brown Boobies supposedly feed 'inshore' around coral atolls and such inshore feeding habits have been used to explain their small populations (Simmons, 1967; Diamond, 1978). However, waters surrounding coral atolls are hundreds of fathoms deep within a few hundred meters offshore. Additional data are clearly needed in order to evaluate the role of 'inshore' and 'pelagic' feeding zones in population regulation mechanisms. Probably the presence of gyres, eddies, current interfaces, and other mechanisms which increase productivity near islands are more important than simple closeness to land as a means of regulating bird populations.

The diet of Blue-footed and Abbott's Boobies remain poorly known. The Anchoveta is the main food of the Peruvian Booby but it can feed on other fish when the Anchoveta population crashes following an El Niño event. Peruvian Brown Pelicans *Pelecanus thagus* and Guanay Cormorants apparently cannot easily utilize other food sources and thus their populations fluctuate more than do those of the booby. The El Niño situation in

Peru, Chile, and Ecuador (and especially on the Galapagos Islands) deserves more careful study (Barber & Chavez, 1983), primarily because the oceanographic conditions (physical, chemical, and biological) are well known and actively being studied. Similar complete and detailed ornithological data for the region would provide us with a model study of a pelecaniform community.

Pelicans Pelecanidae

White pelicans are mainly inland forms that usually feed communally. They often cooperatively herd fish into shallow water and synchronously plunge their bills into the water while swimming (Cottam, Williams & Sooter, 1942; Din & Eltringham, 1974), although Gunter (1958) found this communal feeding rare in American White Pelicans on the Gulf of Mexico. Populations of all species of white pelicans spend varying amounts of time on salt water, usually during the non-breeding season. Feeding ecology of two species is well known, the American White (Clapp *et al.*, 1982 and references therein; Smith, Steinbach & Pampush, 1984) and the Great White *Pelecanus onocrotalus* (Crawford, Cooper & Shelton, 1981; Brown *et al.*, 1982, and references therein), although habits at sea are poorly understood. Virtually no data on diet exist for the other white pelican species.

The Brown Pelican *P. occidentalis* is the only pelecaniform for which actual food capture methods have been documented (Schreiber, Woolfenden & Curtsinger, 1975). They feed primarily by plunge-diving, usually in small groups but also in large flocks when bait fish are concentrating on the surface. They also occasionally feed while sitting on the water, and are efficient scavengers. Such scavenging frequently leads to their becoming entangled in fish hooks and monofilament line which causes high mortality. They are entirely marine but rarely venture far from land (beyond 80 km) although a population is established on the Galapagos, and individuals are occasionally found inland in North America.

Relations between Brown Pelican fledging success and food availability have been studied in California (Anderson, Gress & Mais, 1982; Northern Anchovy Fishery Management Plan, 1983) where data on reproductive success and foraging range of the birds have been collected for over a decade and reasonably complete data are available on the abundance and distribution of the primary food source, Northern Anchovy *Engraulis mordax*. If feeding behaviour of the birds and localized areas of anchovy abundance directly associated with breeding colonies are considered, a strong relation exists between fish abundance and bird production. Only a weak relation exists between large-scale indices of anchovy abundance and pelican fledging rates. Large-scale weather patterns are thus likely to affect reproduction through influence on prey levels. Fledging rates are a suitable indicator of local food conditions because the pelicans are restricted to local feeding areas during breeding (as are most species of birds). High levels of annual variation in food supply result in high annual variation in pelican reproductive performance.

Anderson *et al.* (1982) further suggest that breeding colonies are probably sited where the appropriate nesting substrate and a consistent, available food supply co-occur. With this species there are promising opportunities for investigating relations between food availability during the non-breeding season and its effect on survival, especially for juveniles, and patterns of movements.

Other aspects of feeding ecology

Color

Pelecaniformes vary from essentially white (gannets, tropicbirds, white-phase Red-footed Boobies), through intermediate colors (many boobies, female and sub-adult frigates, juvenile Brown Pelicans), to all dark (cormorants, anhingas, male frigatebirds, adult Brown Pelicans). Juvenile and immature Sulidae, Phaethontidae, and Pelecanidae are generally darker than adults. Bellies of juvenile Brown Pelicans are pure white and juvenile cormorants tend to be speckled or usually pale, resembling the females, and young frigatebirds have white on the belly, breast, and head–neck. Craik (1944) suggested and Phillips (1962) provided data and detailed analysis (see also Simmons, 1972) that white undersides contrast less against the sky and thus should allow seabirds colored in such a way to approach closer to prey items. Their work has been widely quoted. However, Cowan (1972) showed, at least on photographic film, no clear 'simple relationship between partial white and white coloration and contrast in sea-bird mounts'. While it is interesting to speculate that the white coloration on the bellies of juveniles of many Pelecaniformes (i.e. Brown Pelicans, frigatebirds) gives them some 'advantage' in feeding, why should these become black in adult plumage and why are gannet and tropicbird young dark? Probably the explanation is not directly related to just color and feeding and one needs to consider temperature regulation (e.g. Siegfried *et al.*, 1975) and social behaviour both at the breeding colony and at sea. For instance, the 'flash' of black and white of a female frigatebird or Masked Booby as it dives and twists is particularly visible to our eyes and more so than if just a single color were involved. May coloration also serve to notify other individuals that food has been found?

Kleptoparasitism

Kleptoparasitism (see Chapter 5) deserves special mention here since members of the Pelecaniformes serve both as parasites and hosts (Brockman & Barnard, 1979). Frigatebirds resort to kleptoparasitism frequently and other pelecaniform species do so occasionally (Bildstein, 1980; Duffy, 1980). Pelicans, boobies, and cormorants are kleptoparasitized by non-Pelecaniformes, primarily members of the Laridae (Maxson & Bernstein, 1982; Schnell, Woods & Ploger, 1983). Most studies have concentrated on the beneficial effects of the behavior for the predator species but there has been little study of the negative effects of being kleptoparasitized. Great Frigatebirds virtually eliminated Sooty Tern colonies on Christmas Island in 1967 and again in 1983 (Schreiber & Ashmole, 1970; Schreiber & Schreiber, unpublished data) and our data indicate that only a few individual frigatebirds are involved. What are the long-term effects of such predation?

Flightlessness

Flightless carinate species are not common (van Tyne & Berger, 1976) and only one marine form exists: the Flightless Cormorant *Phalacrocorax harrisi* of the Galapagos Islands. This is a very sedentary species, no individuals having been seen more than 1 km away from the breeding colonies. The population exhibits unusual breeding patterns, apparently in relation to the abundance of food in the vicinity of its nesting sites (Snow, 1966; Harris, 1979). Individuals attempt to breed several times in succession if eggs are lost and females breed more frequently than males, who attend fledged young. Breeding occurs in all months and individuals change mates, nest sites, and nesting areas from one attempt to the next; this is in marked contrast to the behavior of most marine birds. The 1972 El Niño provided valuable data in this regard but the extensive and severe El Niño of 1982–83 should provide even greater insights (Barber & Chavez, 1983; F. Ortiz, personal communication).

Age and sex in relation to foraging success

Very little information is available on the influence of age and sex on foraging success in Pelecaniformes. Orians (1969) showed that adult Brown Pelicans foraged more effectively than young birds and Schnell, Woods & Ploger (1983) confirmed the earlier study and showed that

Laughing Gulls *Larus atricilla* were attracted to the successful pelican, regardless of its age. Morrison, Slack & Shanley (1978) found similar results in Olivaceous Cormorants and related the development of capture techniques, manoeuverability, and prey search image to higher mortality in fledglings and the attainment of reproductive maturity. Pelecaniformes with their distinctively plumaged juveniles and adults, and their varied feeding regimes lend themselves to further study of this topic.

Conclusions

The Pelecaniformes provide excellent study subjects for investigation of feeding-niche overlap, segregation between species, and food availability conditions in different areas of the oceans. Gannets are cold-water forms, nesting alone in the North Atlantic but in association with cormorants (and other species) in the South Atlantic and Pacific. Frigatebirds, boobies, and tropicbirds are sympatric in the Pacific and Indian Ocean warm waters and provide instances of overlap not only between families (or genera) but also between species (*Fregata minor* sympatric with *F. ariel* and *F. magnificens*; and with *Sula dactylatra, S. sula*, and *S. leucogaster*). Pelicans and cormorants occur together along the coasts of the eastern Pacific and the central Atlantic.

The main recent studies (e.g. Harris, 1977; Harrison *et al.*, 1983; Diamond, 1984) indicate that variation, but also considerable similarity, exists between regions, communities, and species. Timing of breeding is related to food availability or unavailability in these areas in a complicated fashion and only better methods of measuring the availability of 'prey' (Harrison *et al.*, 1983) will answer the most fundamental questions in tropical seabird biology, of which the Pelecaniformes forms a central group of species.

References

Ainley, D.G., Anderson, D.W. & Kelly, P.R. (1981). Feeding ecology of marine cormorants in Southwestern North America. *Condor*, **83**, 120–31.

Anderson, D.W., Gress, F. & Mais, K.F. (1982). Brown Pelicans: influence of food supply on reproduction. *Oikos*, **39**, 23–31.

Ashmole, N.P. (1971). Seabird ecology and the marine environment. In *Avian biology*, vol. 1, ed. D.S. Farner & J.R. King, pp. 223–86, New York: Academic Press.

Ashmole, N.P. (1975). Pelecaniformes. *New Encyclopaedia Britannica, Macropaedia*, **11**, 15–20.

Barber, R.T. & Chavez, F.P. (1983). Biological consequences of El Niño. *Science*, **222**, 1203–10.

Bartholomew, G.A., Jr (1942). The fishing activities of Double-crested Cormorants on San Francisco Bay. *Condor*, **44**, 13–21.

Bildstein, K.L. (1980). Adult Brown Pelican robs Great Blue Heron of fish. *Wilson Bull.*, **92**, 122–3.

Birkhead, M.E. (1978). Some aspects of the feeding ecology of the Reed Cormorant and Darter on Lake Kariba, Rhodesia. *Ostrich*, **49**, 1–7.

Brockman, H.J. & Barnard, C.J. (1979). Kleptoparasitism in birds. *Anim. Behav.*, **27**, 487–514.

Brown, L.H., Urban, E.K. & Newman, K. (1982). *The birds of Africa*, vol. 1. London: Academic Press.

Burger, A.E. (1978). Functional anatomy of the feeding apparatus of four South African cormorants. *Zool. Afr.*, **13**, 81–102.

Clapp, R.B., Banks, R.C., Morgan-Jacobs, D. & Hoffman, W.A. (1982). *Marine birds of the southeastern United States and Gulf of Mexico. Part 1. Gaviiformes through Pelecaniformes.* Washington, D.C.: U.S. Fish and Wildlife Service, Office of Biological Services. FWS/OBS-82/01. 637 pp.

Cottam, C., Williams, C.S. & Sooter, C.A. (1942). Cooperative feeding of White Pelicans. *Auk*, **59**, 444–5.

Cowan, P.J. (1972). The contrast and coloration of sea-birds: an experimental approach. *Ibis*, **114**, 390–3.

Cramp, S. & Simmons, K.E.L. (1977). *The birds of the western Palearctic*, vol. 1. Oxford: Oxford University Press.

Craik, K.J.W. (1944). White plumage of seabirds. *Nature, Lond.*, **153**, 288.

Crawford, R.J.M., Cooper, J. & Shelton, P.A. (1981). The breeding population of White Pelicans *Pelecanus onocrotalus* at Bird Rock Platform in Walvis Bay, 1949–1978. *Fish. Bull. S. Afr.*, **15**, 67–70.

Crawford, R.J.M. & Shelton, P.A. (1981). Population trends for some southern African seabirds related to fish availability. In *Proceedings of the symposium on birds of the sea and shore*, ed. J. Cooper, pp. 15–41, Cape Town: African Seabird Group.

Diamond, A.W. (1973). Notes on the breeding biology and behavior of the Magnificent Frigatebird. *Condor*, **75**, 200–9.

Diamond, A.W. (1974). The Red-footed Booby on Aldabra Atoll, Indian Ocean. *Ardea*, **62**, 196–218.

Diamond, A.W. (1975*a*). Biology and behaviour of frigatebirds *Fregata* spp. on Aldabra Atoll. *Ibis*, **117**, 302–3.

Diamond, A.W. (1975*b*). The biology of tropicbirds *Phaethon* spp. at Aldabra Atoll, Indian Ocean. *Auk*, **92**, 16–39.

Diamond, A.W. (1978). Feeding strategies and population size in tropical seabirds. *Am. Nat.*, **112**, 215–23.

Diamond, A.W. (1984). Feeding overlap in some tropical and temperate seabird communities. *Stud. Avian Biol.*, no. 8, 24–46.

Din, N.A. & Eltringham, S.K. (1974). Ecological separation between White and Pink-backed Pelicans in the Ruwenzori National Park, Uganda. *Ibis*, **116**, 28–43.

Duffy, D.C. (1980). Patterns of piracy by Peruvian seabirds: a depth hypothesis. *Ibis*, **122**, 521–5.

Duffy, D.C. & Laurenson, L.J.B. (1983). Pellets of Cape Cormorants as indicators of diet. *Condor*, **85**, 305–7.

Dunn, E.H. (1975). Caloric intake of nestling Double-crested Cormorants. *Auk*, **92**, 553–65.

Furness, R.W. (1982). Competition between fisheries and seabird communities. *Adv. Mar. Biol.*, **20**, 225–302.

Furness, R.W. & Cooper, J. (1982). Interactions between breeding seabirds and pelagic fish populations in the Southern Benguela Region. *Mar. Ecol. Prog. Ser.*, **8**, 243–50.

Gould, P.J., King, W.B. & Sanger, G.A. (1974). Red-tailed Tropicbird (*Phaethon rubricauda*). *Smithson. Contrib. Zool.*, no. 158, 206–31.

Gunter, G. (1958). Feeding behavior of Brown and White Pelicans on the Gulf Coast of the United States. *Proc. Louisiana Acad. Sci.*, **21**, 34–9.

Haley, D. (ed.) (1984). *Seabirds of North Pacific and Arctic waters.* Seattle: Pacific Search Press.

Harrington, B.A., Schreiber, R.W. & Woolfenden, G.E. (1972). The distribution of male and female Magnificent Frigatebirds, *Fregata magnificens*, along the Gulf Coast of Florida. *Am. Birds*, **26**, 927–31.

Harris, M.P. (1969). Factors influencing the breeding cycle of the Red-billed Tropicbird in the Galapagos Islands. *Ardea*, **57**, 149–57.

Harris, M.P. (1977). Comparative ecology of seabirds in the Galapagos Archipelago. In *Evolutionary ecology*, ed. B. Stonehouse & C.M. Perrins, pp. 65–76. London: Macmillan.

Harris, M.P. (1979). Population dynamics of the Flightless Cormorant *Nannopterum harrisi*. *Ibis*, **121**, 135–46.

Harrison, C.S., Hida, T.S. & Seki, M.P. (1983). Hawaiian seabird feeding ecology. *Wildl. Monogr.*, **85**, 1–71.

Harrison, P. (1983). *Seabirds. An identification guide.* Boston: Houghton Mifflin.

Hubbs, C.L., Kely, A.L. & Limbaugh, C. (1970). Diversity in feeding by Brandt's Cormorant near San Diego. *Calif. Fish Game*, **56**, 156–65.

Knopf, F.L. & Kennedy, J.L. (1981). Differential predation by two species of piscivorous birds. *Wilson Bull.*, **93**, 554–6.

Lack, D. (1945). The ecology of closely related species with special reference to Cormorant (*Phalacrocorax carbo*) and Shag (*P. aristotelis*). *J. Anim. Ecol.*, **14**, 12–16.

McNally, J. (1957). The feeding habits of cormorants in Victoria. *Fauna Contr. Fish Game Dep. Vict.*, no. 6, 1–36.

Maxson, S.J. & Bernstein, N.P. (1982). Kleptoparasitism by South Polar Skuas on Blue-eyed Shags in Antarctica. *Wilson Bull.*, **94**, 269–81.

Mendall, H.L. (1936). The home-life and economic status of the Double-crested Cormorant. *Maine Bull.*, **39**(3), 1–159.

Miller, B. (1979). Ecology of the Little Black Cormorant *Phalacrocorax sulcirostris* and Little Pied Cormorant *P. melanoleucos* in inland New South Wales. I. Food and feeding habits. *Aust. Wildl. Res.*, **6**, 79–95.

Morrison, M.L., Slack, R.D. & Shanley, E., Jr. (1978). Age and foraging ability relationships of Olivaceous Cormorants. *Wilson Bull.*, **90**, 414–22.

Nelson, J.B. (1976). The breeding biology of frigatebirds: a comparative review. *Living Bird*, **14**, 113–55.

Nelson, J.B. (1978). *The Sulidae: gannets and boobies.* Oxford: Oxford University Press.

Nelson, J.B. (1984). Contrasts in breeding strategies between some tropical and temperate marine Pelecaniformes. *Stud. Avian Biol.*, no. 8, 95–114.

Northern Anchovy Fishery Management Plan. (1983). FMP Amendment 5. Portland, Oregon: Pacific Fishery Management Council.

Orians, G.H. (1969). Age and hunting success in the Brown Pelican (*Pelecanus occidentalis*). *Anim. Behav.*, **17**, 316–19.

Owre, O.T. (1967). Adaptations for locomotion and feeding in the Anhinga and the Double-crested Cormorant. *Orn. Monogr.*, no. 6.

Palmer, R.S. (1962). *Handbook of North American birds*, vol. 1. New Haven, CT: Yale University Press.
Pennycuick, C.J. (1983). Thermal soaring compared in three dissimilar tropical bird species, *Fregata magnificens*, *Pelecanus occidentalis* and *Coragyps atratus*. *J. Exp. Biol.*, **102**, 307–25.
Phillips, G.C. (1962). Survival value of the white coloration of gulls and other seabirds. Unpublished D.Phil. thesis, Oxford University.
Ricklefs, R.E. (1984). Some considerations on the reproductive energetics of pelagic seabirds. *Stud. Avian Biol.*, no. 8, 84–94.
Robertson, I. (1974). The food of nesting Double-crested and Pelagic Cormorants at Mandarte Island, British Columbia, with notes on feeding ecology. *Condor*, **76**, 346–8.
Schnell, G.D., Woods, B.L. & Ploger, B.J. (1983). Brown Pelican foraging success and kleptoparasitism by Laughing Gulls. *Auk*, **100**, 636–44.
Schreiber, R.W. & Ashmole, N.P. (1970). Sea-bird breeding seasons on Christmas Island, Pacific Ocean. *Ibis*, **112**, 363–94.
Schreiber, R.W. & Hensley, D.A. (1976). The diets of *Sula dactylatra*, *Sula sula*, and *Fregata minor* on Christmas Island, Pacific Ocean. *Pacif. Sci.*, **30**, 241–8.
Schreiber, R.W., Woolfenden, G.E. & Curtsinger, W.E. (1975). Prey capture by the Brown Pelican. *Auk*, **92**, 649–54.
Sibley, C.G. & Ahlquist, J.E. (1972). A comparative study of the egg white proteins of non-passerine birds. *Bull. Peabody Mus. Nat. Hist.*, **39**, 276 pp.
Sibley, F.C. & Clapp, R.B. (1967). Distribution and dispersal of central Pacific Lesser Frigatebirds *Fregata ariel*. *Ibis*, **109**, 328–37.
Siegfried, W.R., Williams, A.J., Frost, P.G.H. & Kinahan, J.B. (1975). Plumage and ecology of cormorants. *Zool. Afr.*, **10**, 183–92.
Simmons, K.E.L. (1967). Ecological adaptations in the life history of the Brown Booby at Ascension Island. *Living Bird*, **11**, 187–212.
Simmons, K.E.L. (1972). Some adaptive features of seabird plumage types. *Br. Birds*, **65**, 465–79.
Smith, M., Steinbach, T. & Pampush, G. (1984). *Distribution, foraging relationships and colony dynamics of the American White Pelican* (Pelecanus erythrorhynchos) *in Southern Oregon and Northeastern California*. Portland, Oregon: Nature Conservancy, 56 pp.
Snow, B.K. (1966). Observations on the behaviour and ecology of the Flightless Cormorant *Nannopterum harrisi*. *Ibis*, **108**, 264–80.
Stonehouse, B. (1967). Feeding behaviour and diving rhythms of some New Zealand shags, Phalacrocoracidae. *Ibis*, **109**, 600–5.
van Tyne, J. & Berger, A.J. (1976). *Fundamentals of ornithology*, 2nd edn. London: Wiley.
Williams, A.J. & Burger, A.E. (1978). The ecology of the prey of Cape and Bank Cormorants. *Cormorant*, **4**, 28–9.
Williams, A.J., Siegfried, W.R. & Cooper, J. (1982). Egg composition and hatching precocity in seabirds. *Ibis*, **124**, 456–70.
Winkler, H. (1983). The ecology of cormorants (genus *Phalacrocorax*). In *Limnology of Parakrama Sumudra-Sri Lanka*, ed. F. Schiemer. The Hague: W. Junk.

9

Feeding ecology of Alcidae in the eastern North Pacific Ocean

KEES VERMEER,
Canadian Wildlife Service, P.O. Box 340, Delta, British Columbia, V4K 3Y3, Canada
SPENCER G. SEALY,
Department of Zoology, University of Manitoba, Winnipeg, Manitoba, R3T 2N2, Canada
and
GERALD A. SANGER,
U.S. Fish and Wildlife Service, Alaska Fish and Wildlife Research Center, 1101 East Tudor Road, Anchorage, Alaska 99503, USA

Introduction

The family Alcidae constitutes a group of 22 living species, of which 6 are found in the North Atlantic, 18 in the North Pacific and 3 in the Arctic Ocean (Table 9.1). Storer (1945) divided them into seven morphological groups primarily on the basis of the structure of the hind limb: (1) the auks and murres (*Alca, Pinguinus, Uria*); (2) the puffins (*Cerorhinca, Fratercula*); (3) the guillemots (*Cepphus*); (4) the *Brachyramphus* murrelets; (5) the *Synthliboramphus* murrelets; (6) the Dovekie (*Alle*); and (7) the auklets (*Aethia, Cyclorrhynchus, Ptychoramphus*). We use these groups here, except that the Dovekie is placed with the other auklets, in reviewing the feeding ecology of alcids in the northeastern Pacific Ocean particularly as related to breeding range and nesting habitat. The review is not exhaustive but differences and similarities in use of food resources between closely related or sympatric alcids are emphasized. Some additional comparisons will be made with the more voluminous literature on the diets and foraging alcids in the North Atlantic Ocean (Bradstreet & Brown, 1985), and in the western North Pacific, although little information exists for alcids on the Asian side of the Pacific Ocean.

Breeding range and habitat

Auks and murres

The extinct Great Auk *Pinguinus impennis*, the largest known alcid, which weighed about 5 kg (Coues, 1868), nested on flat tops of bare rocky islands as well as on cliff ledges in the North Atlantic Ocean (Bengtson, 1984). The species' last stronghold in North America was on Funk Island, Newfoundland (Olson, Swift & Mokhiber, 1979) and the last known auks were killed in Iceland in 1844 (Newton, 1861; Grieve, 1885). The Razorbill *Alca torda*, the Great Auk's closest relative, nests on cliff ledges or in crevices in both sub-Arctic and temperate waters of the North Atlantic (see Bédard, 1969*a*), but weighs much less (700 g; Belopolskii, 1957) than did the Great Auk. Common *Uria aalge* and Thick-billed Murres *U. lomvia* nest on cliffs in both the North Atlantic and North Pacific Oceans. The Thick-billed Murre is chiefly an Arctic and sub-Arctic species, but shares some colonies with the Common Murre in temperate waters as far south as Newfoundland in the North Atlantic, north Japan in the western North Pacific (Tuck, 1961; Udvardy, 1963), and Triangle Island, British Columbia (Vallee & Cannings, 1983), in the eastern North Pacific (Fig. 9.1). The Common Murre occurs in Arctic, sub-Arctic and temperate

Table 9.1. *Distribution of alcids*

Species	North Pacific	North Atlantic	Arctic
Razorbill *Alca torda*	—	×	—
Common Murre *Uria aalge*	×	×	—
Thick-billed Murre *U. lomvia*	×	×	×
Tufted Puffin *Fratercula cirrhata*	×	—	—
Horned Puffin *F. corniculata*	×	—	—
Atlantic Puffin *F. arctica*	—	×	—
Rhinoceros Auklet *Cerorhinca monocerata*	×	—	—
Pigeon Guillemot *Cepphus columba*	×	—	—
Black Guillemot *C. grylle*	—	×	×
Spectacled Guillemot *C. carbo*	×	—	—
Marbled Murrelet *Brachyramphus marmoratus*	×	—	—
Kittlitz's Murrelet *B. brevirostris*	×	—	—
Ancient Murrelet *Synthliboramphus antiquus*	×	—	—
Japanese Murrelet *S. wumizusume*	×	—	—
Xantus' Murrelet *Endomychura hypoleuca*	×	—	—
Craveri's Murrelet *E. craveri*	×	—	—
Parakeet Auklet *Cyclorrhynchus psittacula*	×	—	—
Dovekie *Alle alle*	—	×	×
Cassin's Auklet *Ptychoramphus aleuticus*	×	—	—
Crested Auklet *Aethia cristatella*	×	—	—
Whiskered Auklet *A. pygmaea*	×	—	—
Least Auklet *A. pusilla*	×	—	—

environments and is the only murre which breeds along the US west coast from Washington to California (Fig. 9.1) and in northwestern Europe from Scandinavia and the British Isles to Spain (Tuck, 1961; Udvardy, 1963).

Puffins

The Tufted Puffin *Fratercula cirrhata*, Horned Puffin *F. corniculata* and Rhinoceros Auklet *Cerorhinca monocerata*, a misnamed puffin, are restricted to the North Pacific Ocean. Tufted and Horned Puffins range continuously

Fig. 9.1. Breeding range of Common and Thick-billed Murres in the Pacific. (Data from Udvardy, 1963; Campbell, 1976; Campbell & Garrioch, 1979; Sowls *et al.*, 1978, 1980; Vermeer *et al.*, 1976; Varoujean, 1979; Vallee & Cannings, 1983.)

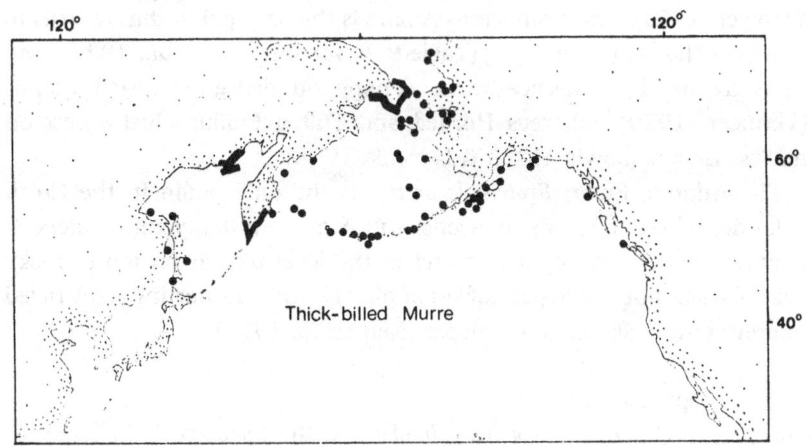

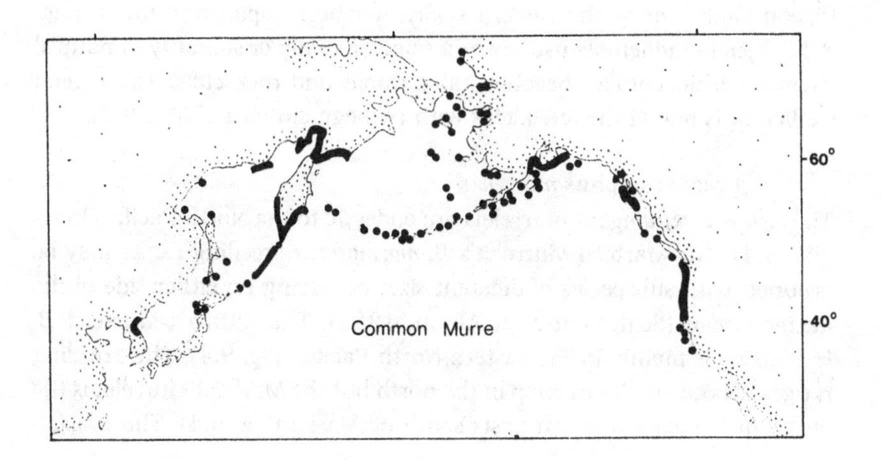

from the northeastern to the northwestern Pacific Ocean, while the distribution of the Rhinoceros Auklet is disjunct (Fig. 9.2). The Horned Puffin is chiefly restricted to sub-Arctic waters, but a few are suspected to nest as far south as British Columbia (Campbell, Carter & Sealy, 1979). Tufted Puffins and Rhinoceros Auklets excavate burrows in which to nest; Horned Puffins generally use natural crevices. Where Horned and Tufted Puffins nest sympatrically, the former use crevices among talus slopes while Tufted Puffins burrow in the earth along cliff tops (Wehle, 1983). Horned and Tufted Puffins use crevices among boulders where they nest in the most northerly part of their breeding ranges, possibly due to the persistence of permafrost (Sealy, 1973*a*). Where Tufted Puffins and Rhinoceros Auklets are sympatric, puffins burrow on steep slopes and cliff tops while Rhinoceros Auklets nest on gradual slopes (Leschner, 1976; Vermeer, 1979). The Rhinoceros Auklet is the only puffin that returns to or leaves the colony at night (Table 9.2, but see Scott *et al.*, 1974), and nests frequently underneath the canopy on densely forested islands (Vermeer, 1979), whereas Horned and Tufted Puffins chiefly nest on treeless islands (Sowls, Hatch & Lensink, 1978).

The Atlantic Puffin *Fratercula arctica* is the only puffin in the North Atlantic, where it breeds in Arctic, sub-Arctic and temperate waters. It burrows in turf-covered slopes and in the level terrain on top of rocky coastal islets, but due to permafrost at high latitudes its nesting is restricted to crevices in cliffs and talus slopes (Nettleship, 1972).

Cepphus *guillemots*

The Pigeon Guillemot *Cepphus columba* and the Spectacled Guillemot *C. carbo* breed in the North Pacific. The Spectacled Guillemot replaces the Pigeon Guillemot in the eastern USSR, northern Japan and Korea (Fig. 9.3). *Cepphus* guillemots usually nest semicolonially or solitarily in natural cavities within boulder beaches, talus slopes and rock cliffs. The Pigeon Guillemot is one of the few alcids with two-egg clutches (Table 9.2).

Brachyramphus *murrelets*

The two *Brachyramphus* murrelets are endemic to the North Pacific Ocean (Fig. 9.4). The Marbled Murrelet's *B. marmoratus* breeding range may be disjunct, with subspecies of different sizes occurring on either side of the Pacific Ocean (Sealy, Carter & Alison, 1982). The Kittlitz's Murrelet *B. brevirostris* is mainly in the eastern North Pacific (Fig. 9.4). The breeding ranges of both species overlap in the north but the Marbled Murrelet is the only *Brachyramphus* which nests south of Alaska (Fig. 9.4). The *Brachy-*

Fig. 9.2. Breeding ranges of Horned and Tufted Puffins and Rhinoceros Auklets in the Pacific. (Data from Udvardy, 1963; Sowls *et al.*, 1978, 1980; Campbell & Garrioch, 1979; Varoujean, 1979; Vermeer, 1979.)

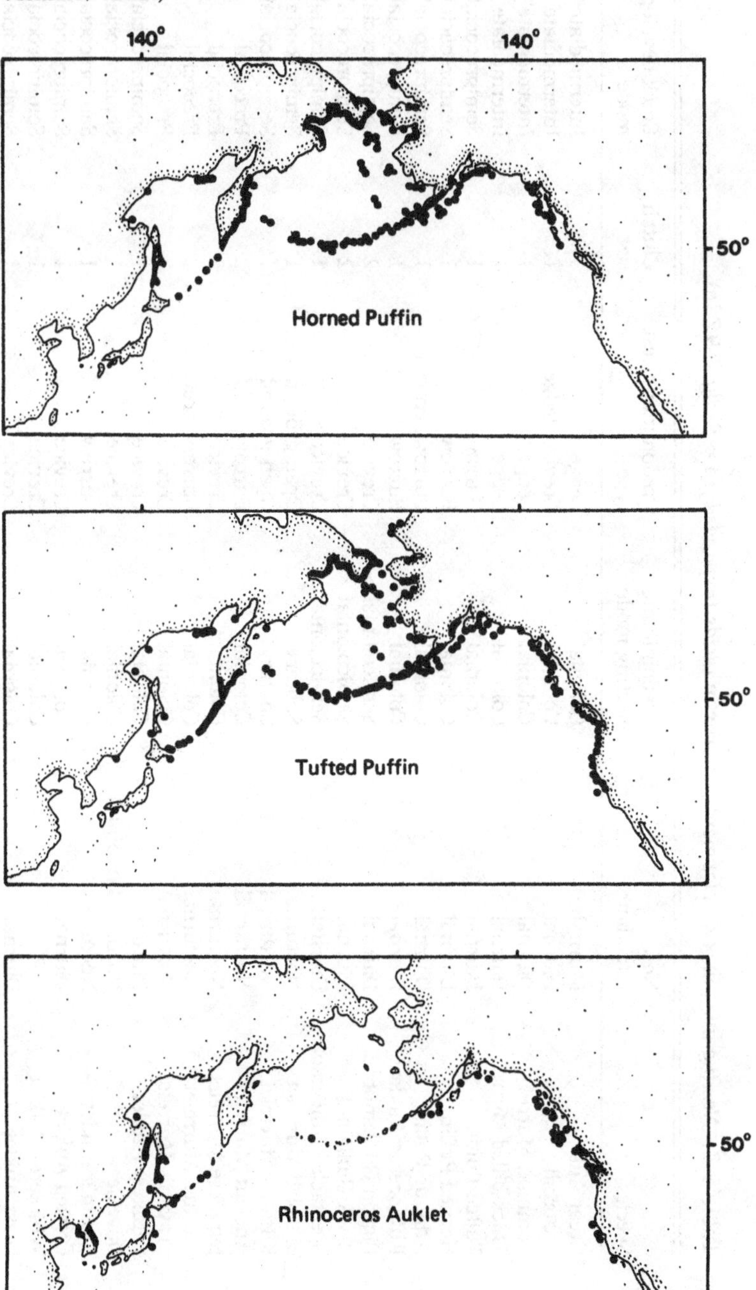

Table 9.2. *Nesting and chick development patterns in alcids (expanded from Sealy, 1973b)*

Species	Nest attendance	Predominant nesting mode	Predominant nest type	Clutch size	Chick development mode
Great Auk	Diurnal(?)	Colonial	Ledge	1	Intermediate[a]
Razorbill	Diurnal	Colonial	Crevice, ledge	1	Intermediate
Common Murre	Diurnal	Colonial	Ledge	1	Intermediate
Thick-billed Murre	Diurnal	Colonial	Ledge	1	Intermediate
Tufted Puffin	Diurnal	Colonial	Burrow	1	Semiprecocial
Horned Puffin	Diurnal	Colonial	Burrow	1	Semiprecocial
Atlantic Puffin	Diurnal	Colonial	Burrow, crevice	1	Semiprecocial
Rhinoceros Auklet	Diurnal	Colonial	Burrow	1	Semiprecocial
Pigeon Guillemot	Diurnal	Semicolonial[b]	Crevice	2	Semiprecocial
Black Guillemot	Diurnal	Semicolonial	Crevice	2	Semiprecocial
Spectacled Guillemot	Diurnal	Semicolonial	Crevice	1–2	Semiprecocial
Marbled Murrelet	Nocturnal	Solitary	Open ground, tree	1	Semiprecocial
Kittlitz's Murrelet	Nocturnal	Solitary	Open ground	1	Semiprecocial
Ancient Murrelet	Nocturnal	Colonial	Burrow	2	Precocial
Japanese Murrelet	Nocturnal	Colonial	Crevice	2	Precocial
Xantus' Murrelet	Nocturnal	Colonial	Burrow, crevice	2	Precocial
Craveri's Murrelet	Nocturnal	Colonial	Crevice	2	Precocial
Parakeet Auklet	Diurnal	Colonial	Crevice	1	Semiprecocial
Dovekie	Nocturnal (mainly)	Colonial	Crevice	1	Semiprecocial
Cassin's Auklet	Nocturnal	Colonial	Burrow	1	Semiprecocial
Crested Auklet	Diurnal	Colonial	Crevice	1	Semiprecocial
Whiskered Auklet	?	Colonial	Crevice	1	Semiprecocial
Least Auklet	Diurnal	Colonial	Crevice	1	Semiprecocial

[a] Intermediate between precocial and semiprecocial.
[b] Small or loose colonies as well as solitary.

ramphus murrelets nest solitarily (Table 9.2). Kittlitz's Murrelets nest on the ground and Marbled Murrelets nest either in trees, crevices or on the ground (Binford, Elliott & Singer, 1975; Kiff, 1981; Johnston & Carter, 1985). Kittlitz's Murrelets nest at higher elevations (average 570 m) than Marbled Murrelets (300 m). Where these murrelets are sympatric, Kittlitz's generally nest twice as far from the sea (average 11 km, up to 75 km) as do Marbled Murrelets (average 6 km, up to 24 km) (Day, Oakley & Barnard, 1983).

Fig. 9.3. Breeding range of Pigeon and Spectacled Guillemots in the Pacific. (Data from Udvardy, 1963; Campbell, 1976; Campbell & Garrioch, 1979; Sowls *et al.*, 1978, 1980; Varoujean, 1979; Vermeer *et al.*, 1983.)

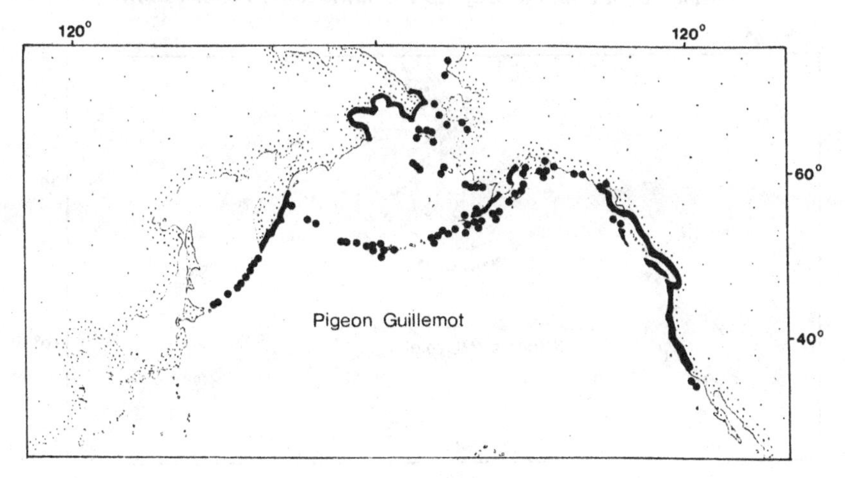

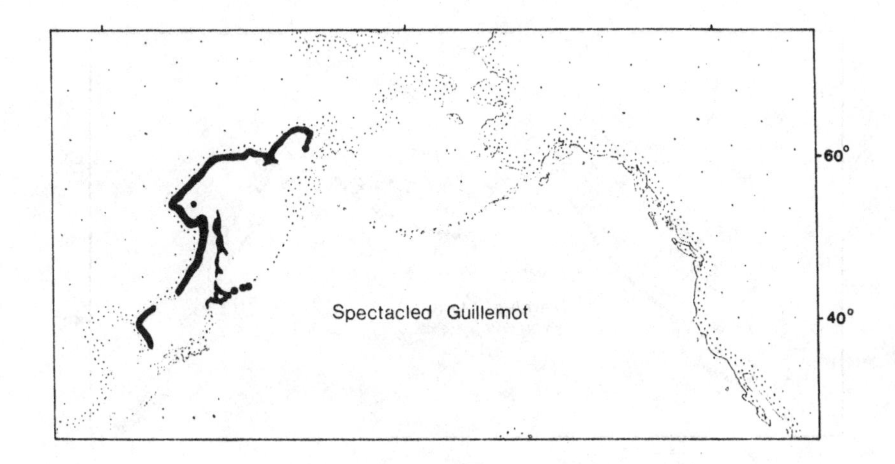

Synthliboramphus (Endomychura) *murrelets*

The genus *Synthliboramphus* (merged *Endomychura*; American Ornithologists' Union, 1982) consists of four species: Ancient Murrelet *S. antiquus*; Japanese Murrelet *S. wumizusume*; Xantus' Murrelet *Endomychura hypoleuca*; and Craveri's Murrelet *E. craveri*. The Ancient Murrelet is most widely distributed and, although it nests in Korea and China (Fig. 9.5), is replaced by the Japanese Murrelet in Japan. The Xantus' and Craveri's Murrelets occur off California and in the Gulf of California, respectively (Fig. 9.5). Most alcids have semiprecocial young that remain in the nest until they fledge (Table 9.2). *Synthliboramphus* is unique among alcids in

Fig. 9.4. Breeding range of Kittlitz's and Marbled Murrelets in the Pacific. (Data from Udvardy, 1963; Sowls *et al.*, 1978, 1980.)

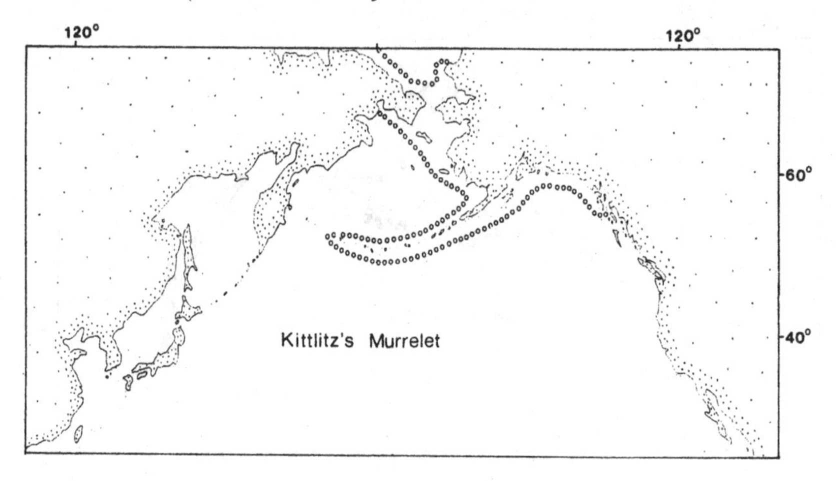

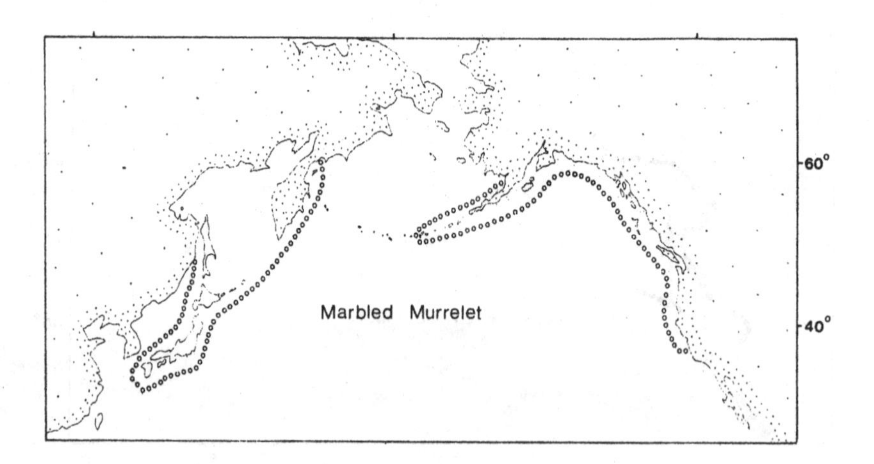

the precocity it exhibits; the young go to sea about 2 days after hatching, accompanied by one or both parents who take them far from land to feed (Drent & Guiguet, 1961; Sealy, 1972, 1976; DeWeese & Anderson, 1976; Sealy & Campbell, 1979; Murray *et al.*, 1983; Eppley, 1984; Vermeer *et al.*, 1984).

Auklets

Five species of auklets, the Parakeet *Cyclorrhynchus psittacula*, Cassin's *Ptychoramphus aleuticus*, Crested *Aethia cristatella*, Whiskered *A. pygmaea*

Fig. 9.5. Breeding range of Ancient, Craveri's, Japanese and Xantus's Murrelets in the Pacific. (Data from Udvardy, 1963; DeWeese & Anderson, 1976; Sowls *et al.*, 1978, 1980; Hasegawa, 1984; Murray *et al.*, 1983.)

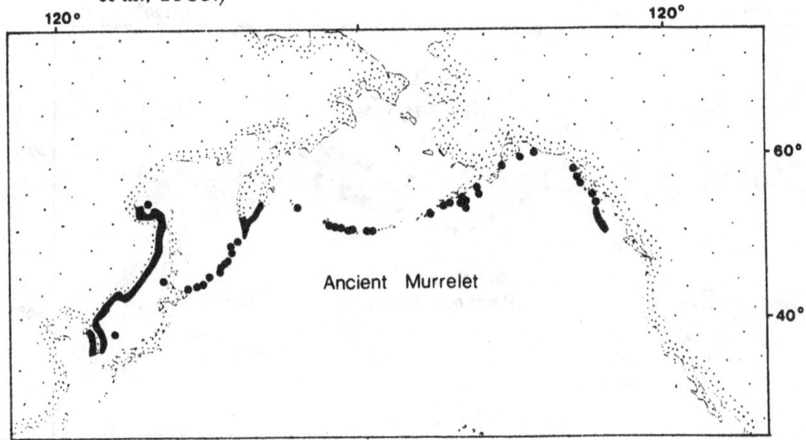

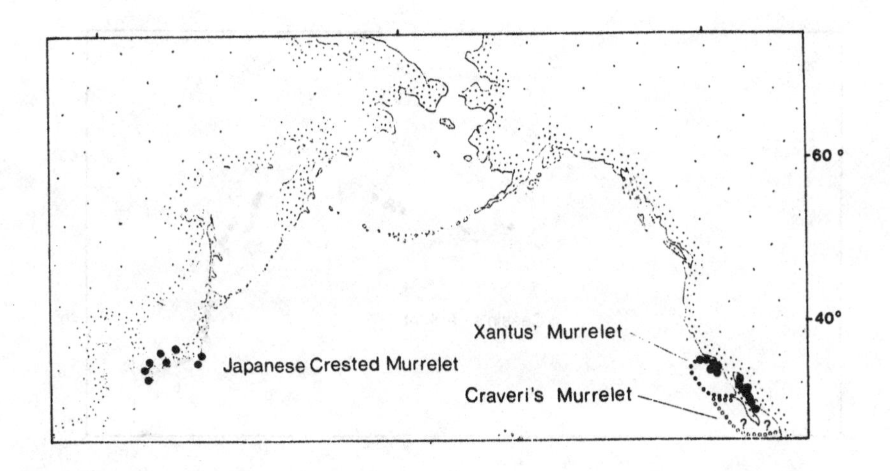

and Least *A. pusilla* Auklets are restricted to the North Pacific (Figs 9.6 and 9.7). Of those, the Cassin's Auklet has the most extensive latitudinal breeding distribution, extending from Baja California to the Gulf of Alaska, and westward to the central Aleutians (Fig. 9.6). All five species may breed in rock crevices but Cassin's Auklet is chiefly a burrow nester (Thoresen, 1964; Manuwal, 1974; Vermeer *et al.*, 1979*a*). Parakeet Auklets are primarily cliff nesters throughout their range (Bédard, 1969*b*; Searing, 1977) but on St Lawrence Island, Alaska, they also nest on talus slopes among weathered outcrops and on peaty slopes where boulders are

Fig. 9.6. Breeding range of Cassin's and Parakeet Auklets in the Pacific. (Data from Udvardy, 1963; Sowls *et al.*, 1978, 1980; Vermeer *et al.*, 1979*b*.)

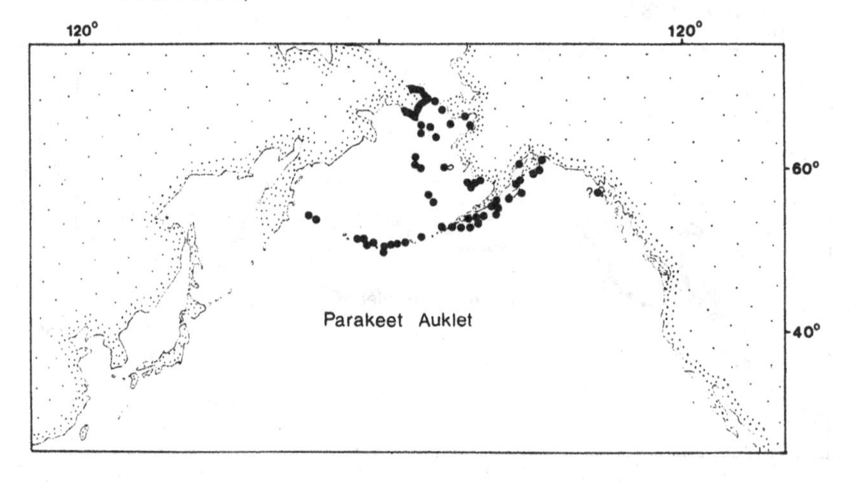

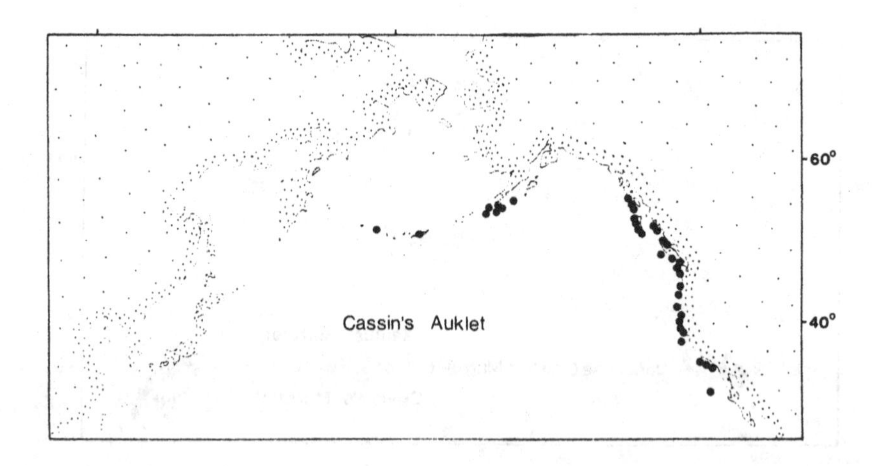

Fig. 9.7. Breeding range of Crested, Least and Whiskered Auklets. (Data from Udvardy, 1963; Sowls *et al.*, 1978, 1980.)

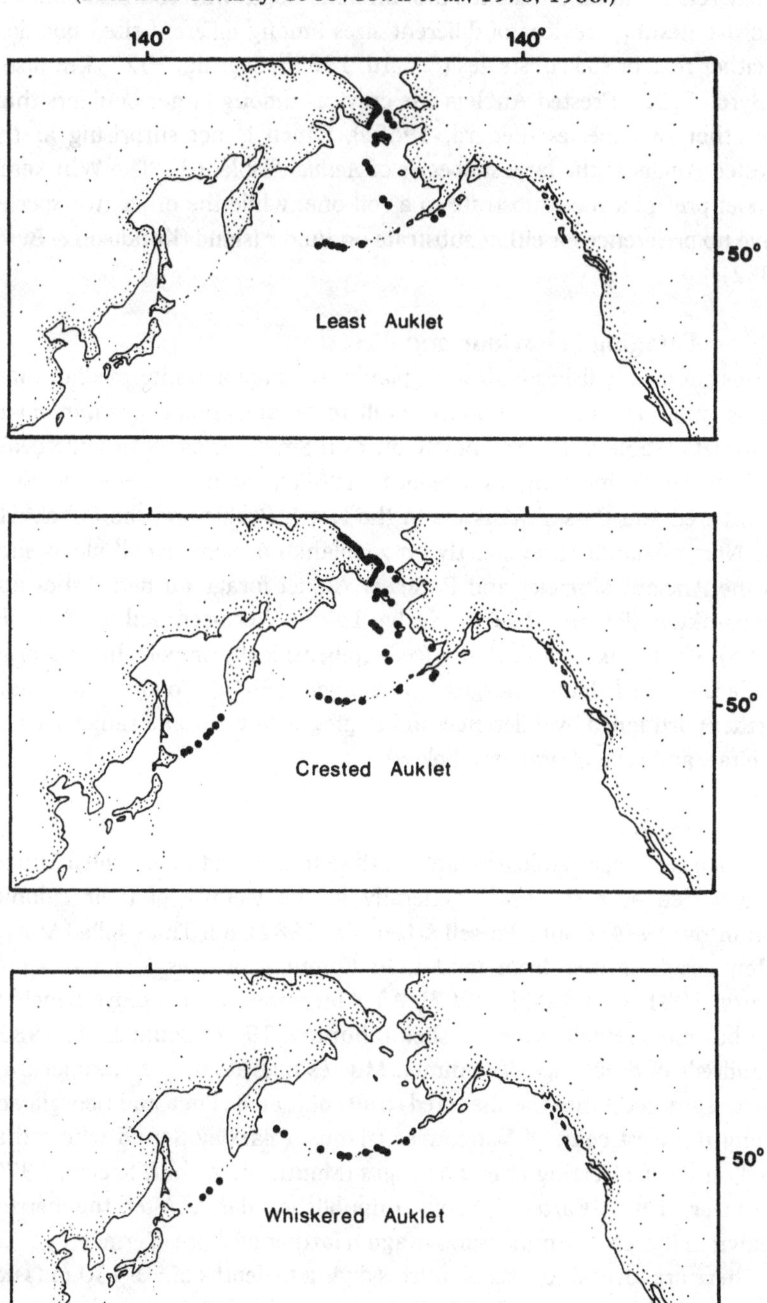

covered with soil and vegetation (Sealy & Bédard, 1973). *Aethia* species on St Lawrence and Buldir Islands nest on talus slopes with rounded boulders and use nesting crevices of different sizes among different-sized boulders in either rock or soil substrates (Bédard, 1969*b*; Searing, 1977; Knudtson & Byrd, 1982). Crested Auklets use crevices among larger boulders than the other two species (Bédard, 1969*b*), which is not surprising as the Crested Auklet is the largest species of *Aethia* (Table 9.1). The Whiskered Auklet prefers a rock substrate to a soil one, while the other two species have no preference for either substrate on Buldir Island (Knudtson & Byrd, 1982).

Foraging behaviour and diet

Alcids eat mostly fishes and/or zooplankton. Large and intermediate-sized alcids such as murres, puffins, guillemots and small *Brachyramphus* murrelets (Table 9.3) prey mostly on fish. Small alcids, with wide beaks and broad fleshy tongues (Bédard, 1969*c*), such as Least, Crested, Whiskered and Cassin's Auklets in the North Pacific and the Dovekie in the North Atlantic, prey mostly on zooplankton. Some small alcids such as the Ancient Murrelet and Parakeet Auklet forage on both fishes and zooplankton (Bédard, 1969*d*; Sealy, 1975*a*; Vermeer, Fulton & Sealy, 1985). The Parakeet Auklet also feeds epibenthically on polychaete worms (Bédard, 1969*d*; Hunt, Burgeson & Sanger, 1981*a*). Food resources are further partitioned by differences in foraging behaviour and range, feeding habitat, and type of prey (see below).

Murres

Foraging behaviour While breeding, both Common and Thick-billed Murres chiefly feed over the shelf, generally in the vicinity of their colonies (Shuntov, 1974; Gould, Forsell & Lensink, 1982), but Thick-billed Murres often forage farther from land than Common Murres (Swartz, 1966; Drury, 1981; Gould *et al.*, 1982). After breeding, murres move widely at sea but most remain over the shelf (Shuntov, 1972; Gould *et al.*, 1982). Hundreds of thousands of Common Murres migrate in late summer from the Oregon coast into the sheltered straits of Juan de Fuca and Georgia and along the west coast of Vancouver Island to Barkley Sound where they feed on Pacific Herring *Clupea harengus* (Manuwal, Wahl & Speich, 1979; Vermeer, 1983; Carter & Sealy, unpublished data). After the herring spawn in March, the murres move again farther offshore (Vermeer, 1983).

There are several reports of murres diving to depths of 50–130 m (Tuck & Squires, 1955; Scott, 1973; Forsell & Gould, 1981), but the present

Table 9.3. *Adult weight of alcids in the eastern North Pacific Ocean*

Species	Average adult body weight in g (*n*)	Location	Source
Large and intermediate-sized alcids			
Thick-billed Murre	976 (76)	Cape Thompson, Alaska	Swartz, 1966
Common Murre	931 (42)	Cape Thompson, Alaska	Swartz, 1966
Tufted Puffin	790 (137)	Alaska	Wehle, 1983
Tufted Puffin	746 (23)	Triangle I., Brit. Col.	Vermeer & Cullen, 1979
Horned Puffin	545 (43)	Alaska	Wehle, 1983
Rhinoceros Auklet	521 (50)	Destruction I., Wash.	Leschner, 1976
Rhinoceros Auklet	520 (48)	Triangle I., Brit. Col.	Vermeer & Cullen, 1982
Pigeon Guillemot	450 (53)	Mandarte I., Brit. Col.	Drent, 1965
Small-sized alcids			
Crested Auklet	300 (84)	St Lawrence I., Alaska	Bédard, 1969*d*
Parakeet Auklet	281 (11)	St Lawrence I., Alaska	Bédard, 1969*d*
Kittlitz's Murrelet	237 (3)	—	Bédard, 1969*c*
Marbled Murrelet	220 (74)	Langara I., Brit. Col.	Sealy, 1975*b*
Ancient Murrelet	206 (154)	Langara I., Brit. Col.	Sealy, 1976
Cassin's Auklet	188 (25)	Triangle I., Brit. Col.	Vermeer & Cullen, 1982
Xantus' Murrelet	167 (375)	Santa Barbara I., California	Murray *et al.*, 1983
Craveri's Murrelet	151 (8)	—	Bédard, 1969*c*
Whiskered Auklet	121 (?)	Buldir I., Alaska	Knudtson & Byrd, 1982
Least Auklet	93 (85)	St Lawrence I., Alaska	Bédard, 1969*d*

record comes from Newfoundland, where fishermen caught murres in stationary gill nets set on the sea floor at depths of up to 180 m (Piatt & Nettleship, 1985). Murres catch slower fish by the tail and faster ones by the head, but both are swallowed at the surface (Swennen & Duiven, 1977). The size of the fish taken by alcids seems to depend on diameter rather than length. Murres mostly eat fishes up to 4 cm in diameter (Swennen & Duiven, 1977).

Adult diet Ogi & Tsujita (1973) examined the food contents of 163 stomachs of murres (not identified to species, but most likely Common Murres as they were from shallow waters) that drowned in gill nets in summer in the eastern Bering Sea and Bristol Bay. The food consisted chiefly of juvenile Walleye Pollock *Theragra chalcogramma*, Pacific Sand-lance *Ammodytes hexapterus* and Capelin *Mallotus villosus*. Where these fish were absent from their diet the euphausiids *Thysanoessa raschii* and *T. longipes* had been eaten (Ogi & Tsujita, 1973). The amphipods *Parathemisto libellula* and *Hyperia medusarum* were also present. Swartz (1966), who examined 84 Common and 176 Thick-billed Murres from Cape Thompson in the Chuckchi Sea in summer, found that both species ate chiefly Arctic Cod *Boregadus saida*, while Thick-billed Murres ate more invertebrates, particularly polychaetes. Hunt *et al.* (1981*a*) found that the summer diet of both murre species near the Pribilof Islands consisted mostly of Walleye Pollock, but Thick-billed Murres ate more amphipods, particularly *P. libellula*.

Sanger (1983), who pooled food samples of 166 Common and 38 Thick-billed Murres from the Gulf of Alaska and adjacent marine regions over all seasons, found that Common and Thick-billed Murres ate at least 23 and 14 species of prey, respectively. By volume, fish were more abundant in the diet of Common (81%) than of Thick-billed Murres (44%). Of the fish taken by the Common Murre, Capelin (30%), Pacific Sandlance (22%) and Walleye Pollock (14%) were the principal prey. The remaining prey were mysids (8%), euphausiids (4%) and pandalid shrimps (4%). Thick-billed Murre prey, in order of importance, were: cephalopods (26%), Capelin (17%), hyperiid amphipods (16%), crangid and pandalid shrimps (12%), unidentified fish (10%) and Walleye Pollock (9%). Ogi (1980) analysed the food of 320 Thick-billed Murres that drowned in salmon gill nets in the oceanic waters ranging from the Kuril Islands to a region east and south of the Aleutians (160° W) from March to June, and reported the importance of food, by average weight, as: squid (73%), fish (17%) and euphausiids (10%). Identified squids were *Gonatopis borealis*, *Berryteuthis magister* and *B. anonychus*, and identified fish were juvenile Atka Mackerel *Pleurogrammus monopterygius* and several species of lanternfish Myctophidae. The predominance of squid and the presence of lanternfish suggest that the murres may feed on them at dusk and dawn. Of the planktonic prey, the euphausiids *Thysanoessa inermis* and *T. longipes* were most common.

Crustaceans may dominate the diet of murres in winter. For example, 12 Common Murre stomachs from the Pribilof Islands in winter contained chiefly amphipods (Preble & McAtee, 1923) and 28 Common Murres from Kachemak Bay, Alaska, taken from November to April, contained mostly mysids (49%) and pandalid shrimps (32%) (Sanger, Jones & Wiswar, 1979). Four Common Murres collected from Kodiak Island during January and February contained mostly pandalid shrimps (50%), Capelin (28%) and gadoid fish (13%) (Krasnow & Sanger, 1982). Most quantitative dietary data for murres are from spring to early fall samples, and more information on their winter diet is desirable.

Common Murres feed primarily on Eulachons *Thaleichthys pacificus*, Northern Anchovies *Engraulis mordax* and rockfishes *Sebastes* sp. during the summer in Oregon (Scott, 1973), although summer diets were variable over the 3 years of this study. Anchovies were important prey in all summers and made up 24% of the food volume in 1969, 23% in 1970, and 79% in 1971. Eulachons (44%) were most abundant in 1969 and rockfishes (22%) were as important as anchovies in 1970. However, in May and June of 1969 euphausiids, mysids and other planktonic crustaceans dominated the diet of murres. Scott (1973) suggested that the largely zooplankton diet in 1969 was abnormal because apparently midwater fishes were scarce in the area at that time. In winter, Pacific Herring predominates in the diet of Common Murres in British Columbia (Robertson, 1972).

In summary, it appears that the food of the two murres feeding over the shelf varies seasonally and between localities. The food of Common Murres consists mostly of fish, and that of the Thick-billed Murre of fish and crustaceans. In winter, however, crustaceans were also important in the diet of Common Murres. Of the crustaceans, euphausiids, hyperiid amphipods and pandalid shrimps were eaten by both murres. Squid is an important food for far-offshore foraging Thick-billed Murres. In the North Atlantic, Common Murres also eat predominantly fish and Thick-billed Murres eat fish and crustaceans (Bradstreet & Brown, 1985). Arctic Cod and Capelin are important prey of both murres in the Atlantic as well as in the Pacific Ocean.

Chick diet Capelin formed the main diet of 120 Common Murre chicks from Ugaiushak Island, Gulf of Alaska (Hatch *et al.*, 1978), with Chum Salmon *Onchorhynchus keta*, Walleye Pollock, Pacific Herring and Pacific Sandlance also present. Drury (1981) reported that Common and Thick-billed Murres at Cape Thompson brought Arctic and Saffron *Eliginus gracilis* Cod to their chicks. Scott (1973) collected five chicks at sea in the company of males, presumably their parents, in Oregon, and found that anchovies (60%) were most abundant in the chick diet. Anchovies (39%) and rockfish (39%) were equally important in the diet of accompanying adults, while in 11 unaccompanied adults, rockfish (52%) predominated. Scott (1973) suggested that these and other observations on parents delivering food to chicks at the nest site, indicated that adults fed their chicks, on average, larger prey (average length of anchovies: 10 cm) than they ate themselves (average length of rockfish: 5 cm). All data indicate that both species feed their chicks chiefly on fish. Bradstreet and Brown (1985) concluded similarly for both murre species in the North Atlantic.

Puffins

Foraging behaviour In winter, Horned and Tufted Puffins feed mostly offshore (Shuntov, 1972; Sanger, 1975; Gould, 1977), but when breeding, Horned Puffins feed primarily inshore, whereas Tufted Puffins, possibly depending on colony location, forage close to colonies as well as over the shelf (Sealy, 1973*a*; Wehle, 1982; Gould *et al.*, 1982; see Chapter 10) and they, but not Horned Puffins are caught in salmon gill nets (Ogi, 1980). During summer, Rhinoceros Auklets feed mostly over the shelf (Vermeer, unpublished data), but little is known of their winter feeding areas. Simons (1981), who observed feeding and diving of Rhinoceros Auklets and Tufted Puffins in the Seattle Aquarium, found that their underwater locomotion was similar, but that puffins stayed submerged significantly longer (average dive 21 s, max. 53 s, $n = 42$) and stayed on the surface between dives significantly longer than did the Rhinoceros Auklet (average dive: 15 s, max. 40 s, $n = 88$). Rhinoceros Auklets also foraged in darkness while the Tufted Puffins did not. In the wild, Rhinoceros Auklets, which commonly feed during the day, have been observed to forage at the onset of dusk (Vermeer, unpublished data; Sanger, unpublished data) and do not arrive at the colony to feed their chicks until dusk (Thoresen, 1964; Leschner, 1976; Vermeer, 1979; but see Scott *et al.*, 1974). In the aquarium, fish were caught by singling one out and driving it to the surface or against the side of the aquarium tank (Simons, 1981). At sea,

Rhinoceros Auklets have been observed to herd sandlance by diving beneath a school, apparently causing it to form a tight cohesive ball and then driving this to the surface (Grover & Olla, 1983). Rhinoceros Auklets and Tufted Puffins catch a fish by either the tail or head but, as in murres, they always swallow fish head first (Simons, 1981). There are indications that Rhinoceros Auklets and Tufted Puffins do not dive as deeply as other alcids, perhaps because of extreme buoyancy. Simons (1981) recorded that 'they are extremely buoyant and must swim constantly to remain under water. At the end of a dive they merely stop flapping [their wings] and glide quickly to the surface.' Atlantic Puffins off Newfoundland are commonly caught in gill nets set 0–30 m below the surface, but have never been recorded from those set at greater than 60m depth (Piatt, Nettleship & Threlfall, 1984; Piatt & Nettleship, 1985).

Adult diet The main prey of 93 Tufted Puffins from the East Kamchatki Current were juvenile fish, euphausiids and squid (Ogi, 1980). That of 58 birds from the Sub-Arctic Current was made up chiefly of fish, of which Atka Mackerel and Northern Smoothtongue *Leuroglossus stilbius*, a bathypelagic species, were identified. Euphausiids made up 37% of the food volume in Tufted Puffins from the East Kamchatka Current, 50% in the Sea of Okhotsk, 31% in the Sub-Arctic Current, 6% in the western Sub-Arctic Gyre and 5.5% in the Alaskan Stream (Ogi, 1980).

Wehle (1982), who examined the food of 73 adult Tufted and 41 adult Horned Puffins on and near their nesting colony on Buldir Island, Alaska, in 1975, found that squid occurred in 97% and fish in 25% of Tufted Puffin stomachs and squid in 88% and fish in 27% of Horned Puffin stomachs. Wehle (1983) also sampled 20 adult Tufted Puffins and 12 adult Horned Puffins at Uguiushak Island, Alaska, in 1976 and 1977. Squid and fish each occurred in six Tufted Puffins and in one and five Horned Puffins, respectively. Three Tufted Puffins and one Horned Puffin had also eaten crustaceans.

There is little information on the diet of adult Rhinoceros Auklets. Fish made up 99% and cephalopods 1% by volume of the food of 21 adult Rhinoceros Auklets from the Gulf of Alaska (Sanger, 1986). Of the fish, Capelin (61%) and Pacific Sandlance (21%) predominated. The Market Squid *Loligo opalescens* occurred in 85%, and Northern Anchovies in 39% of the adults from California (Baltz & Morejohn, 1977). Other fish prey were *Sebastes* spp., *Citharichthys* spp. and *Icichthys lockingtoni*, each in 11.5% of samples. By contrast, Kozlova (1957) stated that adult Rhinoceros

Auklets feed chiefly on crustaceans and partly on fish, but did not provide details.

Chick diet Wehle (1983), studying diets of nestling Horned and Tufted Puffins at the same localities in the Gulf of Alaska, found that Pacific Sandlance and Capelin were the predominant foods of both puffins. The other prey, however, varied. The Tufted Puffin chicks received (in decreasing order of abundance) cephalopods, gadoids, sculpins Cottidae, and greenling *Hexagrammos* sp., while in Horned Puffins the order of importance was greenling, gadoids and cephalopods. In other Alaskan studies, Capelin and Pacific Sandlance were also the major prey delivered to Horned and Tufted Puffins (Manuwal & Boersma, 1977; Baird & Moe, 1978; Moe & Day, 1979). The diet of Tufted Puffin chicks in Alaska (Baird & Moe 1978; Wehle, 1983) and in British Columbia (Vermeer, 1979; Vermeer, Cullen & Porter, 1979) and that of Horned Puffin chicks in Alaska (Moe & Day, 1979; Wehle, 1983) varied in prey composition and species over the summer. The diet of Rhinoceros Auklet chicks has been extensively studied in Washington State (Leschner, 1976; Wilson, 1977), in British Columbia (Vermeer, 1979, 1980; Vermeer & Westrheim, 1984) and in Alaska (Hatch, 1984; A.R. DeGange & L.L. Leschner, personal communication), and varies latitudinally. In Washington State, Pacific Sandlance, Northern Anchovy, Pacific Herring, Night Smelt *Spirinchus starksi* and Surf Smelt *Hypomesus pretiosus* predominate; in British Columbia, Pacific Sandlance, Pacific Saury *Cololabis saira* and rockfish are most important on offshore islands, and Pacific Sandlance, Pacific Herring and rockfish in inshore colonies; while in the Gulf of Alaska Pacific Sandlance and Capelin are the main foods of Rhinoceros Auklet chicks (Vermeer, 1979; Hatch, 1984). Food brought by the parents to the chicks varies drastically over the season and among years (see Vermeer & Westrheim, 1984). Where the diet of Rhinoceros Auklet and Tufted Puffin chicks was investigated simultaneously in British Columbia, Rhinoceros Auklets ate more Pacific Sauries than the puffins, which may relate to the more crepuscular feeding of the Rhinoceros Auklet (Vermeer, 1979). In 1978, the Bluethroat Argentine *Nansenia candida*, a bathypelagic species, was a major food for chicks of both species in British Columbia. The appearance of bathypelagic fishes as a predominant food in the birds' diet may be related to foraging in areas of upwelling (Vermeer, 1979).

Pigeon Guillemot

Foraging behavior Pigeon Guillemots feed near their nest sites, but disperse more widely over inshore waters after the young become independent (Drent, 1965; Scott, 1973; Kuletz, 1983). They feed their chicks diurnally on epibenthic fish (bringing them in one at a time) and invertebrates (Drent, 1965; Follett & Ainley, 1976; Scott, 1973; Krasnow & Sanger, 1982; Kuletz, 1983). The maximum depth to which *Cepphus* guillemots dive appears to be 50 m (Piatt & Nettleship, 1985).

Adult diet Krasnow & Sanger (1986) examined 44 stomachs with food from Kodiak Island and found that shrimp and fish were most important in winter (49% and 45%, respectively, of the volume), crabs (mostly *Cancer oregonensis*) and fish in June (99%, mostly Capelin and Pacific Sandfish *Trichodon trichodon*), fish in July (79%, mostly Capelin), and fish in August (90%, mostly Capelin and gadoids).

Chick diet Drent (1965) assigned 662 food items delivered to chicks on Mandarte Island in southern British Columbia to 10 categories of fish (identified species in brackets): lampreys Petromyzontidae (*Lampetra ayresi*), sticklebacks Gasterosteidae (*Gasterosteus aculeatus*), shiners Embisticidae (*Cymatogaster aggregatus*), sandlance, blennies (including *Pholis laetus*, *Delolepis giganteus* and *Lumpenus anguillaris*), rockfish, sculpins (including *Leptocottus armatus*, *Myxocephalus polyanthocephalus* and *Triglops beani*), gruntfish, Rhamphocottidae, sea-poachers Agaridae and flatfish Bothidae and Pleuronectidae. Benthic fish, such as blennies and sculpins, comprised at least 70% of the prey fed to the young. The chicks had difficulty swallowing sea-poachers (which are armoured) and flatfish. Kuletz (1983) reported that sandlance and blennies made up the main diet of chicks at Naked Island, Alaska. Other prey consisted of herring, smelt, ling cod, cod, sculpins, flatfish, rockfish, squid and shrimp. Follett & Ainley (1976) found that epibenthic fish predominated in chick diets on Southeast Farallon Island, California. Comparing diets of adults and chicks, it seems that adults eat fish and invertebrates, while they feed chiefly fish to their young.

Marbled and Kittlitz's Murrelets

Foraging behavior Marbled Murrelets generally forage in sheltered inshore areas (Sealy, 1975*a*; Forsell & Gould, 1981; Carter, 1984; see Chapter 10) and at Langara Island, British Columbia, they foraged seldom more than 500 m from shore and usually in water less than 30 m deep. At Kodiak

Island, Alaska, Marbled Murrelets usually foraged throughout the water column inshore of the 50 m contour (Krasnow & Sanger, 1982), at least part of the time demersally (Sanger, unpublished observation) and often feeding alone or in pairs. They feed their young at dusk (Table 9.2). Kittlitz's Murrelet also feeds inshore and often in the same bays in Alaska as Marbled Murrelets (Sowls, Hatch & Lensink, 1978).

Adult diet The diet of Marbled Murrelets has been investigated near Langara Island, Queen Charlotte Islands (Sealy, 1975*a*) and in Barkley Sound, Vancouver Island (Carter, 1984) in British Columbia; and at Kodiak Island (Krasnow & Sanger, 1986), and Kachemak Bay (Sanger, 1986), Alaska. At Langara Island, the most important foods present in 75 adult and sub-adult murrelets were Pacific Sandlance and *Thysanoessa spinifera* in April and May, and Pacific Sandlance in June and July. Seaperch *Cymatogaster aggregata*, osmerids Osmeridae, stichaeids Stichaeidae, and *Euphausia pacifica* were less important prey. In Barkley Sound, 43 breeding adults and 26 molting birds from late summer and fall, fed primarily on Pacific Herring and Pacific Sandlance. At Kodiak Island, Capelin (59% by volume), unidentified osmerids (19%) and *Thysanoessa inermis* (16%) dominated the diet of 18 murrelets in one winter, and mysids (85%, *Acanthomysis* and *Neomysis* spp.) were important in the diet of another 18 murrelets from Kachemak Bay the next winter. In April and May, Capelin, Pacific Sandlance and the euphausiids *Thysanoessa inermis* and *T. raschii* were important foods, while from June to August, Pacific Sandlance predominated.

Sixteen prey species, including seven crustaceans and four fish, were observed in 129 stomachs of Marbled Murrelets, pooled from different seasons, from four regions in the Gulf of Alaska (Sanger, 1983). Pacific Sandlance (28% by volume) and Capelin (27%) were most important, followed by unidentified fish (18%), mysids (13%), euphausiids (9%) and Pacific Sandfish (2%). In summary, the diet of adult Marbled Murrelets consisted mostly of Capelin and Pacific Sandlance in Alaska, and of Pacific Sandlance and Pacific Herring in British Columbia. Euphausiids and mysids are important foods in spring and winter.

The food of 16 Kittlitz's Murrelets from the Bering Sea in spring and the Kodiak Island area in summer, consisted (by volume) of fish (70%), mostly unidentified but including Capelin, Pacific Sandlance and Pacific Herring, and euphausiids (29%) such as *Thysanoessa inermis* and *T. spinifera* (Sanger, 1986).

Chick diet There is little information on this because nests are so difficult to find. Simons (1980) observed a Marbled Murrelet on two occasions bringing a single fish (one identified as Capelin) to its chick on East Amatuli Island, Alaska. Other authors mention that the Marbled Murrelet probably makes only one trip each night, carrying one to four fish to its chick (Guiguet, 1956; Cody, 1973; Sealy, 1975*b*; Hatler, Campbell & Dorst, 1978). Carter (1984) observed 144 fish in the bills of Marbled Murrelets during the breeding season in Barkley Sound, which may have been carried to nestlings. Of those fish, 49% were Pacific Sandlance, 11% Pacific Herring, 1% Northern Anchovy and 39% unidentified.

Ancient Murrelet and other Synthliboramphus *species*

Foraging behavior Ancient Murrelets feed over the shelf but more frequently over the shelf break on the British Columbia coast during summer (Vermeer *et al.*, 1984). At the end of the nesting period, parents depart with their chicks (usually 2 days old) from the colony and disperse widely (Sealy, 1975*a*; Sealy & Campbell, 1979; Vermeer *et al.*, 1984). The parents and chicks remain together as a family unit until at least the chicks have grown (Sealy, 1972; Sealy & Campbell, 1979; Vermeer *et al.*, 1984). Similar observations have been made for the closely related Craveri's Murrelet (DeWeese & Anderson, 1976) and Xantus' Murrelet (Murray *et al.*, 1983), while little is known about the foraging behavior of the Japanese Murrelet. There are few observations on Ancient Murrelets in winter except that thousands have been observed feeding in turbulent inshore waters near Victoria (Vermeer, 1983).

Adult diet Sealy (1975*a*) analysed the diets of 46 adults and 22 sub-adult Ancient Murrelets near Langara Island during nesting and found that they had fed mainly on *Euphausia pacifica* and *Thysanoessa spinifera* from early April to late May, but mostly Pacific Sandlance thereafter. Sanger (1986) found that 15 Ancient Murrelets from Alaskan waters fed mostly on *T. inermis* and unidentified *Thysanoessa* spp. (55% by volume), and fish (42%). The fish consisted mostly of gadoids, Walleye Pollock and Capelin.

Auklets

Foraging behavior The plankton-feeding Least, Crested, Whiskered, Parakeet and Cassin's Auklets in the North Pacific, and the Dovekie in the North Atlantic, have different chick-feeding adaptations from the alcids which are predominantly fish-eaters. The plankton-feeders bring food to their young in a neck pouch, and they have relatively wide beaks and broad fleshy tongues which facilitate feeding on zooplankton (Bédard, 1969*c*). Three auklets in Arctic and sub-Arctic regions feed their chicks diurnally, the circadian rhythm of the Whiskered Auklet is at present unclear (Byrd, Day & Knudtson, 1983), while the Cassin's Auklet and perhaps the Dovekie feed their young mainly at night (Table 9.2).

Where Parakeet, Crested and Least Auklets occur together in the eastern Bering Sea, the Crested and Least Auklets feed on zooplankton at middle depths and near the surface, while Parakeet Auklets take a wide variety of zooplankton, invertebrates and fish, some of it demersally or epibenthically (Bédard, 1969*d*; Hunt *et al.*, 1981*a*). In the spring these three species aggregate in dense rafts in leads in the ice while in summer they forage near their colonies (Bédard, 1967, 1969*d*; Hunt *et al.*, 1981*a*). The Crested and Least Auklets fly to and from breeding grounds in flocks of about 20–50 individuals and after landing on the water they drift apart and feed individually, although Least Auklets have also been seen feeding in dense groups of thousands of birds (Bédard, 1969d). Parakeet Auklets leave the colony singly or in small flocks, and feed farther apart than the *Aethia* species (Bédard, 1969*d*). Little is known about the foraging of Whiskered Auklets, but there are indications they may feed earlier as well as later in the day than the other two *Aethia* species, since they leave the colony earlier in the morning and return later in the evening (Byrd *et al.*, 1983). Least, Crested and Parakeet Auklets commonly associate with Grey Whales *Eschrichtius robustus* off Alaska, feeding apparently mainly on gammarid amphipods brought to the surface by the whales (Harrison, 1979).

In the northern Gulf of Alaska, Cassin's Auklets appear to feed mainly over the shelf (Gould *et al.*, 1982; see Chapter 10), while in British Columbia they feed mostly on zooplankton in singles or pairs over the shelf break and seamounts during the summer (Vermeer, Fulton & Sealy, 1985). In October hundreds of Cassin's Auklets have been observed feeding over La Perouse Bank off Vancouver Island (Shepard, 1977).

Adult and chick diet Bédard (1969*d*) reported on the diets of Least, Crested and Parakeet Auklets during the nesting season at the St Lawrence Island in the Bering Sea. The two *Aethia* species had a varied diet of mysids, hyperiids and gammarids during the early stages of nesting. During the chick-rearing period the Least Auklet switched to one main prey, *Calanus* sp., and the Crested Auklet to *Thysanoessa* spp. The Parakeet Auklet maintained a diverse diet throughout the summer, of which the hyperiid amphipod *Parathemisto libellula* was most important. Cephalopods and polychaetes occurred only in the diet of the Parakeet Auklet, which also ate more small fish than the two *Aethia* species. Searing (1977), who studied the diet of Crested and Least Auklets on the same island 10 years later, found that both species ate mostly calanoids during the chick phase. In Bédard's and Searing's studies, calanoid copepods comprised 36% and 97% (by numbers), respectively, which may reflect yearly fluctuations in this prey's availability.

Hunt *et al.* (1981*a*) examined the food brought to chicks of the same three auklets in the Pribilof Islands. The Parakeet Auklet was the most pronounced generalist, feeding on three main prey types, euphausiids (*Thysanoessa inermis* 10% volume, *T. raschii* 19%), small fish (unidentified and *Theragra chalcogramma* 27%) and polychaetes (24%). The Crested Auklet specialized on euphausiids (69%, mostly *Thysanoessa inermis*) and somewhat less on amphipods (30%, mostly *Parathemisto libellula*), while the Least Auklet was highly specialized on calanoid copepods (73.5%, mostly *Calanus marshallae* and *Neocalanus cristatus*). *C. marshallae* was the main food of Least Auklets on St Matthew Island in 1982 (89% volume) and 1983 (84% volume) (Springer & Roseneau, 1985).

Few data exist on the diet of adult Cassin's Auklets. Food categories in neck pouches and stomachs of six Cassin's Auklets, collected at sea off British Columbia during the chick phase, were similar, perhaps reflecting identical adult and chick diets (Vermeer *et al.*, 1985). In British Columbia, Cassin's Auklets feed their chicks mostly large copepods, *Neocalanus cristatus*, as well as *Thysanoessa spinifera*, *T. longipes*, caridean shrimps and larval and juvenile fish (Vermeer, 1981, 1984), while in southern California they feed their young euphausiids, squid and amphipods (Manuwal, 1974). In the northern Gulf of Alaska, eight Cassin's Auklets collected at sea had eaten mostly calanoid copepods and lesser amounts of euphausiids, carideans, crab larvae and unidentified fish (Sanger, 1986).

Nesting distribution and range in relation to diet

Closely related species replace one another geographically, either completely or partially, for example the Rhinoceros Auklet and Tufted Puffin. Although their ranges overlap considerably (Fig. 9.2), their numerical distribution is very different. In the eastern North Pacific most Tufted Puffins breed in Alaska (Sowls *et al.*, 1978), while the main breeding population of Rhinoceros Auklets is in southeastern Alaska and British Columbia (Vermeer, 1979). Differences in range and distribution are linked to nesting differences. Tufted Puffins prefer treeless soil-covered islands such as are relatively abundant in the Aleutians and the Bering Sea but not in British Columbia, where Tufted Puffins are chiefly restricted to a few offshore islands (Vermeer, 1979). Rhinoceros Auklets also nest there with Tufted Puffins, but more frequently breed on soil-covered, forested islands which are relatively numerous in British Columbia.

Nesting distribution may also affect the type of prey fed to nestlings. For example, the rarity of Pacific Herring in the diet of nestling Tufted Puffins in British Columbia (Vermeer, 1979), probably reflects simply that herring is an inshore species and thus largely unavailable to breeding adult Tufted Puffins (Vermeer & Westrheim, 1984).

Besides nesting distribution, feeding behaviour also affects diet. For example, in British Columbia, where the two species nest on the same offshore island (Triangle Island), Rhinoceros Auklets utilized Pacific Sauries significantly more than did Tufted Puffins. This may relate to Rhinoceros Auklets feeding at dusk, when sauries migrate to the water surface (Vermeer, 1979). These two puffins thus appear to partition food during the breeding season both directly, through differences in feeding behavior, and indirectly by nesting distribution.

The diet of alcids also varies with latitude. For example, Capelin is an important food of nestling Rhinoceros Auklets and Tufted Puffins in Alaska, but not farther south (Vermeer, 1979). This undoubtedly relates to Capelin being a species characteristic of more northern, colder waters (Harris & Hartt, 1977; see also Ainley & Sanger, 1979; Vermeer & Westrheim, 1984). Pacific Sauries are an important food of nestling Rhinoceros Auklets in British Columbia and Washington State in August (Vermeer, 1979; U. Wilson, personal communication), but not in northern Alaska. Pacific Sauries are pelagic, warm-water fish and therefore are usually unavailable to alcids in sub-Arctic waters. Similarly, Northern Anchovies are warm-water fish and a major food of nestling Rhinoceros Auklets in Washington State, but not farther north (Leschner, 1976; Vermeer, 1979). A similar replacement of fish was observed in the diet of Common Murres. Arctic Cod, Capelin, Pacific Sandlance and Walleye Pollock are important foods of murres at northern latitudes (Swartz, 1966; Ogi & Tsujita, 1973; Sanger, 1983), while anchovies, eulachons and rockfish prevail in the murre diet at more southern localities (Scott, 1973).

Geographical replacement of plankton prey also occurs. For example, *Thysanoessa inermis* has its centre of abundance in Alaska (Brinton, 1962), and is one of the most common euphausiids in the diet of adult and nestling alcids and other seabirds in the northern Gulf of Alaska and the

Bering Sea (Krasnow & Sanger, 1986; Sanger 1986), but has not been recorded in the diet of alcids or other seabirds in British Columbia (Vermeer, 1984). *T. inermis*, although not uncommon, is much less numerous in British Columbia waters than *T. spinifera*, *T. longipes* and *Euphausia pacifica* (Fulton, Arai & Mason, 1982), which are the only euphausiids taken by alcids in British Columbia (Sealy, 1975*a*; Vermeer, 1984). The predominance of the copepod *Neocalanus cristatus* in the diet of nestling Cassin's Auklets in British Columbia and its absence in the diet of California birds (Manuwal, 1974; R. Boekelheide, personal communication) also reflect prey availability; the species is abundant in British Columbia but scarce or absent in California waters (Vermeer, 1981).

Finally, the size and distribution of breeding colonies of auklets may be determined by the availability and abundance of calanoid copepod prey. Vermeer (1981) suggested that the large breeding concentrations of Cassin's Auklets in British Columbia (62% of the world population) relate to the availability and abundance of the high energy content copepod *Neocalanus cristatus*. Springer & Roseneau (1985) suggest that the distribution of copepod biomass may govern numbers and distribution of nesting Least Auklets in the Bering Sea.

Feeding frequencies, prey and meal sizes and food consumption

Feeding frequencies and meal sizes of nestlings

Feeding frequencies and meal sizes of some North Pacific alcids are shown in Table 9.4. Those of murres in the North Atlantic are included since none has been documented for murres in the North Pacific. Alcids that visit their colonies at night usually feed their chicks once during a 24-h day (e.g. Richardson, 1961; Thoresen, 1964), while alcids that visit their nests during daylight feed their chicks several to many times each day (Bédard, 1969*a*; Norderhaug, 1970, 1980). The feeding frequency in diurnal alcids depends on the proximity of the feeding area. Pigeon Guillemots, which are inshore foragers, visit their colony many times during the day, while murres, which forage further ashore, bring food to their young less frequently. Feeding frequencies of diurnal alcids also vary over the nesting season and among years (e.g. Vermeer, 1979; Vermeer, Cullen & Porter, 1979; Gaston & Nettleship, 1981). A combination of a low feeding frequency and reduced meals may lead to starvation and death. For example, in 1977 Tufted Puffin chicks starved on Triangle Island, British Columbia, as parents appeared unable to provide their chicks with sufficient food, while in 1978 chicks thrived as a result of being fed more

Table 9.4. *Feeding frequencies and meal sizes of some alcid species*

	Feeding frequency per 24 h; mean (range)	Average meal size in g (sample size)	Location, year	Source
Circadian rhythm: nocturnal				
Rhinoceros Auklet	2	30 (2165)	British Columbia, 1976–80	Vermeer & Westrheim, 1984
Cassin's Auklet	2	17·6–19·7 (241)	British Columbia, 1978–79	Vermeer, 1981
Circadian rhythm: diurnal				
Tufted Puffin	1·3 (0–6)	14 (22)	British Columbia, 1977	Vermeer *et al.*, 1979*a*
Tufted Puffin	3·5 (0–6)	22 (35)	British Columbia, 1978	Vermeer, 1979
Tufted Puffin	— (0–6)	9–15 (144)	Alaska, 1976–77	Wehle, 1983
Horned Puffin	— (2–6)	11 (15)	Alaska, 1975	Wehle, 1983
Pigeon Guillemot	16·2 (12–23)	One fish	British Columbia, 1960	Drent, 1965
Parakeet Auklet	6·6	—	Alaska, 1967	Sealy & Bédard, 1973
Common Murre	2·65	5–30, one fish	USSR	Kartaschew, 1960
Common Murre	2·1	9[a], one fish	Labrador	Birkhead & Nettleship, 1982
Common Murre	3	26·5[a], one fish	Newfoundland	Mahoney, 1979
Thick-billed Murre	3·4–4·6 (0–10)	12·5[a], one fish	Canadian eastern Arctic, 1975–77	Gaston & Nettleship, 1981

[a] Estimated.

frequently and with larger meals than the previous year (Vermeer, 1979; Vermeer, Cullen & Porter 1979)

The number of fish parents bring their chicks varies among fish-feeding alcids. Murres and Pigeon Guillemots usually bring one fish at a time, which in murres is held lengthwise in the bill with the tail forward (e.g. Tuck & Squires, 1955; Birkhead, 1976; Mahoney, 1979), and in Pigeon Guillemots is grasped by the head and dangles from the bill (e.g. Drent, 1965). Puffins and Rhinoceros Auklets carry fish crosswise in their bills. The number of fish carried depends on their depth or width. The smaller the depth or width, the more fish are carried. In Rhinoceros Auklet bill loads containing only one prey species, the number of first-year Pacific Sandlances averaged 12 (max. 26) and that of second-year and older sandlances averaged 3 (max. 7) (Vermeer & Devito, in press). Slender fish, like sandlance, are carried in the bill depth-or widthwise, while herring, sauries and rockfish, whose depth is at least twice their width, are usually carried widthwise. The larger the fish, the smaller the number that was carried by the auklets. The longest fish made up the largest meals in biomass and provided most energy to nestlings. Small fish of different species were positively associated with one another in Rhinoceros Auklet meals, as well as certain small fish with larger ones, but not vice versa. By complementing small fish of one species with fish of another species, foraging Rhinoceros Auklets boost the energy content of a meal (Vermeer & Devito, in press).

The amount of food which plankton-feeding alcids bring to their chicks depends on the amount of prey encountered during feeding, and the size of the gullet and the neck pouch. Bédard (1969*d*) found that the gullet of the Least Auklet, which is the smallest alcid (90 g), can hold 2.5 cm^3 of food and its neck pouch 11 cm^3 of food. In the closely related Crested Auklet (290 g), the gullet and neck pouch can hold 15 and 32 cm^3 of food, respectively. The quantity of food delivered to chicks by plankton-feeding alcids is usually less than the gullet and pouch capacity, but larger alcids deliver more food to their chicks than smaller ones. For example in Alaska, filled neck pouches of Least Auklets contained on the average 6 g and those of Crested and Parakeet Auklets 20–21 g of food (Bédard, 1969*d*), while in British Columbia, Cassin's Auklets (188 g) regurgitated on average a 19 g meal to their chicks (Vermeer, 1984).

Prey size

The prey of large and intermediate-sized alcids (Table 9.3) is generally larger than that of small alcids, many of whom feed predominantly on zooplankton. For example, the prey of the Cassin's Auklet is smaller than that of the Rhinoceros Auklet and Tufted Puffin (Table 9.5). Rhinoceros Auklets and Tufted Puffins bring similar-sized prey to their chicks (Table 9.5); information on the prey size of murres, the largest living alcids (see Scott, 1973; Gaston & Nettleship, 1981; Birkhead & Nettleship, 1982), also indicates that they feed on similar-sized fishes to Tufted Puffins. However, the average estimated meal size of single fish fed to murre chicks (Table 9.4), as well as Scott's (1973) observation that murres feed their

Table 9.5. *Size of prey (with sample size in parenthesis) in the diet of Rhinoceros Auklet, Tufted Puffin and Cassin's Auklet chicks (Vermeer, 1979; recalculated from Vermeer & Westrheim, 1984; Vermeer, unpublished data)*

	Length[a] (mm)	
Principal prey species	Mean	Range
Rhinoceros Auklet (520 g)		
Pacific Sandlance *Ammodytes hexapterus* (1st yr)	81 (4692)	35–105
Pacific Sandlance (2nd yr and older)	130 (1353)	105–185
Pacific Herring *Clupea harengus* (1st yr)	70 (1198)	40–100
Pacific Herring (2nd yr)	145 (516)	100–185
Pacific Saury *Cololabis saira*	164 (383)	30–230
Sockeye Salmon *Oncorhynchus nerka*	124 (116)	95–170
Yellowtail Rockfish *Sebastes flavidus*	54 (156)	25–60
Widow Rockfish *S. entomelas*	64 (976)	45–85
Bluethroat Argentine *Nansenia candida*	101 (382)	75–125
Tufted Puffin (746 g)		
Pacific Sandlance (1st yr)	85 (80)	65–105
Pacific Sandlance (2nd yr and older)	135 (32)	105–175
Yellowtail Rockfish	54 (14)	50–59
Widow Rockfish	67 (16)	55–80
Bluethroat Argentine	100 (47)	89–115
Cassin's Auklet (188 g)		
Neocalanus cristatus (adults)	7·6 (842)	6·8–8·6
Thysanoessa spinifera (adults)	26·0 (544)	18·0–33·6
T. spinifera (juveniles)	14·5 (282)	10·3–19·1
T. longipes (adults)	15·9 (486)	11·2–19·0
T. longipes (juveniles)	10·0 (464)	7·3–12·5
Euphausia pacifica (adults)	22·0 (357)	16·3–26·4
Pacific Sandlance (1st yr)	40·0 (5)	35·0–45·0
Irish Lord (1st yr)	20·0 (6)	17·0–25·0

[a] Fish were measured from tip of snout to fork of tail and crustaceans from tip of rostrum to tip of telson.

young generally larger fish than they eat themselves, suggests that murre chicks receive larger prey fish on average, albeit within the same size range, than puffins.

Food consumption of adult alcids

Consumption by adult alcids can be estimated from food intake of birds in captivity or indirectly from estimates of daily consumption on basis of body weight. For example, it has been estimated that murres eat the equivalent of about 10–30% of their body weight daily (Uspenski, 1956; Johnson & West, 1975), while the daily intake of fish by captive murres varied over the course of a year from 90 to 300 g per bird or about 10–33% of the body weight (Swennen & Duiven, 1977). Food intake is dependent upon metabolic requirements of the birds, as well as on caloric values and lipid content of the prey. Consumption is greater when the requirements of murres are high and the lipid content of the prey is low (Marsault, 1975). A small alcid eats more food relative to its body size than a larger bird because of greater energy loss in birds with higher body surface/body weight ratios (Kendeigh, 1970; see also Vermeer & Cullen, 1982; Vermeer & Devito, in press). Nilsson & Nilsson (1976) considered

Fig. 9.8. Composition of breeding populations of seabirds in the eastern North Pacific. (Data from Varoujean, 1979; Sowls *et al.*, 1978, 1980; Vermeer *et al.*, 1983.)

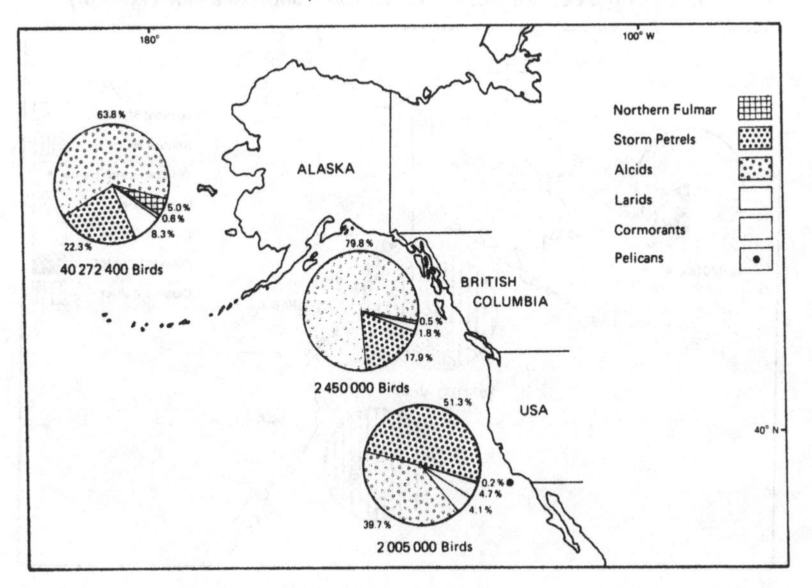

Fig. 9.9. Composition of breeding population of fish-eating alcids in the eastern North Pacific. (Data sources as for Fig. 9.8.)

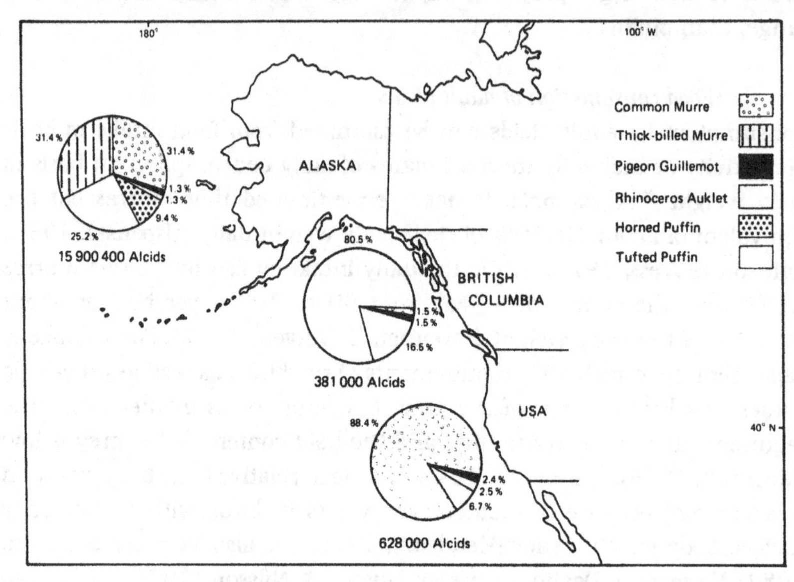

Fig. 9.10. Composition of breeding population of plankton-feeding alcids in the eastern North Pacific. (Data sources as for Fig. 9.8.)

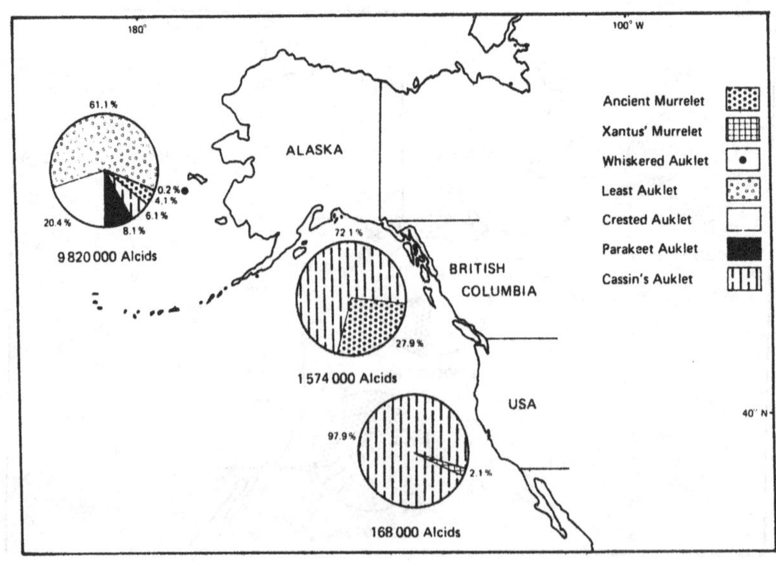

the latter factor when estimating daily food consumption of wild piscivorous birds in a fresh-water environment with the formula $\log F = -0.293 + 0.85 \log W$, where F = food consumption in g day^{-1} and W = weight of bird in g. There is no similar equation for piscivorous birds in the marine environment, but the above equation may approximate the daily food consumption of alcids and other seabirds, at least on a relative scale. With known breeding populations and body weights of alcids (Table 9.3), the daily consumption of alcids in the eastern North Pacific can be approximately established. Alcids make up about 64% of a breeding population of 45 million seabirds in the eastern North Pacific (Fig. 9.8). Of those alcids about 17 million eat predominantly fish (Fig. 9.9) and about 12 million eat mostly plankton (Fig. 9.10). Fish-eating alcids are calculated to consume about 2675 tonnes and plankton-feeding alcids 413 tonnes, while all other seabirds consume about 866 tonnes of food per day (Fig. 9.11).

Conclusions

Resource partitioning occurs among alcids as indicated by their diverse nesting and feeding strategies. Some alcids nest colonially or semicolonially, on cliff tops, ledges, in natural crevices or earthen burrows on bare or forested islands, while others nest alone in pairs on hillsides or in trees, frequently far from the sea. Related species partition nesting resources and prey, either by having partially or completely different breeding ranges or by nesting in different habitats. Alcids partition food by foraging on fish or zooplankton in different areas or at different depths and times of day. They differ in feeding frequencies, meal and prey sizes, and age classes.

This brief review shows that there are many information gaps which presently hinder a thorough understanding of the feeding ecology of Pacific alcids. Some of the most important gaps could be addressed by investigations of the topics suggested below (several of which were also identified by Bradstreet & Brown (1985)).

1. The feeding ecology of Whiskered Auklet, and of Craveri's, Kittlitz's and Xantus' Murrelets in the eastern North Pacific is virtually unknown. Diets of adult Cassin's and Rhinoceros Auklets and young Ancient Murrelets are poorly known. 2.The feeding behavior and diet of entire communities of alcids needs studying to determine whether food and nest partitioning is a result of interspecific competition or inherent differences in behaviour. 3.The type and size of prey occurring in the diet of adults and nestlings should be correlated with the distribution and abundance of prey found in the sea to determine prey availability and selection. The nutritional

Fig. 9.11. Daily food consumption of breeding populations of fish-eating and plankton-feeding alcids in the eastern North Pacific (based on Nilsson & Nilsson's (1976) method).

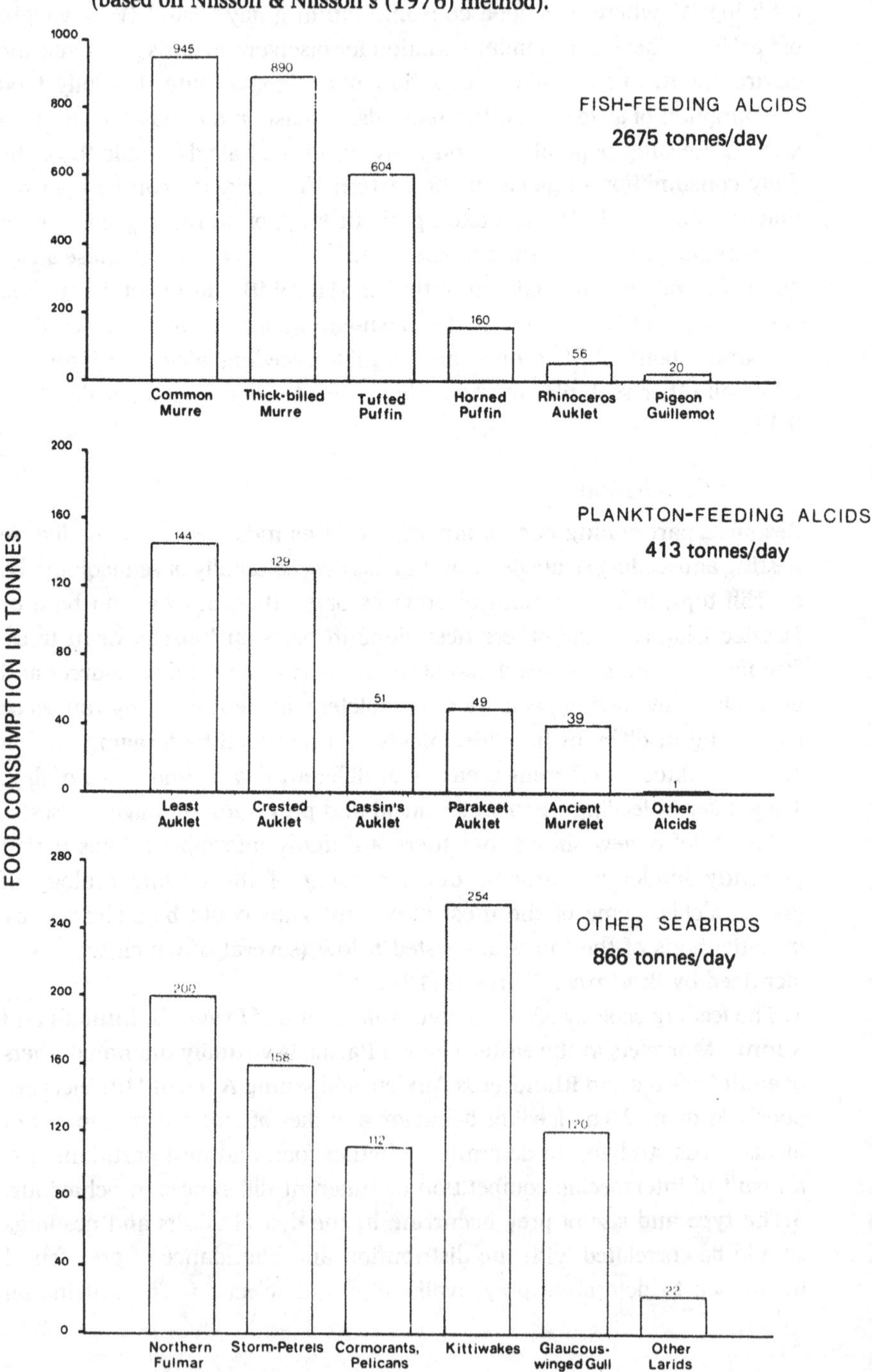

composition of major prey, as well as that of the most abundant non-prey items in the water column, should be compared. 4.Studies of the diving and feeding behavior and energetics of birds are required. Even the depths and duration of dives are poorly known, as is their prey. 5.More data are needed on how diets of adults and chicks compare and on relations between food consumption and metabolic and growth requirements. 6.More information is needed on the life cycles of the main prey species, which should involve cooperative programs with oceanographers and fisheries biologists, in order better to understand alcid–prey interactions.

Acknowledgements The authors thank M.S.W. Bradstreet for providing a copy of an unpublished manuscript on the feeding ecology of Atlantic alcids, K. Devito and L. Rankin for drafting maps on alcid distribution, R.W. Butler and P.G.H. Evans for reviewing a preliminary, and J.P. Croxall the final version of this manuscript.

References

Ainley, D.G. & Sanger, G.A. (1979). Trophic relationships of seabirds in the northeastern Pacific Ocean and Bering Sea. In *Conservation of marine birds of northern North America*, ed. J.C. Bartonek & D.N. Nettleship, pp. 95–122. Washington, D.C.: U.S. Fish & Wildlife Service, Wildlife Research Report 11.

American Ornithologists' Union (1982). Thirty-fourth supplement to the American Ornithologists' Union Check-list of North American birds. *Auk*, **99**, Suppl.

Baird, P.A. & Moe, R.A. (1978). The breeding biology and feeding ecology of marine birds in Sitkalidak Strait area, Kodiak Island, 1977. In *Environmental assessment of the Alaskan Continental Shelf. Annual reports of principal investigators*, vol. 3, pp. 313–524. Boulder, Colorado: NOAA, Environ. Res. Lab.

Baltz, D.M. & Morejohn, G.V. (1977). Food habits and niche overlap of seabirds wintering on Monterey Bay, California. *Auk*, **94**, 526–43.

Bédard, J. (1967). Ecological segregation among plankton-feeding Alcidae (*Aethia* and *Cyclorrhynchus*). Ph.D. thesis, University of British Columbia.

Bédard, J. (1969*a*). Histoire naturelle du Gode, *Alca torda* L., dans le golfe Saint-Laurent, province de Quebec, Canada. *Etude de Service Canadien de la Faune*, no. 7, 79 pp.

Bédard, J. (1969*b*). The nesting of the Crested, Least, and Parakeet Auklets on St. Lawrence Island, Alaska. *Condor*, **71**, 386–98.

Bédard, J. (1969*c*). Adaptive radiation in Alcidae. *Ibis*, **111**, 189–98.

Bédard, J. (1969*d*). Feeding of the Least, Crested, and Parakeet Auklets around St. Lawrence Island, Alaska. *Can. J. Zool.*, **47**, 1025–50.

Belopolskii, I.O. (1957). *Ecology of sea colony birds of the Barents Sea*. (Transl. 1961 by Israel Program for Scientific Translations, Jerusalem.) Springfield, Virginia: Clearinghouse for Federal Scientific and Technical Information.

Bengtson, S.A. (1984). Breeding ecology and extinction of the Great Auk *Pinguinus impennis*: anecdotal evidence and conjectures. *Auk*, **101**, 1–12.

Binford, L.C., Elliott, B.G. & Singer, S.W. (1975). Discovery of a nest and the downy young of the Marbled Murrelet. *Wilson Bull.*, **87**, 305–19.

Birkhead, T.R. (1976). Breeding biology and survival of Guillemots (*Uria aalge*). D.Phil. thesis, Oxford University. 205 pp.

Birkhead, T.R. & Nettleship, D.N. (1982). Studies of alcids breeding at the Gannet Clusters, Labrador. Manuscript report. Canadian Wildlife Service, Dartmouth, Nova Scotia. 144 pp.

Bradstreet, M.S.W. & Brown, R.G.B. (1985). Feeding ecology of the Atlantic Alcidae. In *The Atlantic Alcidae*, ed. D.N. Nettleship & T.R. Birkhead, pp. 263–318. London & New York: Academic Press.

Brinton, E. (1962). The distribution of Pacific euphausiids. *Bull. Scripps Inst. Oceanog. Univ. Calif.*, **8**, 51–270.

Byrd, G.V., Day, R.H. & Knudtson, E.P. (1983). Patterns of colony attendance and censusing of auklets at Buldir Island, Alaska. *Condor*, **85**, 274–80.

Campbell, R.W. (Compiler) (1976). Seabird colonies of Vancouver Island. Map issued by the British Columbia Prov. Mus., Victoria, B.C.

Campbell, R.W., Carter, H.R. & Sealy, S.G. (1979). Nesting of Horned Puffins in British Columbia. *Can. Field-Nat.*, **93**, 84–6.

Campbell, R.W. & Garrioch, H.M. (Compilers) (1979). Seabird colonies of the Queen Charlotte Islands. Map issued by the British Columbia Prov. Mus., Victoria, B.C.

Carter, H.R. (1984). At-sea biology of the Marbled Murrelet (*Brachyramphus marmoratus*) in Barkley Sound, British Columbia. M.Sc. thesis, Univ. Manitoba, Winnipeg.

Cody, M.L. (1973). Coexistence, coevolution and convergent evolution in seabird communities. *Ecology*, **54**, 31–44.

Coues, E. (1868). A monograph of the Alcidae. *Proc. Acad. Nat. Sci. Philad.*, **20**, 2–81.

Day, R.H., Oakley, K.L. & Barnard, D.R. (1983). Nest sites and eggs of Kittlitz's and Marbled Murrelets. *Condor*, **85**, 265–73.

DeWeese, L.R. & Anderson, D.W. (1976). Distribution and breeding biology of Craveri's Murrelet. *Trans. San Diego Soc. Nat. Hist.*, **18**, 155–68.

Drent, R.H. (1965). Breeding biology of the Pigeon Guillemot, *Cepphus columba*. *Ardea*, **53**, 99–160.

Drent, R.H. & Guiguet, C.J. (1961). A catalogue of British Columbia seabird colonies. *Occ. Paper B.C. Prov. Mus.*, no. 12, 1–173.

Drury, W.H. (1981). Ecological studies in the Bering Strait region. In *Environmental assessment of the Alaskan Continental Shelf. Final reports of principal investigators*, vol. 3, pp. 175–87. Boulder, Colorado: NOAA, Environ. Res. Lab.

Eppley, Z.A. (1984). Development of thermoregulatory abilities in Xantus' Murrelet chicks *Synthliboramphus hypoleucus*. *Physiol. Zool.*, **57**, 307–17.

Follett, W.I. & Ainley, D.G. (1976). Fishes collected by Pigeon Guillemots, *Cepphus columba* Pallas, nesting on Southeastern Farallon Island, California. *Calif. Fish & Game*, **62**, 28–31.

Forsell, D.J. & Gould, P.J. (1981). *Distribution and abundance of marine birds and mammals wintering in the Kodiak area of Alaska*. U.S. Dept Interior, U.S. Fish and Wildlife Service. FWS/OBS-81/13, 72 pp.

Fulton, J., Arai, M.N. & Mason, J.C. (1982). Euphausiids, coelenterates, ctenophores, and other zooplankton from the Pacific Coast Ichthyoplankton Survey, 1980. *Can. Tech. Rep. Fish. Aquat. Sci.*, no. 1125, 75 pp.

Gaston, A.J. & Nettleship, D.N. (1981). The Thick-billed Murres of Prince Leopold Island. *Can. Wild. Serv. Monogr. Ser.*, no. 6, 349 pp.

Gould, P.J. (1977). Shipboard surveys of marine birds. In *Environmental assessment of the Alaskan Continental Shelf, Annual reports of principal investigators, vol. 3, pp. 193–284. Boulder, Colorado: NOAA, Environ. Res. Lab.*

Gould, P.J., Forsell, D.J. & Lensink, C.J. (1982). Pelagic distribution and abundance of seabirds in the Gulf of Alaska and eastern Bering Sea. FWS/OBS-82/48. U.S. Fish and Wildl. Ser., Biol. Serv. Prog. Anchorage, Alaska. 294 pp.

Grieve, S. (1885). *The Great Auk or Garefowl Alca impennis, its history, archeology and remains.* Edinburgh: Grange Publishing Works.

Grover, J.J. & Olla, B.L. (1983). The role of the Rhinoceros Auklet (*Cerorhinca monocerata*) in mixed-species feeding assemblages of seabirds in the Strait of Juan de Fuca, Washington. *Auk*, **100**, 979–82.

Guiguet, C.J. (1956). Enigma of the Pacific. *Audubon Mag.*, **58**, 164–7, 174.

Harris, C.K. & Hartt, A.C. (1977). Assessment of pelagic and nearshore fish in three bays on the east and south coast of Kodiak Island, Alaska. *Env. Ass. Alaska. Cont. Shelf Nat. Oc. Atm. Admin./Outer Cont. Shelf Env. Ass. Prog.* Wo. 485. Quart. Rep. Apr.–June, 1, 483–688.

Harrison, C.S. (1979). The association of marine birds and feeding Gray Whales. *Condor*, **81**, 93–5.

Hasegawa, H. (1984). Status and conservation of seabirds in Japan, with special attention to the Short-tailed Albatross. In *The status and conservation of the world's seabirds*, ed. J.P. Croxall, P.G.H. Evans & R.W. Schreiber, pp. 489–500. Cambridge: ICBP.

Hatch, S.A. (1984). Nestling diet and feeding rate of Rhinoceros Auklets in Alaska. In *Marine birds: their feeding ecology and commercial fisheries relationships*, ed. D.N. Nettleship, G.A. Sanger & P.F. Springer, pp. 106–15. Ottawa: Can. Wild. Serv. Spec. Publ.

Hatch, S.A., Nysewander, D.R., DeGange, A.R., Peterson, M.R., Baird, P.A., Wohl, K.D. & Lensink, C.J. (1978). Population dynamics and trophic relationships of marine birds in the Gulf of Alaska and Southern Bering Sea. In *Environmental assessment of the Alaskan Continental Shelf, Annual reports of principal investigators*, vol. 3, pp. 1–908. Boulder, Colorado: NOAA, Environ. Res. Lab.

Hatler, D.F., Campbell, R.W. & Dorst, A. (1978). Birds of the Pacific Rim National Park. *Occ. Pap. B.C. Prov. Mus.*, **20**, 192 pp.

Hunt, G.L., Jr., Burgeson, B. & Sanger, G.A. (1981*a*). Feeding ecology of seabirds in the eastern Bering Sea. In *The eastern Bering Sea shelf: oceanography and resources*, vol. 2, ed. D.W. Hood & J.A. Calder, pp. 629–48. Seattle: Univ. of Washington Press.

Hunt, G.L., Jr, Eppley, Z., Burgeson, B. & Squibb, R. (1981*b*). Reproductive ecology, foods, and foraging areas of seabirds nesting on the Pribilof Islands, 1975–1979. In *Environmental assessment of the Alaskan Continental Shelf, Final reports of principal investigators*, vol. 12, pp. 1–257. Washington, D.C.: NOAA.

Johnson, S.R. & West, G.C. (1975). Growth and development of heat regulation in nestlings, and metabolism of adult Common and Thick-billed Murres. *Ornis Scand.*, **6**, 109–15.

Johnston, S. & Carter, H.R. (1985). Cavity-nesting Marbled Murrelets. *Wilson Bull.* **97**, 1–3.

Kartaschew, N.N. (1960). *Die Alkenvögel des Nordatlantiks.* Wittenberg Lutherstad: Die Neue Brehm-Bucherei.
Kendeigh, S.C. (1970). Energy requirements for existence in relation to size of bird. *Condor*, **72**, 60–5.
Kiff, L. (1981). Eggs of the Marbled Murrelet. *Wilson Bull.*, **93**, 400–3.
Knudtson, E.P. & Byrd, G.V. (1982). Breeding biology of Crested, Least, and Whiskered Auklets on Buldir Island, Alaska. *Condor*, **84**, 197–202.
Kozlova, E.V. (1957). Charadriiformes, Suborder Alcae. In *Fauna of USSR: birds*, vol. 3, no. 2, 140 pp. (Transl. 1961 by Israel Program for Scientific Translations, Jerusalem.) Springfield, Virginia: Clearinghouse for Federal Scientific and Technical Information.
Krasnow, L.D. & Sanger, G.A. (1986). Feeding ecology of marine birds in the near-shore waters of Kodiak Island. In *Environmental assessment of the Alaskan continental shelf. Final reports of the principal investigators*, vol. 45, pp. 505–630, Anchorage, Alaska: NOAA, National Ocean Service.
Kuletz, K.J. (1983). Mechanisms and consequences of foraging behavior in a population of breeding Pigeon Guillemots. M.Sc. thesis, University of California, Irvine. 79 pp.
Leschner, L.L. (1976). The breeding biology of the Rhinoceros Auklet on Destruction Island. M.Sc. thesis, Univ. Washington, Seattle. 77 pp.
Mahoney, S.P. (1979). Breeding biology and behaviour of the Common Murre (*Uria aalge aalge. Pont.*) on Gull Island, Newfoundland. M.Sc. thesis, Memorial University, Newfoundland.
Manuwal, D.A. (1974). The natural history of Cassin's Auklet *Ptychoramphus aleuticus*. *Condor*, **76**, 421–31.
Manuwal, D.A. & Boersma, D. (1977). Dynamics of marine bird populations in the Barren Islands, Alaska. In *Environmental assessment of the Alaskan Continental Shelf. Annual reports of principal investigators*, vol. 4, pp. 294–420. Boulder, Colorado: NOAA, Environ. Res. Lab.
Manuwal, D.A., Wahl, T.R. & Speich, S.N. (1979). THe seasonal distribution and abundance of marine bird populations in the Strait of Juan de Fuca and Northern Puget Sound in 1978. *NOAA Tech. Mem.* ERL MESA-44. Boulder, Colorado: NOAA. 391 pp.
Marsault, B.M. (1975). Auks breeding in captivity. *Bird Study*, **22**, 44–6.
Moe, R.A. & Baird, P.A. (1978). Some notes on the feeding ecology of the Tufted Puffin *Lunda cirrhata* in the Sitkalidak Strait Region of Kodiak Island, Alaska, during the 1977 breeding season. *Pacific Seabird Group Bull.*, **5**, 42.
Moe, R.A. & Day, R.H. (1979). Populations and ecology of seabirds of the Koniuji Group, Sumagin Islands, Alaska. In *Environmental assessment of the Alaskan Continental Shelf. Annual reports of principal investigators*, vol. 6, pp. 395–491. Boulder, Colorado: NOAA, Environ. Res. Lab.
Murray, K.G., Winnett-Murray, K., Eppley, Z.A., Hunt, G.L., Jr & Schwartz, D.B. (1983). Breeding biology of the Xantus' Murrelet, *Condor*, **85**, 12–21.
Nettleship, D.N. (1972). Breeding success of the Common Puffin *Fratercula arctica* on different habitats at Great Island. Newfoundland. *Ecol. Monogr.*, **42**, 239–68.
Newton, A. (1861). Abstract of Mr J. Wolley's researches in Iceland respecting the Garefowl or Great Auk *Alca impennis*, Linn. *Ibis*, **3**, 374–99.
Nilsson, S.G. & Nilsson, I.N. (1976). Numbers, food consumption, and fish predation by birds in Lake Möckeln, southern Sweden. *Ornis. Scand.*, *7*, 61–70.
Norderhaug, M. (1970). The role of the Little Auk *Plautus alle* (L.) in arctic

ecosystems. In *Antarctic ecology*, vol. 1, ed. M.W. Holdgate, pp. 558–60. London: Academic Press.

Norderhaug, M. (1980). Breeding biology of the Little Auk *Plautus alle* in Svalbard. *Norsk Polarinstitutt Skrifter*, no. 173, 1–45.

Ogi, H. (1980). The pelagic feeding ecology of Thick-billed Murres in the North Pacific, March–June. *Bull. Fac. Fish., Hokkaido Univ.*, **31**, 50–72.

Ogi, H. & Tsujita, T. (1973). Preliminary examination of stomach contents of murres (*Uria* spp.) from the eastern Bering Sea and Bristol Bay, June–August, 1970 and 1971. *Jap. J. Ecol.*, **23**, 201–9.

Olson, S.L., Swift, C.C. & Mokhiber, C. (1979). An attempt to determine the prey of the Great Auk *Pinguinus impennis*. *Auk*, **96**, 790–2.

Piatt, J.F. & Nettleship, D.N. (1985). Diving depths of four alcids. *Auk*, **102**, 293–7.

Piatt, J.F., Nettleship, D.N. & Threlfall, W. (1984). Mortality of Common Murres and Atlantic Puffins in Newfoundland, 1951–1981. In *Marine birds: their feeding ecology and commercial fisheries relationships*, ed. D.N. Nettleship, pp. 196–206. Ottawa: Can. Wildl. Serv. Spec. Publ.

Preble, E.A. & McAtee, W.L. (1923). A biological survey of the Pribilof Islands, Alaska. *N. Am. Fauna*, no. 46.

Richardson, F. (1961). Breeding biology of the Rhinoceros Auklet on Protection Island, Washington. *Condor*, **63**, 456–73.

Robertson, I. (1972). Studies on fish-eating birds and their influence on stocks of Pacific Herring, *Clupea harengus*, in the Gulf Islands of British Columbia. *Herring investigations, Pac. Biol. Stn Nanaimo, B.C. Unpubl. Rep.* 28 pp.

Sanford, R.C. & Harris, S.W. (1967). Feeding behavior and food consumption rates of a captive California Murre. *Condor*, **69**, 298–302.

Sanger, G.A. (1975). Observations on the pelagic biology of the Tufted Puffin. *Pacific Seabird Gr. Bull.*, **2**, 30–1.

Sanger, G.A. (1986). Diets and food web relationships of seabirds in the Gulf of Alaska and adjacent marine regions. In *Environmental assessment of the Alaskan continental shelf. Final reports of the principal investigators*, vol. 45, pp. 631–711, Anchorage, Alaska: NOAA, National Ocean Service.

Sanger, G.A., Jones, R.D., Jr & Wiswar, D.W. (1979). The winter feeding habits of selected species of marine birds in Kachemak Bay, Alaska. In *Environmental assessment of the Alaskan Continental Shelf, Annual reports of principal investigators*, pp. 309–47. Boulder, Colorado: NOAA, Environ. Res. Lab.

Scott, J.M. (1973). Resource allocation in four syntopic species of marine diving birds. Ph.D. thesis, Oregon State University, Corvallis.

Scott, J.M., Hoffman, W., Ainley, D. & Zeillemaker, C.F. (1974). Range expansion and activity patterns in Rhinoceros Auklets. *West. Birds*, **5**, 13–20.

Sealy, S.G. (1972). Adaptive differences in breeding biology in the marine bird family Alcidae. Ph.D. thesis, University of Michigan, Ann Arbor.

Sealy, S.G. (1973*a*). Breeding biology of the Horned Puffin on St. Lawrence Island, Bering Sea, with zoogeographical notes on the North Pacific puffins. *Pacific. Sci.*, **27**, 99–119.

Sealy, S.G. (1973*b*). Adaptive significance of post-hatching developmental patterns in the Alcidae. *Ornis Scand.*, **4**, 113–21.

Sealy, S.G. (1975*a*). Feeding ecology of the Ancient and Marbled Murrelets near Langara Island, British Columbia. *Can. J. Zool.*, **53**, 418–33.

Sealy, S.G. (1975*b*). Aspects of the breeding biology of the Marbled Murrelet in British Columbia. *Bird-Banding*, **46**, 141–54.

Sealy, S.G. (1976). Biology of nesting Ancient Murrelets. *Condor*, **78**, 294–306.

Sealy, S.G. & Bédard, J. (1973). Breeding biology of the Parakeet Auklet *Cyclorrhynchus psittacula* on St. Lawrence Island, Alaska. *Astarte*, **6**, 59–68.
Sealy, S.G. & Campbell, R.W. (1979). Post-hatching movements of young Ancient Murrelets. *West. Birds*, **10**, 25–30.
Sealy, S.G., Carter, H.R. & Alison, D. (1982). Occurrences of the Asiatic Marbled Murrelet *Brachyramphus marmoratus perdix* (Pallas) in North America. *Auk*, **99**, 778–81.
Searing, G. (1977). Some aspects of cliff-nesting sea-birds at Kongkok Bay, St. Lawrence Island, Alaska, during 1976. In *Environmental assessment of the Alaskan Continental Shelf, Annual reports of principal investigators*, vol. 5, pp. 263–412. Boulder, Colorado: OCSEAP, NOAA/BLM.
Shepard, M.G. (1977). Pelagic birding trips – fall 1977. *Discovery*, **6**, 90–1.
Shuntov, V.P. (1974). *Seabirds and the biological structure of the ocean.* (Transl. from Russian.) Bur. Sport Fish. Wildl. and Natl. Sci. Found., Washington, D.C. Nat. Tech. Inf. Serv. TT 74-55032. 566 pp.
Simons, T.R. (1980). Discovery of a ground-nesting Marbled Murrelet. *Condor*, **82**, 1–9.
Simons, T.R. (1981). Feeding behaviour of captive alcids. *U.S. Dept Int. Nat. Park Serv. Mauii, Hawaii. Unpubl. Rep.* 8 pp.
Sowls, A.L., Degange, A.R., Nelson, J.W. & Lester, G.S. (1980). *Catalog of California Seabird Colonies.* U.S. Dept Int. Biol. Serv. Prog. FWS/OBS-80/37, 371 pp.
Sowls, A.L., Hatch, S.A. & Lensink, C.J. (1978). *Catalog of Alaskan seabird colonies.* Washington, D.C.: U.S. Fish and Wildlife Service.
Springer, A.M. & Roseneau, D.G. (1985). Copepod-based food webs: auklets and oceanography in the Bering Sea. *Mar. Ecol.-Prog. Ser.*, **21**, 229–37.
Storer, R.W. (1945). Structural modifications in the hind limb in the Alcidae. *Ibis*, **87**, 433–56.
Swartz, L.G. (1966). Sea-cliff birds. In *Environment of the Cape Thompson Region, Alaska*, ed. N.J. Wilimovsky & J.N. Wolfe, pp. 611–78. Oak Ridge, Tennessee: U.S. Atomic Energy Commission.
Swennen, C. & Duiven, P. (1977). Size of food objects of three fish-eating seabird species: *Uria aalge, Alca torda* and *Fratercula arctica* (Aves, Alcidae). *Neth. J. Sea Res.*, **11**, 92–8.
Thoresen, A.C. (1964). The breeding behavior of the Cassin Auklet. *Condor*, **66**, 456–76.
Tuck, L.M. (1961). The Murres. Their distribution, populations and biology. *Can. Wildl. Ser.*, no. 1. Ottawa.
Tuck, L.M. & Squires, H.J. (1955). Food and feeding habits of Brunnich's Murre *Uria lomvia lomvia* on Akpatok Island. *J. Fish. Res. Board Can.*, **12**, 781–92.
Udvardy, M.D.F. (1963). Zoogeographical study of the Pacific Alcidae. In *Pacific basin biogeography*, ed. J.L. Gressitt, pp. 85–111. Honolulu: B.P. Bishop Museum.
Uspenski, S.M. (1956). *The bird bazaars of Novaya Zemlya.* USSR Academy of Sciences popular series, Moscow (Translated by Canada Department of Northern Affairs and Natural Resources, 1958).
Vallee, A. & Cannings, R.J. (1983). Nesting of the Thick-billed Murre, *Uria lomvia*, in British Columbia. *Can. Field-Nat.*, **97**, 450–1.
Varoujean, D.H. (1979). *Seabird colony catalogue. Washington, Oregon and California.* U.S. Fish and Wildl. Rep. 456 pp.
Vermeer, K. (1979). Nesting requirements, food and breeding distribution of

Rhinoceros Auklets, *Cerorhinca monocerata*, and Tufted Puffins, *Lunda cirrhata*. *Ardea*, **67**, 101–10.

Vermeer, K. (1980). The importance of timing and type of prey to reproductive success of Rhinoceros Auklets *Cerorhinca monocerata*. *Ibis*, **122**, 343–50.

Vermeer, K. (1981). The importance of plankton to breeding Cassin's Auklets. *J. Plankton Res.*, **3**, 315–29.

Vermeer, K. (1983). Marine bird populations in the Strait of Georgia; comparison with the west coast of Vancouver Island. *Can. Tech. Rep. Hydr. Ocean Sci.*, no. 19, 18 pp.

Vermeer, K. (1984). The diet and food consumption of nestling Cassin's Auklets during summer, and a comparison with other plankton-feeding alcids. *Murrelet*, **65**, 65–77.

Vermeer, K. & Cullen, L. (1979). Growth of Rhinoceros Auklets and Tufted Puffins, Triangle Island, British Columbia. *Ardea*, **67**, 22–7.

Vermeer, K. & Cullen, L. (1982), Growth comparison of a plankton- and fish-feeding alcid. *Murrelet*, **63**, 34–9.

Vermeer, K., Cullen, L. & Porter, M. (1979*a*). A provisional explanation of the reproductive failure of Tufted Puffins *Lunda cirrhata* on Triangle Island, British Columbia. *Ibis*, **121**, 348–54.

Vermeer, K. & Devito, K. (In press). Size, caloric content and association of prey fishes in meals of nestling Rhinoceros Auklets, *Murrelet*.

Vermeer, K., Fulton, J.B. & Sealy, S.G. (1985). Differential use of zooplankton by Ancient Murrelets and Cassin's Auklets in the Queen Charlotte Islands. *J. Plankton Res.*, **7**, 443–59.

Vermeer, K., Manuwal, D.A. & Bingham, D.S. (1976). Seabirds and pinnipeds of Sartine Island, Scott Island Group, British Columbia. *Murrelet*, **57**, 14–16.

Vermeer, K., Robertson, I., Campbell, R.W., Kaiser, G. & Lemon, M. (1983). *Distribution and densities of marine birds on the Canadian west coast*. Vancouver, B.C.: Can. Wildl. Serv. Rep. 73 pp.

Vermeer, K., Sealy, S.G., Lemon, M. & Rodway, M. (1984). Predation and potential environmental perturbances on Ancient Murrelets nesting in British Columbia. In *Status and conservation of the world's seabirds*, ed. J.P. Croxall, P. Evans, & R.W. Schreiber, pp. 757–70. Cambridge: ICBP.

Vermeer, K., Vermeer, R.A., Summers, K.R. & Billings, R.R. (1979*b*). Numbers and habitat selection of Cassin's Auklet breeding on Triangle Island, British Columbia. *Auk*, **96**, 143–51.

Vermeer, K. & Westrheim, S.J. (1984). Fish changes in diets of nestling Rhinoceros Auklets and their implications. In *Marine birds: their feeding ecology and commercial fisheries relationships*, ed. D.N. Nettleship, G.A. Sanger & P.F. Springer, pp. 96–105. Ottawa: Can. Wildl. Serv. Spec. Publ.

Wehle, D.H.S. (1980). The breeding biology of the puffins: Tufted Puffin *Lunda cirrhata*, Horned Puffin *Fratercula corniculata*, Common Puffin *F. arctica*, and Rhinoceros Auklet *Cerorhinca monocerata*. Ph.D. thesis, University of Alaska, Fairbanks.

Wehle, D.H.S. (1982). Food of adult and subadult Tufted and Horned Puffins. *Murrelet*, **63**, 51–8.

Wehle, D.H.S. (1983). The food, feeding, and development of young Tufted and Horned Puffins in Alaska. *Condor*, **85**, 427–42.

Wilson, U.W. (1977). A study of the biology of the Rhinoceros Auklet on Protection Island, Washington. M.S. thesis, University of Washington, Seattle.

10

Trophic levels and trophic relationships of seabirds in the Gulf of Alaska

GERALD A. SANGER
U.S. Fish and Wildlife Service, Alaska Fish and Wildlife Research Center,
1101 East Tudor Road, Anchorage, Alaska 99503, USA.

Introduction

Recently, studies of the feeding ecology of seabirds in the sub-Arctic North Pacific and adjacent seas (e.g. Bedard, 1969; Ogi & Tsujita, 1973; Sealy, 1975; Ainley & Sanger, 1979; Ogi, Kudobera & Nakamura, 1980) have increased dramatically, along with those of other aspects of their biology (e.g. Sanger, 1972; Shuntov, 1972; Bartonek & Nettleship, 1979). This trend has been particularly evident for waters off the coasts of Alaska, where investigators associated with the interdisciplinary Alaskan Outer Continental Shelf Environmental Assessment Program (OCSEAP) of the United States National Oceanic and Atmospheric Administration (NOAA) and the Mineral Management Service have begun to produce a profusion of reports on the enormous seabird resources in Alaskan waters (e.g. Sowls, Hatch & Lensink, 1978; Forsell & Gould, 1981; Hunt *et al.*, 1981*a,b,c*; Gould, Forsell & Lensink, 1982; Hatch, 1984; Baird & Gould, 1986; Sanger, 1986).

Despite this flurry of activity, however, there has been little attempt to relate seabirds more directly to the marine ecosystem. As main objectives, this paper aims to show more clearly how seabirds may relate to lower trophic levels of the ecosystem of the Gulf of Alaska, and how related species compare trophically.

The general approach of using a system of prey trophic levels better to understand the trophic relationships between seabirds and their prey is not new (e.g. Knox, 1970; Sanger, 1972; Ainley & Sanger, 1979; Sanger & Jones, 1984); the concept of numerical average trophic levels for marine predators that feed at more than one level was recently introduced (Mearns *et al.*, 1981). This paper attempts to bring these two concepts together for a specific community of seabirds.

The Gulf of Alaska and its seabird community

The Gulf of Alaska (Fig. 10.1) is located in the northeastern corner of the North Pacific Ocean. Its geographic limits are considered to extend from the south central coast of Alaska south to 52° N, west to 165° W, and east to 132° W (Gould *et al.*, 1982). The city of Anchorage (pop. 250 000) lies at the head of Cook Inlet, a 300 km fjord, and there are several smaller, isolated towns along the coast. The region remains essentially pristine in character, however. Great-circle shipping lanes between the Pacific coast of North America and the Orient traverse oceanic areas of the Gulf. The continental shelf supports a rapidly growing international trawl fishery. Major oil wells have existed in Cook Inlet since the 1950s, and the outer continental shelf has recently been the site of a moderate amount of petroleum exploration, although there have been no major finds outside Cook Inlet to date.

Water temperatures in the surface layer of the Gulf of Alaska are mild for these latitudes, generally averaging about 12–14° C by late summer (Favorite, Dodimead & Nasu, 1976; Xiong & Royer, 1984). Sea ice forms only at the heads of the most protected bays and inlets in winter. Water circulation over the continental shelf occurs in three regimes (Royer, 1981*a*) – a coastal jet, a mid-shelf doldrum, and the Alaska Current at the shelf break. Seaward of the continental shelf, water circulates in the counterclockwise Alaskan Gyre (Royer, 1981*b*), which originates with the eastward-flowing, trans-Pacific Sub-Arctic Current (Favorite *et al.*, 1976).

Fig. 10.1. The Gulf of Alaska, eastern North Pacific Ocean.

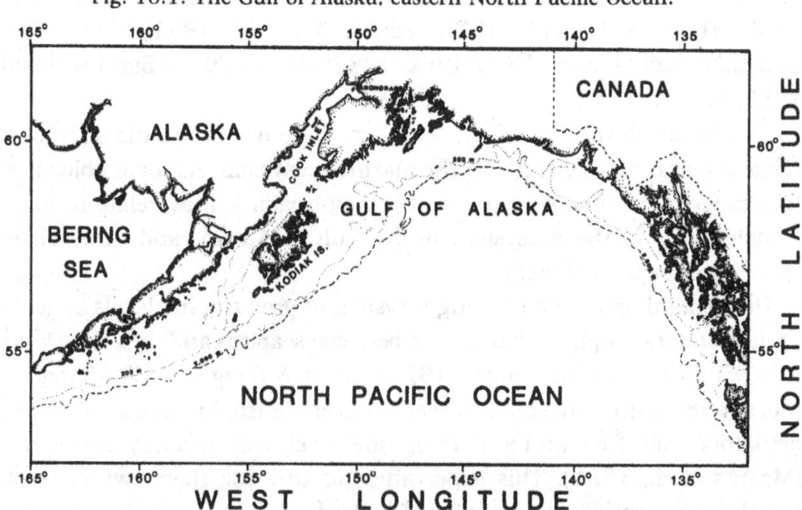

Large populations of zooplankton (Dunn, Kendall & Wolotira, 1981; Rogers *et al.*, 1983), fish (Macy *et al.*, 1978), seabirds (Sowls *et al.*, 1978; Gould *et al.*, 1982), and marine mammals (Calkins & Pitcher, 1983; Pitcher & Calkins, 1983; Rice & Wolman, 1983; Kajimura, 1984) attest to a high biological productivity, but very little is known about the primary productivity that drives the system.

Sowls *et al.* (1978) documented the locations and species composition and estimated population sizes of seabirds at breeding colonies in the Gulf. Numbers are dominated by Northern Fulmars *Fulmarus glacialis*, Fork-tailed Storm Petrels *Oceanodroma furcata*, Black-legged Kittiwakes *Rissa tridactyla*, and Tufted Puffins *Fratercula cirrhata*. During extensive pelagic surveys from 1975 to 1978, mostly over the shelf, Gould *et al.* (1982) recorded 65 species of marine birds, including loons Gaviidae, grebes Podicipedidae and sea ducks. In summer, numbers were dominated by Sooty and Short-tailed Shearwaters *Puffinus griseus* and *P. tenuirostris*, visitors from the southern hemisphere. The most commonly seen Alaskan breeders were Northern Fulmars, Fork-tailed Storm Petrels, Black-legged Kittiwakes, murres *Uria* spp., and Tufted Puffins. Diets and food web relationships of 39 species were described (Sanger, 1986), and the breeding biology of the most abundant species was documented (Baird & Gould, 1986).

Methods

This account deals with the 19 species of seabirds for which dietary information is the most complete. They include the eight species listed above and 11 others: Pelagic Cormorant *Phalacrocorax pelagicus*, Glaucous-winged Gull *Larus glaucescens*, Arctic and Aleutian Terns *Sterna paradisea* and *S. aleutica*, Pigeon Guillemot *Cepphus columba*, Marbled, Kittlitz's, and Ancient Murrelets *Brachyramphus marmoratus*, *B. brevirostris* and *Synthliboramphus antiquus*, Cassin's and Rhinoceros Auklets *Ptychoramphus aleuticus* and *Cerorhincha monocerata*, and Horned Puffin *Fratercula corniculata*.

Food samples from birds collected at sea during the whole season, and from nestlings of a few species on breeding colonies, came from three main areas or periods: (1) from the Kodiak Island area (Fig. 10.1) in spring and summer of 1977 and 1978 and during the intervening winter (Krasnow & Sanger, 1986); (2) from birds collected at sea during OCSEAP cruises from 1975 through 1978; and (3) from birds drowned in research salmon gillnets of the National Marine Fisheries Service (NMFS) from 1969 to 1971, and from NMFS marine mammal research cruises in 1973 and 1974. In total, about 93% of the samples were from the Gulf of Alaska

(74% from the Kodiak Archipelago), 6% from the Bering Sea, and 1% from south of the Aleutian Islands. Seasonally, about 71% were sampled between June and August, 10% between November and March, 10% in April and May, and 9% in September and October. Thus, data are largely from the Kodiak Island area during the summer. Cases of possible geographic or seasonal bias to the general conclusions are noted in the results.

Except for netted birds, birds were collected at sea by shotgun. Usually within 5 minutes of collection, specimens were weighed (to the nearest g) with a spring (Pesola) balance and their stomachs were injected with 10% buffered formalin to stop digestion. Standard measurements were recorded, age and sex was detemined, and the digestive tract was removed and preserved in formalin. Some specimens were frozen intact in the ship's freezer and processed at a later date. Ashore in the laboratory, prey items were counted and identified to the lowest possible taxon, and the volume of each was estimated visually as a percentage of the total.

Data analysis

Diets of prey organisms were determined from the literature as far as possible, and they were assigned trophic levels (TLs) after Mearns *et al.* (1981). It was often necessary, however, to make assumptions about species composition and TLs in a prey's diet. For example, euphausiids are known to eat phytoplankton, small copepods and other microzooplankton (Mauchline, 1980; Parsons & LeBrasseur, 1970). By assuming that the euphausiids' diet is comprised of equal volumes of diatoms (TL 1) and herbivores (TL 2), the average TL of their diet is 1.5.

Diets of prey were generally assumed to be comprised of equal portions of their prey organisms (invariably in two TLs). An exception was for Capelin *Mallotus villosus*, which were shown by Vesin, Leggett & Able (1981) to eat varying proportions of two of three main kinds of foods (diatoms, small copepods, and euphausiids) as they grew through three size classes. Weighted average trophic levels for each size class (< 80 mm, 80–140 mm, > 140 mm) were calculated after Mearns *et al.* (1981), as follows:

$$\mathrm{ATL} = (\mathrm{TL}_{\mathrm{p1}} \cdot \%V_{\mathrm{p1}}) + (\mathrm{TL}_{\mathrm{p2}} \cdot \%V_{\mathrm{p2}})$$

where ATL = weighted average trophic level, $\mathrm{TL}_{\mathrm{p}n}$ = trophic level of prey n, and $\%V_{\mathrm{p}n}$ = % volume of prey n in diet.

Generalized diets for each bird species (Sanger, 1986), based on data pooled from all regions, seasons and years, were expressed in aggregate percentage volume (%*V*; cf. Martin, Gensch & Brown, 1946; Swanson *et al.*, 1974), aggregate percent numbers (%*N*), percent frequency of occurrence (%*F*), and an Index of Relative Importance (IRI, after Pinkas, Oliphant & Iverson (1971)). By combining the other three parameters, the IRI attempts to overcome the shortcomings of using any of them alone as representing a predator's diet. The IRI is calculated as follows:

$$\text{IRI} = \%F\,(\%V + \%N)$$

The IRI itself has shortcomings, but it is a useful tool with which to evaluate the importance of a prey species to a predator, and to compare diets among predators.

Weighted average TLs of the birds were calculated similarly by adding the products of trophic levels times the percentage of total IRI (Sanger, 1986) for each prey of a bird species, for however many additional TLs existed in the diet. Average TLs of each prey were usually fractional, resulting in as many as five fractional TLs in the diets of some bird species. The bird's average TL was considered to be one higher than the average of its prey (Mearns *et al.*, 1981).

Insight into the nature of trophic relationships between bird species was gained by examining the influence of the birds' feeding behaviour and foraging habitats. The average distributions of the 19 species of seabirds across each of the four general habitats (bay and fjord, continental shelf, shelf break, and oceanic) were adapted from Gould *et al.* (1982).

Results

Assignments of trophic levels to prey

A nominal food chain of five TLs (Table 10.1) was used to calculate or assign TLs to prey organisms of the birds (Table 10.2). The birds ate many

Table 10.1. *Nominal food chain for prey of seabirds in coastal and shelf waters of the Gulf of Alaska*

Trophic level	Trophic order	Representative organisms
1	Primary producers	Nanoplankton, diatoms
2	Herbivores	Calanoid copepods
3	Primary carnivores	Sandlance
4	Secondary carnivores	Adult Walleye Pollock
5	Third-order Carnivores	Squid (adults)

Table 10.2. *Assignments of trophic levels to prey of seabirds in the Gulf of Alaska*

Prey	Assigned or calculated trophic level	Primary food(s)	Reference(s)
Nereid polychaetes	3·0	small fauna	Implied, Hedgpeth *et al.*, 1968
Squid (larval)	3·0	larval copepods & euphausiids	Nesis, 1965
Squid (juvenile)	4·0	adult copepods; amphipods chaetognaths & larval squid	Nesis, 1965; Naito *et al.*, 1977
Gastropods	2·5	algae, periphyton, clams	Meglitsch, 1972
Clams	2·0	phytoplankton, detritus	Meglitsch, 1972
Isopods	2·5	detritus, periphyton	Implied, Kozloff, 1973
Copepods	2·0	phytoplankton	Meglitsch, 1972
Amphipods (gammarid)	2·5	detritus, phytoplankton carrion	Meglitsch, 1972
Amphipods (hyperiid)	3·0	phytoplankton, copepods, amphipods	Dunbar, 1946; Wing, 1976
Mysids	2·5	detritus, phytoplankton copepods	Mauchline, 1980
Euphausiids	2·5	phytoplankton, microzooplankton, copepods	Parsons & LeBrasseur, 1970; Mauchline, 1980

Crabs	3·0	herbivores, carrion	Meglitsch, 1972
Shrimp	3·0	copepods, etc.	Omori, 1974
Insects	2·0	kelp detritus	Implied
Salmonidae (juvenile)	3·2	euphausiids, copepods	Feller & Kaczynski, 1975
Pacific Herring	3·2	copepods, euphausiids	Wespestad & Barton, 1981
Capelin (post-larval)	2·8	diatoms (20%) copepods (80%)	Vesin *et al.*, 1981
Capelin (juveniles)	3·1	copepods (75%) euphausiids (25%)	Vesin *et al.*, 1981
Capelin (adults)	3·2	euphausiids (60%) copepods (40%)	Vesin *et al.*, 1981
Walleye Pollock (juvenile)	3·5	euphausiids, copepods	Smith, 1981
Gadids (other)	3·5	shrimp, gammarid amphipods	Rogers *et al.*, 1983
Pacific Sandfish	3·0	copepods, gammarid amphipods	Rogers *et al.*, 1983
Hexagrammidae	3·5	crabs, shrimp, gammarid amphipods	Rogers *et al.*, 1983
Steichaeidae	3·0	copepods, gammarid amphipods?	Rogers *et al.*, 1983
Pacific Sandlance	3·0	copepods, barnacle larvae, gammarid amphipods	Rogers *et al.*, 1983
Cottidae	3·5	smaller fish	Rogers *et al.*, 1983
Other fish	3·0[a]	—	—
	3·2[b]	—	—

[a] When mostly capelin and sandlance in diet.
[b] When other fish dominant, or for birds that weigh more than 100 g.

additional kinds of prey (Sanger, 1986), but Table 10.2 includes only those comprising at least 0.1% of the total IRI of any one bird species. I used references on prey diets from the Gulf of Alaska as much as possible (e.g. Wing, 1976; Rogers *et al.*, 1983), but I often had to rely on studies from elsewhere in the sub-Arctic Pacific (e.g. Omori, 1974; Feller & Kaczynski, 1975), from other oceans (Vesin *et al.*, 1981), or a general account of invertebrates (Meglitsch, 1972).

Oceanic waters of the sub-Arctic Pacific are likely to have one more TL than coastal areas (Parsons & LeBrasseur, 1970). Nanoplankton (phytoplankton less than 20 μm) and radiolarians are, respectively, the main primary producers and herbivores in oceanic waters. Euphausiids eat radiolarians in oceanic areas, and thus function as primary carnivores. Along the coast, euphausiids eat microphytoplankton directly (Parsons & LeBrasseur, 1970; see also Mauchline, 1980), and thus function as herbivores. I assume that euphausiid diets are made up of equal parts of phytoplankton and microzooplankton (Table 10.2), and that euphausiids thus have a TL of 2.5.

According to Nesis (1965; cited by Clarke, 1966) the smallest individuals (presumably larvae) of *Gonatus fabricii*, a common boreal squid, had eaten larval euphausiids and copepods, and larger juveniles had eaten adult copepods, amphipods and chaetognaths. Naito, Murakami & Kobayashi (1977) state that the common sub-Arctic Pacific squid *Berryteuthis magister* eats planktonic crustaceans when juvenile, but switches to fish and squid by the time it is adult. I thus assigned a TL of 3 to larval squid, and 4 to juveniles (Table 10.2). It seems unlikely that any of the bird species considered here eat adults of the largest species of squid.

The diets of the two main fish prey, Pacific Sandlance *Ammodytes hexapterus* and Capelin *Mallotus villosus*, are fairly well known. With a diet mostly of calanoid copepods, Sandlance should have a TL of 3.0. Trophic levels of three size classes of Capelin were calculated from Vesin *et al.* (1981) as noted above, and were found to be 2.8, 3.1, and 3.2, respectively, for fish < 80 mm, 80–140 mm, and > 140 mm.

Procellariiforms In the diet of 43 Northern Fulmars, unidentified squid (presumably juveniles) made up 95% of the total IRI, and six other kinds of prey in small amounts completed their diet (Fig. 10.2). Prey TLs ranged from 2 for copepods to 4 for squid, and averaged 4.0 for all prey, the highest among the 19 bird species studied.

Unidentified squid (presumably larval) accounted for 60% of the total IRI of eight Fork-tailed Storm Petrels (Fig. 10.2). Four other kinds of prey

Fig. 10.2. Composition, percentage of total Index of Relative Importance, and the trophic level spectra of prey in the diets of Northern Fulmars, Fork-tailed Storm Petrels, and Sooty and Short-tailed Shearwaters. The large triangle indicates the average trophic level of prey.

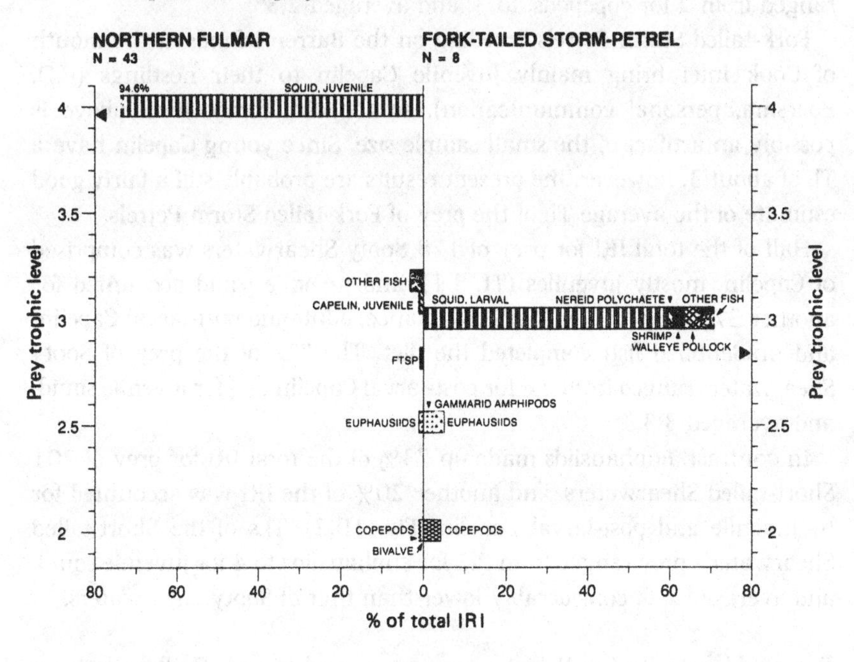

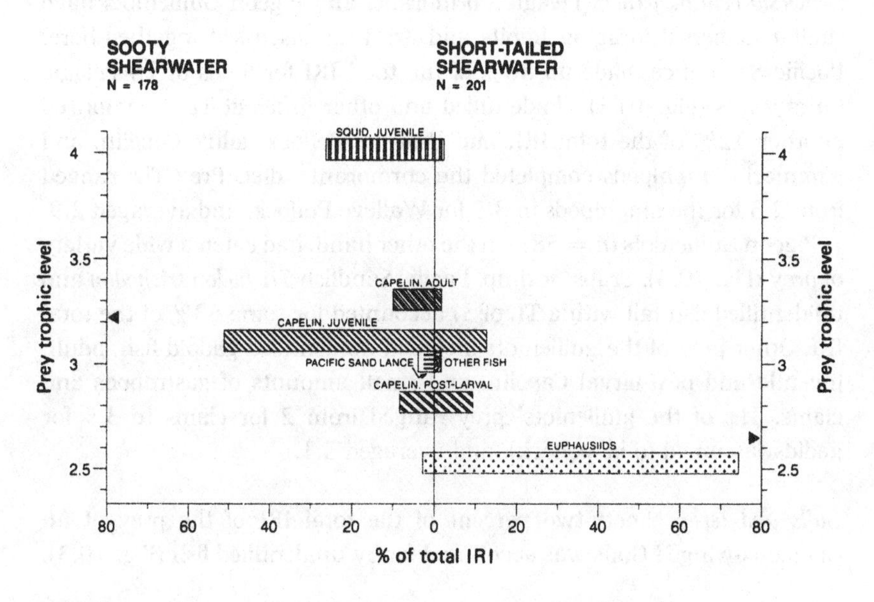

with a TL of 3 (polychaetes, shrimp, juvenile Walleye Pollock *Theragra chalcogramma*, and unidentified fish) contributed an additional 10% to the IRI, while gammarids and euphausiids together comprised another 7%, and copepods accounted for 5% (Fig. 10.2). TLs of the Storm Petrel's prey ranged from 2 for copepods, to 3, and averaged 2.8.

Fork-tailed Storm Petrels breeding on the Barren Islands at the mouth of Cook Inlet bring mainly juvenile Capelin to their nestlings (P.D. Boersma, personal communication), so the diet composition above is possibly an artifact of the small sample size. Since young Capelin have a TL of about 3, however, the present results are probably still a fairly good estimate of the average TL of the prey of Fork-tailed Storm Petrels.

Half of the total IRI for prey of 178 Sooty Shearwaters was comprised of Capelin, mostly juveniles (TL 3.1), and juvenile squid accounted for another 27% (Fig. 10.2). Pacific Sandlance, adult and post-larval Capelin, and unidentified fish completed the diet. The TLs of the prey of Sooty Shearwaters ranged from 2.8 for post-larval Capelin to 4 for juvenile squid, and averaged 3.3.

In contrast, euphausiids made up 73% of the total IRI for prey of 201 Short-tailed Shearwaters, and another 20% of the IRI was accounted for by juvenile and post-larval Capelin (Fig. 10.2). TLs of the Short-tailed Shearwater's prey ranged from 2.5 for euphausiids to 4 for juvenile squid, and averaged 2.6, considerably lower than that of Sooty Shearwaters.

Demersal-benthic feeders Pelagic Cormorants and Pigeon Guillemots have similar demersal foraging habits and are thus discussed together here. Pacific Sandlance made up 86% of the total IRI for foods of 16 Pelagic Cormorants (Fig. 10.3). Unidentified and other fishes at TL 3 comprised another 12% of the total IRI, and Walleye Pollock, adult Capelin, and gammarid amphipods completed the cormorants' diet. Prey TLs ranged from 2.5 for the amphipods to 3.5 for Walleye Pollock, and averaged 2.9.

Pigeon Guillemots (n = 58), on the other hand, had eaten a wide variety of prey (Fig. 10.3). Crabs, shrimp, Pacific Sandfish *Trichodon trichodon* and unidentified fish (all with a TL of 3) accounted for some 53% of the total IRI. Other prey of the guillemots included unidentified gadoid fish, adult, juvenile and post-larval Capelin, and small amounts of gastropods and clams. TLs of the guillemots' prey ranged from 2 for clams to 3.5 for gadids (assumed to be pollock), and averaged 3.1.

Gulls and terns Ninety-two percent of the total IRI of the prey of 66 Glaucous-winged Gulls was accounted for by unidentified fish (Fig. 10.3).

Fig. 10.3. Composition, % of total IRI, and the trophic level spectra of prey in the diets of Pelagic Cormorants, Pigeon Guillemots, Glaucous-winged Gulls and Black-legged Kittiwakes. The large triangle indicates the average trophic level of prey.

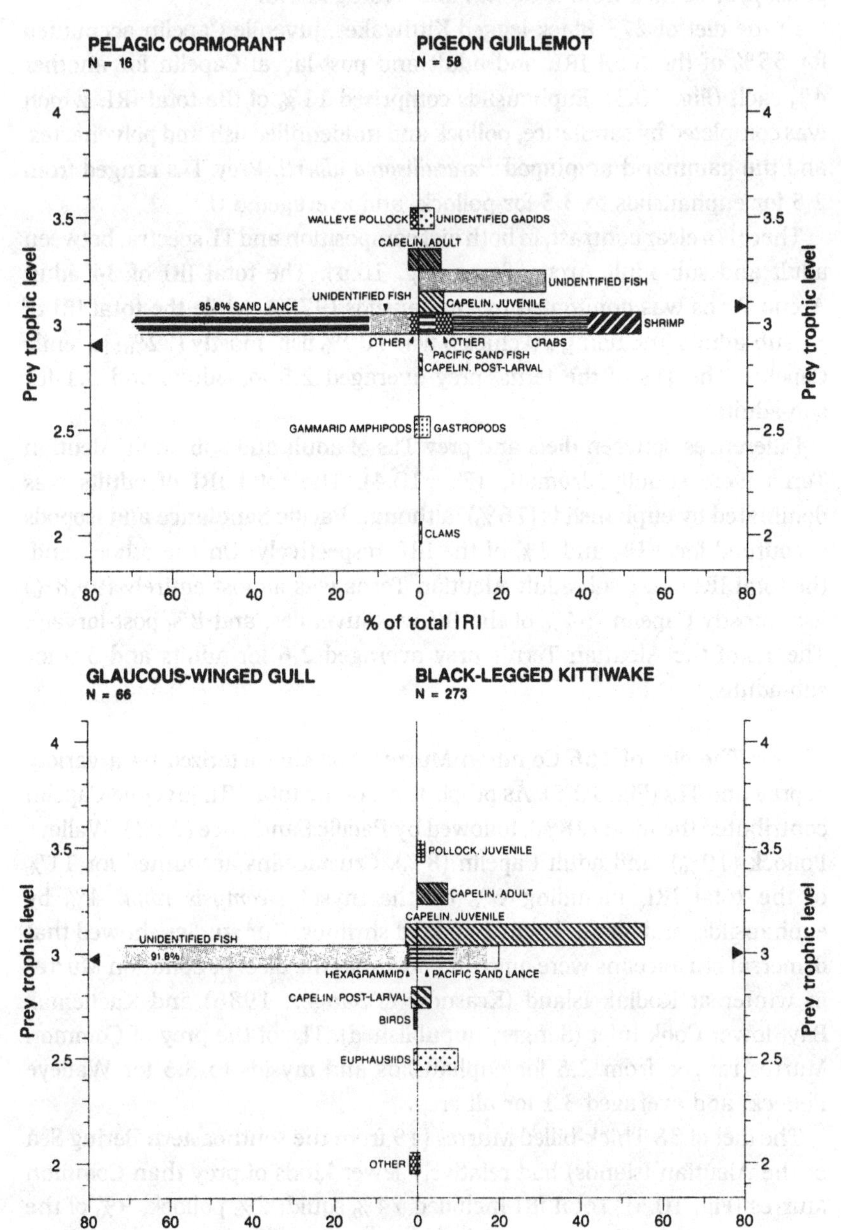

Other prey included juvenile and post-larval Capelin, Pacific Sandlance, greenling Hexagrammidae, birds, euphausiids, and a variety of other invertebrates that were present in trace amounts. Trophic levels of the gulls' prey ranged from 2 to 3.1, and averaged 3.0.

In the diet of 273 Black-legged Kittiwakes, juvenile Capelin accounted for 55% of the total IRI, and adult and post-larval Capelin for another 4% each (Fig. 10.3). Euphausiids comprised 11% of the total IRI, which was completed by sandlance, pollock and unidentified fish and polychaetes, and the gammarid amphipod *Paracallisoma alberti*. Prey TLs ranged from 2.5 for euphausiids to 3.5 for pollock, and averaged 3.0.

There is a clear contrast, in both diet composition and TL spectra, between adult and sub-adult Arctic Terns (Fig. 10.4). The total IRI of 34 adult Arctic Terns was dominated by euphausiids (97%), while the total IRI of 31 sub-adults (including 20 chicks) was 100% fish, mostly (72%) juvenile Capelin. The TLs of the terns' prey averaged 2.5 for adults and 3.1 for sub-adults.

Differences between diets and prey TLs of adult and sub-adult Aleutian Terns were equally dramatic (Fig. 10.4). The total IRI of adults was dominated by euphausiids (76%), although Pacific Sandlance and isopods accounted for 12% and 2% of the IRI, respectively. On the other hand, the total IRI of 47 sub-adult Aleutian Terns was almost entirely (99.8%) fish, mostly Capelin (54% of the IRI was juveniles, and 8% post-larvae). The TL of the Aleutian Tern's prey averaged 2.6 for adults and 3.0 for sub-adults.

Murres The diet of 166 Common Murres was characterized by a variety of prey and TLs (Fig. 10.5). As proportions of the total IRI, juvenile Capelin contributed the most (28%), followed by Pacific Sandlance (22%), Walleye Pollock (19%), and adult Capelin (8%). Crustaceans accounted for 11% of the total IRI, including 6% by the mysid *Neomysis rayii*, 4% by euphausiids, and about 1% by pandalid shrimps. Our studies showed that demersal crustaceans were quite important in the diets of Common Murres in winter at Kodiak Island (Krasnow & Sanger, 1986) and Kachemak Bay, lower Cook Inlet (Sanger, unpublished). TLs of the prey of Common Murres ranged from 2.5 for euphausiids and mysids to 3.5 for Walleye Pollock, and averaged 3.1 for all prey.

The diet of 38 Thick-billed Murres (19 from the southeastern Bering Sea or the Aleutian Islands) had relatively fewer kinds of prey than Common Murres (Fig. 10.5). Total IRI included 74% squid, 7% pollock, 9% of the large (20–40 mm body length) hyperiid amphipod *Parathemisto libellula*,

Fig. 10.4. Composition, % of total IRI, and the trophic level spectra of prey in the diets of adult and sub-adult Arctic Terns and Aleutian Terns. The large triangle indicates the average trophic level of prey.

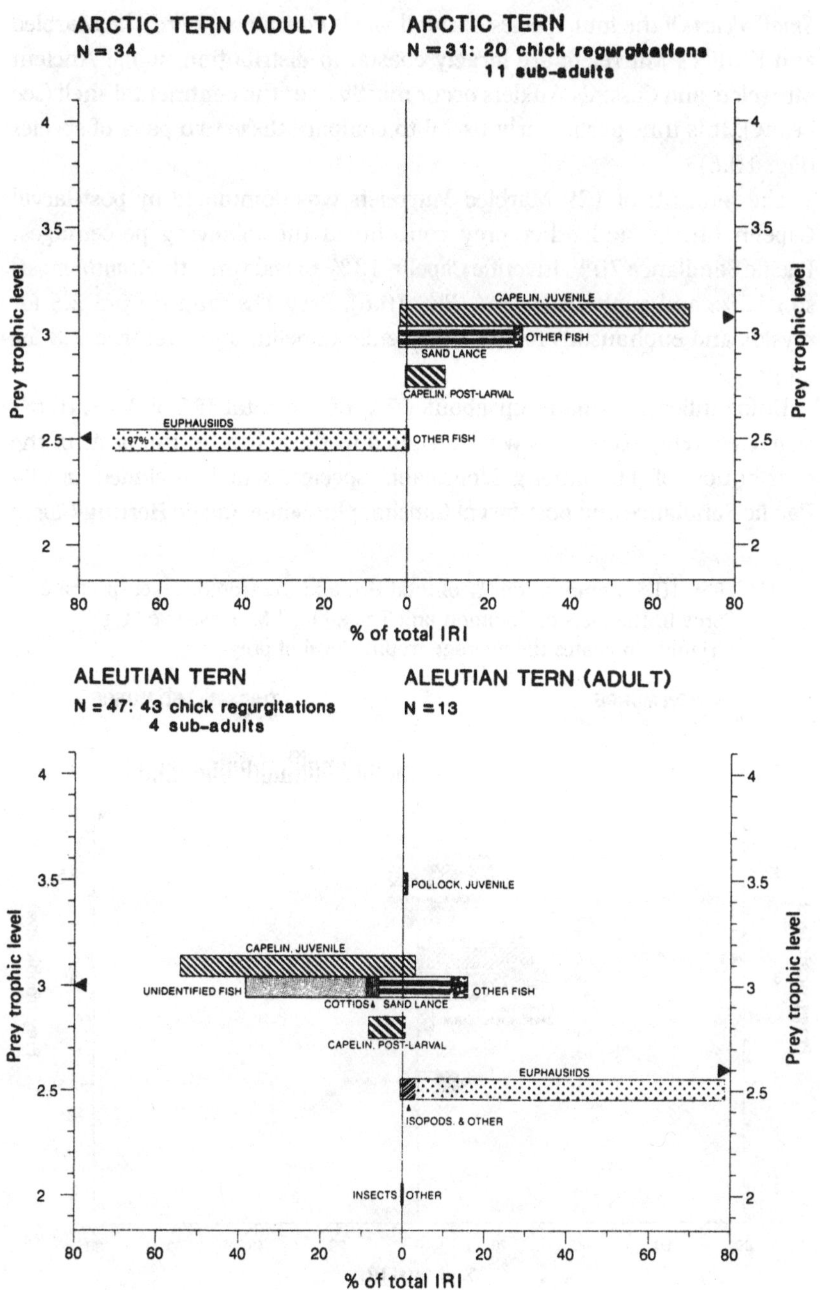

and 1% shrimp. Prey TLs ranged from 2.5 for euphausiids to 4 for squid, and averaged 3.8.

Small alcids Of the four species of small alcids considered here, the Marbled and Kittlitz's Murrelets are largely coastal in distribution, while Ancient Murrelets and Cassin's Auklets occur mostly over the continental shelf (see below). It is thus particularly useful to compare these two pairs of species (Fig. 10.6).

The total IRI of 129 Marbled Murrelets was dominated by post-larval Capelin (40%), and other prey contributed the following percentages: Pacific Sandlance 20%, juvenile Capelin 13%, mysids (mostly *Acanthomysis* sp.) 12%, and euphausiids 6% (Fig. 10.6). Prey TLs ranged from 2.5 for mysids and euphausiids to 3.1 for juvenile Capelin, and averaged 2.8 for all prey.

Unidentified fish made up about 62% of the total IRI of 15 Kittlitz's Murrelets (Fig. 10.6). TLs were assigned to these fish on the basis of the distribution of TLs among identifiable species, which included mostly Pacific Sandlance and post-larval Capelin, plus some Pacific Herring *Clupea*

Fig. 10.5. Composition, % of total IRI, and the trophic level spectra of prey in the diets of Common and Thick-billed Murres. The large triangle indicates the average trophic level of prey.

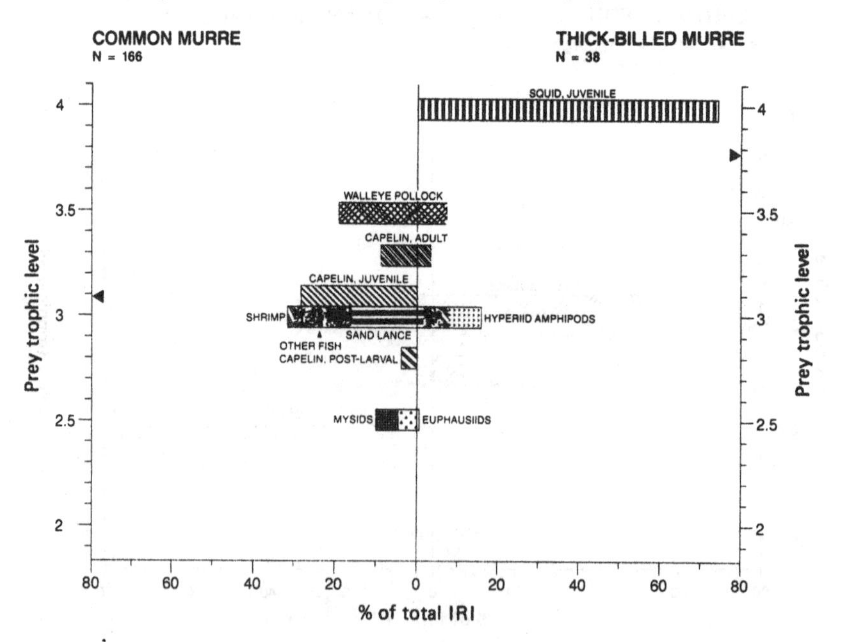

Fig. 10.6. Composition, % of total IRI, and the trophic level spectra of prey in the diets of Marbled, Kittlitz's and Ancient Murrelets, and of the Cassin's Auklet. The large triangle indicates the average trophic level of prey.

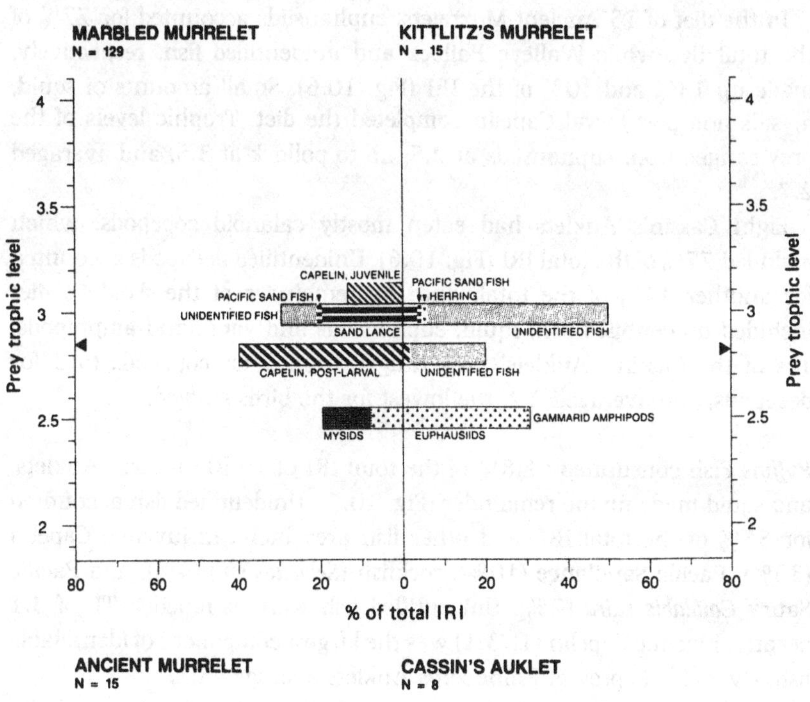

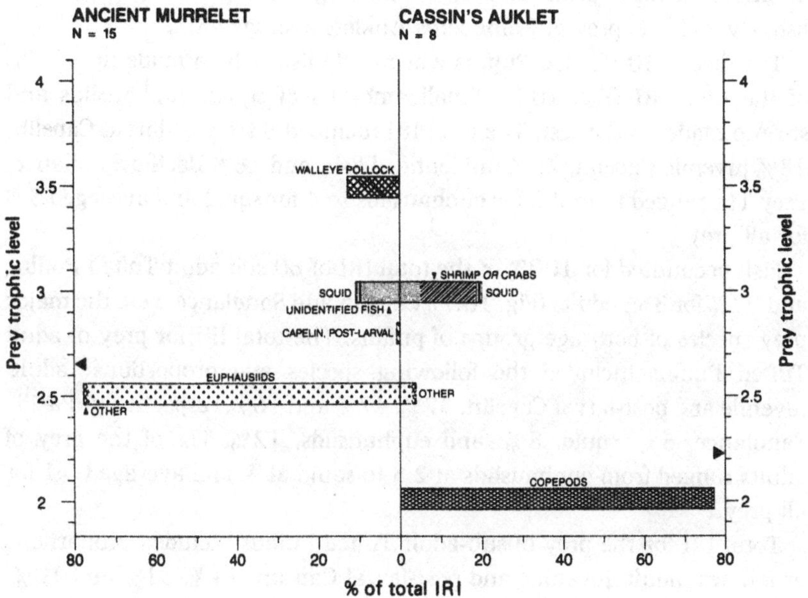

harengus and Pacific Sandfish (Fig. 10.6). Euphausiids made up 30% of the total IRI and gammarid amphipods were present in trace amounts. Prey TLs ranged from 2.5 for euphausiids to 3 for Sandlance, and averaged 2.8 for all prey.

In the diet of 15 Ancient Murrelets, euphausiids accounted for 77% of the total IRI, while Walleye Pollock and unidentified fish, respectively, made up 13% and 10% of the IRI (Fig. 10.6). Small amounts of squid, mysids and post-larval Capelin completed the diet. Trophic levels of the prey ranged from euphausiids at 2.5, up to pollock at 3.5, and averaged 2.7.

Eight Cassin's Auklets had eaten mostly calanoid copepods, which included 77% of the total IRI (Fig. 10.6). Unidentified decapods accounted for another 14% of the total, and the remainder of the Auklet's diet included unidentified fish, squid, euphausiids and gammarid amphipods. TLs of the Cassin's Auklets' prey ranged from 2 for copepods to 3 for decapods, and averaged 2.2, the lowest for the birds studied.

Puffins Fish constituted 98.8% of the total IRI of 16 Rhinoceros Auklets, and squid made up the remainder (Fig. 10.7). Unidentified fish accounted for 55% of the total IRI, and other fish prey included juvenile Capelin (30%), Pacific Sandlance (10%), rockfish (*Sebastes* sp.) (4%), and Pacific Saury *Cololabis saira* (2%). Unidentified fish were assigned a TL of 3.1 because juvenile Capelin (TL 3.1) was the largest component of identifiable fish. Overall, the prey of Rhinoceros Auklets averaged 3.1.

The diet of 40 Horned Puffins was mostly fish, which made up 97.7% of the total IRI (Fig. 10.7). Small amounts of squid, euphausiids and shrimp made up the rest. The total IRI included 33% post-larval Capelin, 28% juvenile Capelin, 20% unidentified fish, and 16% Pacific Sandlance. Prey TLs ranged from 2.5 for euphausiids to 4 for squid and averaged 3.0 for all prey.

Fish accounted for 100% of the total IRI of 60 sub-adult Tufted Puffins and 92% for 364 adults (Fig. 10.7). Capelin and Sandlance were the major prey species of both age groups of puffins. The total IRI for prey of adult Tufted Puffins included the following species and proportions: adult, juvenile and post-larval Capelin, 37%, 47% and 16%, respectively; Pacific Sandlance, 5%; squid, 8%; and euphausiids, 12%. TLs of the prey of adults ranged from euphausiids at 2.5 to squid at 4, and averaged 3.1 for all prey.

Total IRI for the prey of sub-adult Tufted Puffins included proportions as follows: adult, juvenile, and post-larval Capelin, 24%, 31% and 10%,

Fig. 10.7. Composition, % of total IRI, and the trophic level spectra of prey in the diets of Rhinoceros Auklets, and of Horned and Tufted Puffins. The large triangle indicates the average trophic level of prey.

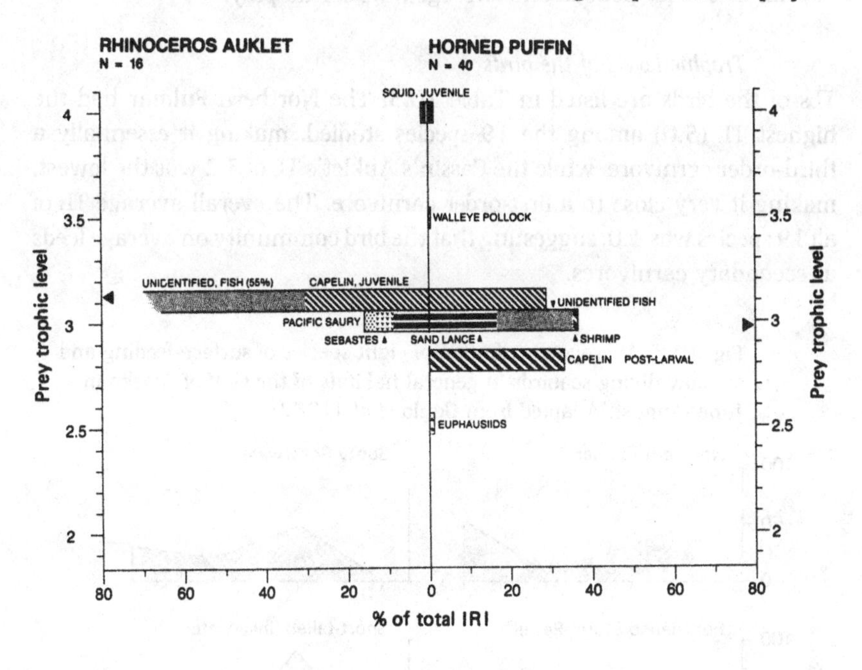

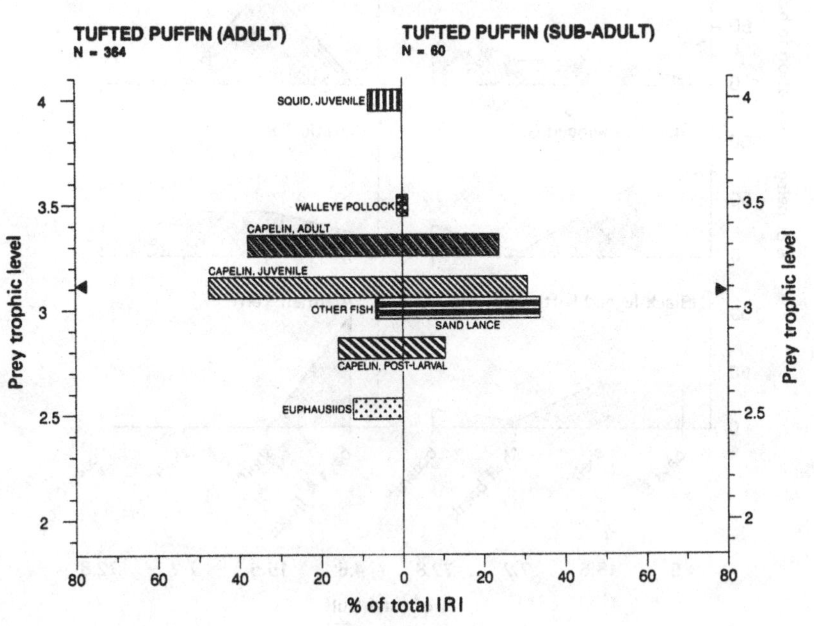

respectively; Pacific Sandlance, 33%; Walleye Pollock, 2%; plus traces of Pacific Sandfish and nereid polychaetes. TLs ranged from 2.8 for post-larval Capelin to 3.5 for pollock and averaged 3.1 for all prey.

Trophic levels of the birds

TLs of the birds are listed in Table 10.3. The Northern Fulmar had the highest TL (5.0) among the 19 species studied, making it essentially a third-order carnivore, while the Cassin's Auklet's TL of 3.2 was the lowest, making it very close to a first-order carnivore. The overall average TL of all 19 species was 4.0, suggesting that the bird community on average feeds as secondary carnivores.

Fig. 10.8. Average distribution of eight species of surface-feeding and shallow-diving seabirds in general habitats of the Gulf of Alaska in June–August. Adapted from Gould *et al.* (1982).

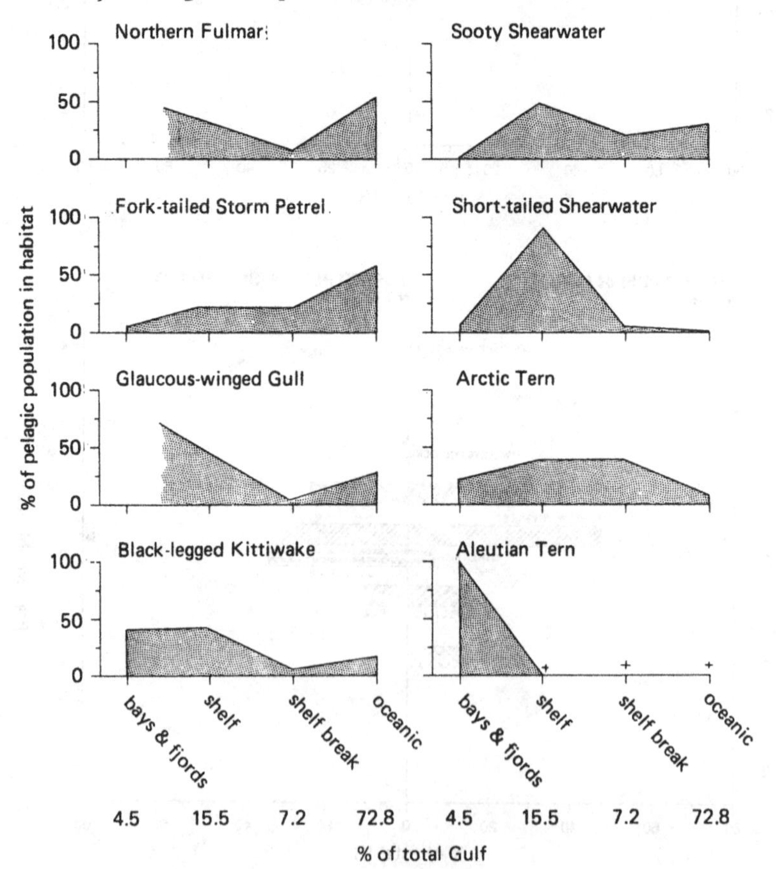

Evaluation of foraging habitats and feeding behaviour

The distribution of surface feeders and shallow divers (Fig. 10.8) suggests that these species are generally more scattered across nearshore, shelf, shelf-break and oceanic habitats than are deep-diving species (Fig. 10.9). Of the former group, essentially all Aleutian Terns were seen in the bay–fjord habitat, 93% of the Short-tailed Shearwaters were over the shelf, 73% of the Glaucous-winged Gull population was in the bay–fjord and shelf habitats combined, and 80% of the Black-legged Kittiwakes were equally divided between the bay–fjord and shelf areas. The remaining four species of surface feeders or shallow divers were more evenly distributed; 20% or more of the populations of Northern Fulmars, Sooty Shearwaters, Fork-tailed Storm Petrels and Arctic Terns were seen in each of three of the four habitats (Fig. 10.8).

In contrast, all 10 species of deep divers appeared to be more restricted in their distributions (Fig. 10.9). Essentially, all Pelagic Cormorants and

Table 10.3. *Summary of trophic levels of seabirds and their prey in the Gulf of Alaska, ranked from highest to lowest*

Species	No. of birds	Trophic levels: For birds' prey – Mean	Min.	Max.	Average for bird species
Northern Fulmar	43	4·0	2·0	4·0	5·0
Thick-billed Murre	38	3·8	2·5	4·0	4·8
Sooty Shearwater	178	3·3	2·5	4·0	4·3
Rhinoceros Auklet	16	3·1	3·0	4·0	4·1
Tufted Puffin	364	3·1	2·5	4·0	4·1
Common Murre	166	3·1	2·5	3·2	4·1
Pigeon Guillemot	58	3·1	2·0	3·5	4·1
Black-legged Kittiwake	373	3·0	2·0	3·5[a]	4·0
Glaucous-winged Gull	66	3·0	2·0	3·1	4·0
Horned Puffin	40	3·0	2·5	4·0	4·0
Pelagic Cormorant	16	2·9	2·0	3·5	3·9
Marbled Murrelet	129	2·8	2·5	3·1	3·8
Fork-tailed Storm Petrel	8	2·8	2·0	3·0	3·8
Kittlitz's Murrelet	15	2·8	2·5	3·0	3·8
Ancient Murrelet	15	2·7	2·5	3·5	3·7
Short-tailed Shearwater	201	2·6	2·5	4·0	3·6
Aleutian Tern	13	2·6	2·0	3·5	3·6
Arctic Tern	34	2·5	2·5	3·1	3·5
Cassin's Auklet	8	2·2	2·0	3·0	3·2

[a] TL 4·0 prey in diet in trace amount.

Kittlitz's Murrelets were confined to the bay–fjord habitat; Pigeon Guillemots, Marbled Murrelets and Rhinoceros Auklets (Sanger, unpublished data) were strongly oriented to the bay–fjord habitat (70% each of the guillemots and auklets, and 58% of the murrelets). Distributions of murres and the other five species of divers were centered over shelf waters, with 66% each of the murres and Ancient Murrelets, 91% of the Cassin's

Fig. 10.9. Average distribution of 11 species of diving seabirds in general habitats of the Gulf of Alaska in June–August. Adapted from Gould *et al.* (1982).

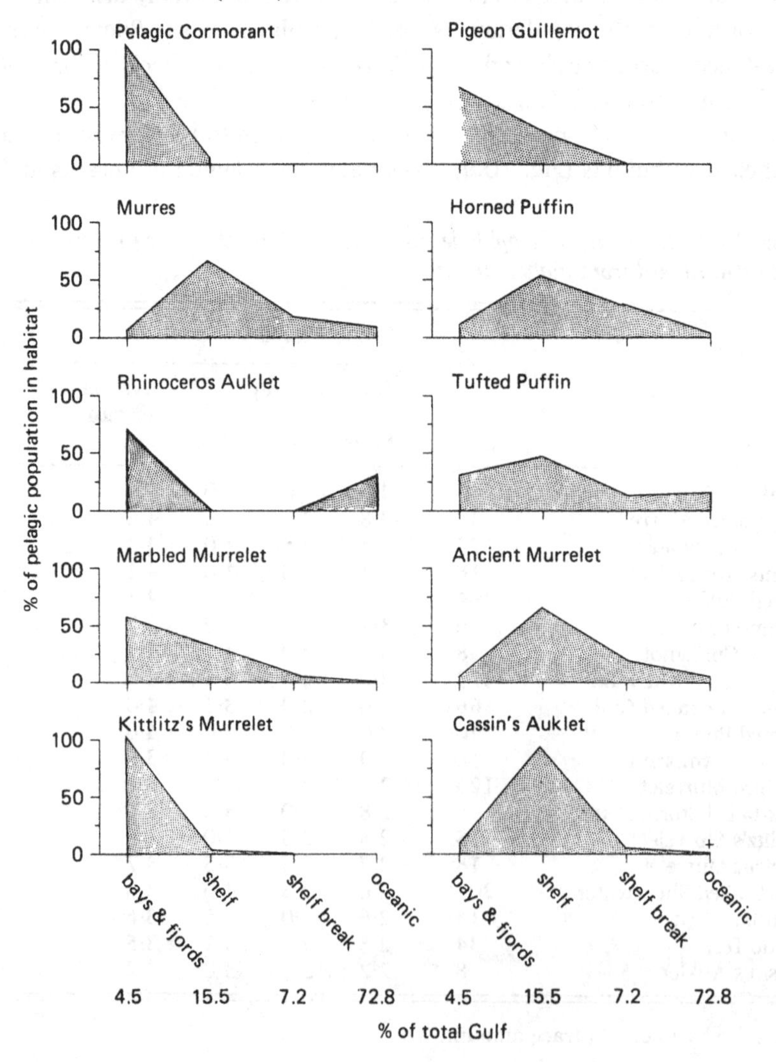

Auklets, 52% of the Horned Puffins and 45% of the Tufted Puffins seen there. Twenty percent or more of the estimated numbers of all species of deep divers were seen in no more than two habitats (Fig. 10.9).

A portrayal of average distribution patterns of birds across general habitats (cf. Figs 10.8 and 10.9) gives a general idea of where populations of birds are located at sea for discussion of TLs. For breeding birds, it is likely that numbers using the nearshore habitats are greater than indicated in Figs 10.8 and 10.9, due to an unknown amount of turnover between birds at sea and birds on colonies. Also, most of the data (Gould *et al.*, 1982) from which these figures were adapted are from nearshore areas, and values for areas with the lowest coverage (oceanic) could be inflated.

Discussion

The indication that, on average, most seabirds in the Gulf of Alaska feed as secondary carnivores (Table 10.3) differs from a previous belief that most sub-Arctic Pacific seabirds fed as second- and third-order carnivores (Ainley & Sanger, 1979). While the present results do not refute this, they suggest that much more feeding by seabirds in the Gulf occurs at the lower TL. The results also give a more precise view of the spectra of TLs in seabird diets, and allow trophic relationships of species to be compared on firmer quantitative footing. They also allow speculation about the origin and nature of lower trophic levels in the birds' food webs.

Comparison of TL spectra and diets between closely related species is especially relevant to assessing how they relate trophically, and how they partition food resources. Sooty and Short-tailed Shearwaters are both abundant in coastal waters of Alaska (Gould *et al.*, 1982); they often occur in mixed flocks, they are highly similar in appearance, there are similarities in the composition and TL spectra of their diets, and they consequently are sometimes thought to be ecologically similar. Despite this, the Sooty Shearwaters' average TL of 4.3 was about 0.7 higher than that of Short-tailed Shearwaters (Fig. 10.2). This resulted from juvenile squid and Capelin dominating the Sooties' diet, while Short-tails ate many euphausiids.

For Arctic and Aleutian Terns, the diets and TL spectra were more similar between the same age group of both species than between adults and sub-adults of the same species (Fig. 10.4). Adult diets of both species were composed largely of euphausiids, which resulted in average TLs of about 2.5–2.6, while adults had fed their chicks almost exclusively fish, resulting in average TLs about 0.5 higher than for adults of both species. This undoubtedly maximizes foraging efficiency for parents, who bring large prey of a relatively high TL and weight to their chicks, thus maximizing food energy fed to chicks per foraging trip. Adults eat smaller prey of lower TLs themselves, which maximizes the trophic efficiency of their food chain. It appears that the two species minimize foraging competition by feeding in different areas: nearly all Aleutian Terns were seen in bays and fjords, while Arctic Terns were dispersed more widely, all the way to the shelf break (Fig. 10.8).

Average TLs of Common and Thick-billed Murres were further apart (about 0.7) than one might have expected for these similar species (Fig. 10.5), although this difference could be a result of half of the latter being collected outside the Gulf (see p. 240). Depending on where Thick-billed Murres occur, they are known to prey heavily on squid (Ogi, 1980), as shown in this study, or on fish (e.g. Tuck & Squires, 1955). Simultaneous collection of food samples of both species from the same area and time period are needed.

Marbled and Kittlitz's Murrelets are nearly identical in size and similar in appearance (in summer); the distribution of Kittlitz's Murrelets overlaps almost entirely that of Marbled Murrelets and their tundra nesting habitats are similar (notwithstanding tree-nesting Marbled Murrelets). Their average TLs are nearly identical, and the TL spectra and the diet composition of their prey overlap (Fig. 10.6). How, then, do these two species co-exist? Crustaceans of TL 2.5 were more important to Kittlitz's Murrelets than Marbled Murrelets (30% *versus* 20% of the total IRI), although unidentified fish comprised most of the diet of the 15 birds sampled. Mysids (12.5% of the IRI) were eaten by Marbled Murrelets in winter (Sanger, unpublished), while euphausiids (6% of the IRI) were eaten in summer (Krasnow & Sanger, 1986). Euphausiids were more important to Kittlitz's Murrelets, and accounted for 30% of the IRI. Thus, despite nearly identical TLs, Marbled and Kittlitz's Murrelet appear to avoid competition by the Kittlitz's eating more crustaceans in summer and foraging closer to shore, while the Marbled Murrelet forages across a broader range of habitats (Fig. 10.9) and eats mostly fish.

Foraging habitats of Ancient Murrelets and Cassin's Auklets are nearly identical, concentrated over the shelf (Fig. 10.9). Small sizes of stomach samples permit only tentative conclusions about diets and TLs, however. While both species relied heavily on crustaceans, Ancient Murrelets had eaten mostly euphausiids (TL 2.5, 77% of IRI), while copepods (TL 2) accounted for 77% of the IRI of the diet of Cassin's Auklets. Elsewhere in the eastern North Pacific, however, Cassin's Auklets eat both copepods and euphausiids during the breeding season (Thoreson, 1964; Manuwal,

1974; Vermeer, 1981). Thus, while their foraging ranges are highly similar, these species feed about 0.5 TL apart (Table 10.3), with the Cassin's Auklet eating mainly planktonic crustaceans, while Ancient Murrelets eat a wider variety of fish and crustaceans (Fig. 10.6). More data are needed to corroborate this.

The average TLs of the three puffins are similar, about 3.1 for Rhinoceros Auklets and adult and sub-adult Tufted Puffins, and about 3.0 for Horned Puffins. The diet of Rhinoceros Auklets was almost all fish (99% of the IRI; Fig. 10.7), and the species occurs mainly in coastal waters of the southern Gulf (Sanger, A.W. Sowls & J. Nelson, unpublished data). Sub-adult Tufted Puffins eat only fish. In contrast, breeding Horned Puffins forage very near their colonies (Wehle, 1983), although the population at large may occur mostly over the shelf (Fig. 10.9). About 2.5% of the IRI of Horned Puffins was crustaceans and the remainder mostly fish (Fig. 10.7). Adult Tufted Puffins occurred mostly in the bay–fjord and shelf habitats of the northern part of the Gulf (Fig. 10.9; see also Wehle, 1983), and they ate prey of a wider spectra of TLs, that included crustaceans as 20% of the total IRI (Fig. 10.7).

What are the likely lower and upper limits of TLs of seabirds in the Gulf? It seems that Cassin's Auklets, with a TL of 3.2, are approaching the lower limits attainable by seabirds in the sub-Arctic Pacific region. It is likely that other small alcids not treated here, i.e. the Parakeet Auklet *Cyclorrhynchus psittacula*, Crested, Least and Whiskered Auklets *Aethia cristatella*, *A. pusilla* and *A. pygmaea* would have similarly low average TLs. The latter two species, with diets largely of copepods (Bédard, 1969; R.H. Day, personal communication), are likely to be the closest to primary carnivores among sub-Arctic Pacific seabirds.

At the upper end of the trophic scale, Short-tailed, Black-footed and Laysan Albatrosses *Diomedea albatrus*, *D. nigripes* and *D. immutabilis* undoubtedly have the largest average TLs among sub-Arctic Pacific seabirds. Their body and bill size allow them to capture relatively large prey, and their oceanic distribution in effect adds one TL compared with a coastal distribution (cf. Parsons & LeBrasseur, 1970). The Laysan Albatross, which eats mostly squid (65% by volume; Harrison, Hida & Seki, 1983), which themselves have high TLs for their size (Table 10.2), and the Short-tailed Albatross probably have the highest average TLs among sub-Arctic Pacific seabirds. The Black-footed Albatross, with a diet mostly of fish and squid (50% and 32% by volume, respectively, in study of Harrison *et al.* (1983)), would presumably have a TL somewhat lower.

In general, it would seem that birds that feed at lower average TLs are in a better position to exploit blooms of phytoplankton due to less time lag and more efficient transfer of energy to lower TLs. They would also seem less susceptible to perturbations in food supply such as those caused by such activities as commercial fishing. Indeed, they may actually benefit from removal from the ecosystem of competitors that feed near their trophic levels (cf. Furness, 1982).

Although the vast majority of biological production in the ocean begins with phytoplankton in the water column (e.g. Strickland, 1970; Steele, 1974), the role of organic detritus may be more important in coastal areas than previously recognized (e.g. Tenore, 1977; Lees & Driskell, 1981). Food-chain pathways that include detritus may result in a more stable food supply than non-detrital food chains. This could be reflected in demersal-benthic feeders like Pelagic Cormorants and Pigeon Guillemots showing stable productivity over the years, compared with midwater and surface feeders. Winter survival of species like Common Murres and Marbled Murrelets may be enhanced by their ability to alter their 'normal' diet of pelagic fishes to include demersal crustaceans, thus seasonally linking themselves to a detrital-based food chain.

These attempts at quantifying seabird trophic levels rely heavily on untested assumptions. Hopefully, however, they provide a better understanding of how seabirds could relate to lower trophic levels of the Gulf of Alaska. The biggest obstacle to developing this concept further at present is the scarcity of data on annual and seasonal variability in the diet of birds, and on diets of the seabirds' prey.

Although it may be premature for ecosystem modelers to introduce such concepts into their current efforts, as additional data and studies clarify the nature of lower trophic levels in marine food webs this information will have to be accounted for. Regulatory bodies such as the North Pacific Fisheries Management Council routinely state in their environmental impact statements that fisheries will be managed for the health of the entire ecosystem; yet, as this paper has shown, very little is known about the lowest TLs of the ecosystem of the Gulf of Alaska. Whether the concern is about stocks of commercial fishes or about seabirds, all life in the sea begins with primary productivity. As data and ideas about the lower trophic levels of marine ecosystems improve, modelers must begin to take them into account to keep their models as realistic as possible. With such models, managers of seabirds, fisheries, and other marine resources will be in a much better position than at present to predict the effect of commercial fisheries on all trophic levels of the ecosystem.

Acknowledgements Bird specimens and food samples were collected by several personnel of the U.S. Fish and Wildlife Service (see Sanger, 1986). Prey were identified by Alan Fukuyama, Valerie Hironaka and David Wiswar, who were assisted by various taxonomic specialists, in particular George Mueller and his staff at the Marine Sorting Center, University of Alaska, Fairbanks. To all of these people I express my thanks.

Special thanks go to Craig Harrison and John Croxall for encouraging me to write and contribute to this volume. I am indebted to Kenneth Briggs for bringing to my attention the very pertinent reference of Mearns *et al.* (1981), whose approach to calculating fractional trophic levels was nearly identical, although completely independent from mine. Harry Ohlendorf gave helpful advice on the presentation of data portrayed in Figs 10.2–10.7. Lynne Krasnow provided length-weight data for capelin, which I used to estimate volumetric composition of size classes of capelin in the birds' diets.

An earlier draft of this paper benefitted from reviews by John Piatt, William Pearcy, David Irons, Scott Hatch, L. Krasnow and Calvin Lensink.

References

Ainley, D.A. & Sanger, G.A. (1979). Trophic relations of seabirds in the northeastern Pacific Ocean and Bering Sea. In *Conservation of marine birds of northern North America*, ed. J.C. Bartonek & D.N. Nettleship, pp. 95–122. U.S. Fish & Wildlife Service, Wildlife Research Report 11.

Baird, P.A. & Gould, P.J. (eds) (1986). The breeding biology and feeding ecology of marine birds in the Gulf of Alaska. In *Environmental assessment of the Alaskan continental shelf. Final reports of the principal investigators*, vol. 45, pp. 121–504. Anchorage, Alaska: NOAA, National Ocean Service.

Bartonek, J.C. & Nettleship, D.N. (eds) (1979). *Conservation of marine birds of northern North America.* U.S. Fish & Wildlife Service, Wildlife Research Report, 11, 319 pp.

Bédard, J. (1969). Feeding of the Least, Crested, and Parakeet Auklets around St. Lawrence Island, Alaska. *Can. J. Zool.*, **47**, 1025–50.

Calkins, D.G. & Pitcher K.W. (1983). Population assessment, ecology, and trophic relationships of Steller Sea Lions in the Gulf of Alaska. In *Environmental assessment of the Alaska continental shelf. Final reports of the principal investigators*, vol. 19, pp. 445–56. Juneau, Alaska: NOAA, National Ocean Service.

Clarke, M.R. (1966). A review of the systematics and ecology of oceanic squid. *Adv. Mar. Biol.*, **4**, 91–300.

Dunbar, M.J. (1946). On *Themisto libellula* in Baffin Island coastal waters. *J. Fish. Res. Bd Can.*, **6**, 419–34.

Dunn, J.R., Kendall A.W. & Wolotira R.J. (1981). Seasonal composition and food web relationships of marine organisms in the nearshore zone – including components of the ichthyoplankton, meroplankton, and holoplankton. In *Environmental assessment of the Alaskan continental shelf.*

Final reports of the principal investigators, vol. 13 *Biological studies*, pp. 357–776. Juneau, Alaska: NOAA, Office of Marine Pollution Assessment.

Favorite, F., Dodimead, A.J. & Nasu, K. (1976). Oceanography of the subarctic Pacific Region. *Bull. Int. N. Pacif. Fish. Comm.*, **33**, 187 pp.

Feller, R.J. & Kaczynski, V.W. (1975). Size selective predation by juvenile Chum Salmon *Onchorhynchus keta* on epibenthic prey in Puget Sound. *J. Fish. Res. Bd Can.*, **32**, 1419–29.

Forsell, D.J. & Gould, P.J. (1981). *Distribution and abundance of marine birds and mammals wintering in the Kodiak area of Alaska.* U.S. Department of the Interior, Fish & Wildlife Service, FWS/OBS – 81/13. 72 pp.

Furness, R.W. (1982). Competition between fisheries and seabird communities. *Adv. Mar. Biol.*, **20**, 225–307.

Gould, P.J., Forsell, D.J. & Lensink, C.J. (1982). *Pelagic distribution and abundance of seabirds in the Gulf of Alaska and eastern Bering Sea.* U.S. Department of the Interior, Fish & Wildlife Service, FWS/OBS – 82/48. 294 pp.

Harrison, C.S., Hida, T.S. & Seki, M.P. (1983). Hawaiian seabird feeding ecology. *Wildl. Monogr.*, no. 85, 71 pp.

Hatch, S.A. (1984). Nestling diets and feeding rates of rhinoceros auklets in Alaska. In *Marine birds: their feeding ecology and commerical fisheries relationships*, ed. D.N. Nettleship, G.A. Sanger & P.F. Springer, pp. 106–15. Ottawa: Can. Wild. Serv., Spec. Publ.

Hedgpeth, J.W. (revisor), Ricketts, E.F. & Calvin, J. (1968). *Between Pacific tides*, 4th edn. Stanford: Stanford University Press.

Hunt, G.L., Jr, Burgeson, B. & Sanger, G.A. (1981*a*). Feeding ecology of seabirds of the eastern Bering Sea. In *The eastern Bering Sea shelf: oceanography and resources*, vol. 2, ed. D.A. Hood & J.A. Calder, pp. 629–47. NOAA, Office of Marine Pollution Assessment. Seattle: University of Washington Press.

Hunt, G.L., Jr, Eppley, Z. & Drury, W.H. (1981*b*). Breeding distribution and reproductive biology of marine birds in the eastern Bering Sea. In *The eastern Bering Sea shelf: oceanography and resources*, vol. 2, ed. D.A. Hood & J.A. Calder, pp. 649–88. NOAA, Office of Marine Pollution Assessment. Seattle: University of Washington Press.

Hunt, G.L., Jr, Gould, P.J., Forsell, D.J. & Peterson, H. (1981*c*). Pelagic distribution of marine birds in the eastern Bering Sea. In *The eastern Bering Sea shelf: oceanography and resources*, vol. 2, ed. D.A. Hood & J.A. Calder, pp. 689–718. NOAA, Office of Marine Pollution Assessment. Seattle: University of Washington Press.

Kajimura, H. (1984). Opportunistic feeding of the Northern Fur Seal, *Callorhinus ursinus*, in the eastern North Pacific Ocean and eastern Bering Sea. *NOAA Technical Report*, NMFS SSRF – 779, 49 pp.

Knox, G.A. (1970). Antarctic marine ecosystems. In *Antarctic ecology*, vol. 1, ed. M.W. Holdgate, pp. 69–96. London & New York: Academic Press.

Kozloff, E.N. (1973). *Seashore life of Puget Sound, the Strait of Georgia and the San Juan Archipelago.* Seattle & London: University of Washington Press.

Krasnow, L.D. & Sanger, G.A. (1986). Feeding ecology of marine birds in the nearshore waters of Kodiak Island. In *Environmental assessment of the Alaskan continental shelf. Final reports of the principal investigators*, vol. 45, pp. 505–630. Anchorage, Alaska: NOAA, National Ocean Service.

Lees, D.C. & Driskell, W.B. (1981). Investigations on shallow subtidal habitats and assemblages in lower Cook Inlet, Alaska. In *Environmental assessment of the Alaskan continental shelf. Final reports of the principal investigators*, vol. 14, pp. 419–610. Boulder, Colorado: NOAA/BLM.

Macy, P.T., Wall, J.M., Lampsakis, N.D. & Mason, J.E., (1978). *Resources of non-salmonid pelagic fishes of the Gulf of Alaska and eastern Bering Sea*, parts 1 & 2. Seattle, Washington: National Marine Fisheries Service.

Manuwal, D.A. (1974). The natural history of Cassin's Auklet *Ptychoramphus aleuticus*. *Condor*, **76**, 421–31.

Martin, A.C., Gensch, R.H. & Brown, C.P. (1946). Alternative methods in upland gamebird food analysis. *J. Wildl. Mgmt*, **10**, 8–12.

Mauchline, J. (1980). The biology of mysids and euphausiids. *Adv. Mar. Biol.*, **18**, 1–680.

Mearns, A.J., Young, D.R., Olson, R.J. & Schafer, H.A. (1981). Trophic structure of the cesium–potassium ratio in pelagic ecosystems. *Calif. Coop. Ocean. Fish. Invest. Rep.*, **22**, 99–110.

Meglitsch, P.A. (1972). *Invertebrate zoology* (2nd edn). London: Oxford University Press.

Naito, M., Murakami, K. & Kobayashi, T. (1977). Growth and food habit of oceanic squids (*Ommastrephes bartrami*, *Onychoteuthis boreali-japonicus*, *Berryteuthis magister*, and *Gonatopsis borealis*) in the western subarctic Pacific region. In *Fisheries biology productivity in the Subarctic Pacific region*, pp. 339–51. University of Hokkaido, Hakodate: Special Volume, Research Institute of North Pacific Fisheries.

Nesis, K.N. (1965). The distribution and nutrition of the young of the squid *Gonatus fabricii* (Licht.) in the northwest Atlantic and the Norwegian Sea. *Okeanologiya*, **5**, 134–41. (In Russian.)

Ogi, H.(1980). The pelagic feeding ecology of Thick-billed Murres in the North Pacific, March–June. *Bull. Faculty Fish., Hokkaido University*, **31**, 50–72.

Ogi, H., Kudobera, T. & Nakamura K. (1980). The pelagic feeding ecology of the Short-tailed Shearwater *Puffinus tenuirostris* in the subarctic Pacific region. *J. Yamashina Inst. Ornithol.*, **12**, 157–82.

Ogi, H. & Tsujita, T. (1973). Preliminary examination of stomach contents of murres (*Uria* spp.) from the eastern Bering Sea and Bristol Bay, June–August, 1970 and 1971. *Jap. J. Ecol.*, **23**, 201–9.

Omori, M. (1974). The biology of pelagic shrimp in the ocean. *Adv. Mar. Biol.*, **12**, 233–436.

Parsons, T.R. & LeBrasseur, R.J. (1970). The availability of food to different trophic levels in the marine food chain. In *Marine food chains*, ed. J.H. Steele, pp. 325–43. Berkeley & Los Angeles: University of California Press.

Pinkas, L., Oliphant, M.S. & Iverson, I.L.K. (1971). Food habits of Albacore, Bluefin Tuna, and Bonito in California waters. *Fish. Bull. Calif.*, **152**, 105 pp.

Pitcher, K.W. & Calkins, D.G. (1983). Biology of the Harbor Seal *Phoca vitulina richardsoni* in the Gulf of Alaska. In *Environmental assessment of the Alaskan continental shelf. Final reports of the principal investigators*, vol. 19, pp. 231–310. Juneau, Alaska: NOAA, National Ocean Service.

Rice, D.W. & Wolman, A.A. (1983). Summer distribution and abundance of Fin, Humpback, and Grey Whales in the Gulf of Alaska. In *Environmental assessment of the Alaskan continental shelf. Final reports of the principal investigators*, vol. 20, pp. 1–45, Juneau, Alaska: NOAA, National Ocean Service.

Rogers, D.E., Rabin, B.J., Garrison, K. & Wangerin, M. (1983). Seasonal composition and food web relations of marine organisms in the nearshore zone of Kodiak Island – including ichthyoplankton, zooplankton, and fish. In –Environmental assessment of the Alaskan continental shelf. Final reports of the principal investigators, vol. 17, pp. 541–658. Juneau, Alaska: NOAA, National Ocean Service.

Royer, T.C. (1981*a*). Baroclinic transport in the Gulf of Alaska, Part II. A fresh water driven coastal current. *J. Mar. Res.*, **39**, 251–66.

Royer, T.C. (1981*b*). Baroclinic transport in the Gulf of Alaska, Part I. Seasonal variations in the Alaska Current. *J. Geophys. Res.*, **39**, 239–50.

Sanger, G.A. (1972). Preliminary standing stock and biomass estimates of seabirds in the subarctic Pacific region. In *Biological oceanography of the northern North Pacific Ocean*, ed. A.Y. Takenouti *et al.*, pp. 589–611. Tokyo: Idemitsu Shoten.

Sanger, G.A. (1986). Diets and food web relationships of seabirds in the Gulf of Alaska and adjacent marine regions. In *Environmental assessment of the Alaskan continental shelf. Final reports of the principal investigators*, vol. 45, pp. 631–771. Anchorage, Alaska: NOAA, National Ocean Service.

Sanger, G.A. & Jones, R.D., Jr (1984). Winter feeding ecology and trophic relationships of Oldsquaws and White-winged Scoters on Kachemak Bay, Alaska. In *Marine birds: their feeding ecology and commercial fisheries relationships*, ed. D.N. Nettleship, G.A. Sanger & P.F. Springer, pp. 20–8. Ottawa: Can. Wild. Serv., Spec. Publ.

Sealy, S. (1975). Feeding ecology of the Ancient and Marbled Murrelets near Langara Island, British Columbia. *Can. J. Zool.*, **53**, 418–33.

Shuntov, V.P. (1972). *Seabirds and the biological structure of the ocean.* Vladivostok: Far East Book Publishers (In Russian). English translation: TT-74-55032, Washington, D.C.: U.S. Department of Commerce, Nat. Tech Inf. Serv.

Smith, G.B. (1981). The biology of Walleye Pollock. In *The eastern Bering Sea shelf: oceanography and resources*, vol. 1, ed. D.W. Hood & J.A. Calder, pp. 527–52. NOAA. Seattle: University of Washington Press.

Sowls, A.L., Hatch, S.A. & Lensink, C.J. (1978). *Catalog of Alaskan seabird colonies.* U.S. Department of the Interior, Fish & Wildlife Service, FWS/OBS – 78/78. 32 pp., + appendices and 153 maps.

Steele, J.H. (1974). *The structure of marine ecosystems.* Cambridge, Mass.: Harvard Press.

Strickland, J.D.H. (1970). Introduction to recycling of organic matter. In *Marine food chains*, ed. J.H. Steele, pp. 3–5. Berkeley & Los Angeles: University of California Press.

Swanson, G.A., Krapu, G.L., Bartonek, J.C., Serie, J.R. & Johnson, D.H. (1974). Advantages in mathematically weighting waterfowl food habits data. *J. Wildl. Mgmt*, **38**, 302–7.

Tenore, K.R. (1977). Food chain pathways in detrital feeding benthic communities: a review, with new observations on sediment resuspension and detritial recycling. In *Ecology of marine benthos*, ed. B.C. Coull, pp. 37–53. Columbia: University of South Carolina Press.

Thoreson, A.C. (1964). The breeding behaviour of the Cassin Auklet. *Condor*, **66**, 456–76.

Tuck, L.M. & Squires, H.J. (1955). Food and feeding habits of Brunnich's Murre (*Uria lomvia lomvia*) on Akpatok Island. *J. Fish. Res. Bd Can.*, **12**, 781–92.

Vermeer, K. (1981). The importance of plankton to Cassin's Auklets during breeding. *J. Plankt. Res.*, **3**, 315–29.

Vesin, J.-P., Leggett, W.C. & Able, K.W. (1981). Feeding ecology of Capelin *Mallotus villosus* in the estuary and western Gulf of St. Lawrence and its multispecies implications. *Can. J. Fish. Aquat. Sci.*, **38**, 257–67.

Wehle, D.H.S. (1983). The food, feeding, and development of young Tufted and Horned Puffins in Alaska. *Condor*, **85**, 427–42.

Wespestad, V.G. & Barton, L.H. (1981). Distribution, migration and status of Pacific Herring. In *The eastern Bering Sea shelf: oceanography and resources*, vol. 1, ed. D.A. Hood & J.A. Calder, pp. 509–25. NOAA, Office of Marine Pollution Assessment. Seattle: University of Washington Press.

Wing, B.L. (1976). *Ecology of* Parathemisto libellula *and* P. pacifica (*Amphipoda: Hyperiidea*) *in Alaskan coastal waters*. Seattle: National Marine Fisheries Service.

Xiong, Q. & Royer, T.C. (1984). Coastal temperature and salinity in the northern Gulf of Alaska. *J. Geophys. Res.*, **89**, 8061–8068.

11

Energy flux to pelagic birds: a comparison of Bristol Bay (Bering Sea) and Georges Bank (Northwest Atlantic)

D. C. SCHNEIDER

Department of Ecology and Evolutionary Biology, University of California, Irvine, California 92717, USA. Current address: Newfoundland Institute for Cold Ocean Science, Memorial University, St. John's, Newfoundland, A1B 3X7, Canada.

G. L. HUNT, Jr

Department of Ecology and Evolutionary Biology, University of California, Irvine, California 92717, USA.
and

K. D. POWERS

Manomet Bird Observatory, Manomet, Massachusetts 02345, USA.

Introduction

Recent estimates of the food requirements of pelagic birds have shown that bird populations are capable of removing prey at rates sufficient to have an impact on local prey stocks (Wiens & Scott, 1975; Furness, 1978) and at rates comparable to marine mammals (Sissenwine, 1984) and commercial fisheries (Hunt, Burgeson & Sanger, 1981; Furness & Cooper, 1982). The potential for competitive interaction among and within these three consumer groups is indicated by the degree of dietary overlap on schooling prey such as krill, or clupeid and gadoid fish. However, foraging activity by birds, mammals, and fisheries is highly localized and hence population interaction depends on the degree of spatial and temporal overlap by these predators. Estimates of spatial variation in energy flux are needed to further our understanding of population interactions among top predators in marine ecosystems.

Estimates of spatial variation in energy flux are also needed to increase our understanding of prey mortality. Most attempts at rational management of fish stocks have focused on the relation of recruitment of stock and on the availability of food to larval fish. Substantial mortality of fish occurs before recruitment to commercially important size classes and hence a better understanding of this mortality is needed to predict the consequences of fisheries practices. Seabirds, because of their visibility, are a promising group for investigating spatial heterogeneity in losses from prey populations of krill, fish, and squid.

Most estimates of energy flux to marine bird populations have been based on colony studies (cf. Furness, 1982). Population sizes are readily determined at colonies, life-history information can be obtained, and behavioural observations can be used to construct time–energy budgets. Despite these advantages, colony studies cannot be used to investigate spatial variation in energy flux to birds. Substantial flux can occur in areas without breeding colonies (Powers & Backus, in press), non-breeding birds can account for a substantial fraction of energy flux to birds in the vicinity of colonies (Wiens & Scott, 1975; Wiens, 1984), and the distribution of foraging effort is not uniform around colonies (Kinder *et al.*, 1983).

The commonest method of investigating spatial variation in pelagic bird abundance is by direct counts from ships. The major advantage of direct counts from moving ships is that the locations of a large number of individuals are recorded. The major disadvantage is that the data cannot be used to estimate numbers and biomass on an absolute scale (Powers, 1982). However, if standard methods are used, the resulting data, which are on a relative scale, can be used to construct and test models describing important features of the spatial dynamics of pelagic birds (Ford *et al.*, 1982). Data on a relative scale can also be used to test hypotheses about spatial variation in energy flux (Schneider & Hunt, 1982).

Consistent patterns of spatial variation in energy flux to birds have been reported in relation to meso-scale (100 km) physical features in Bristol Bay in the southeastern Bering Sea (Schneider, Hunt & Harrison, in press) and on Georges Bank in the northwest Atlantic (Powers & Backus, 1986). Both ecosystems occupy relatively wide continental shelves at middle latitudes. Currents across the two shelves are primarily tidal (Allen *et al.*, 1983), resulting in low rates of lateral advection and resupply of nutrients. Algal production is highly seasonal and energy flux to birds is structured in relation to a system of fronts, or water mass boundaries, that follow topographic contours. These and other similarities in the structure of the two ecosystems (Allen *et al.*, 1983) led us to expect little variation in energy flux between the two regions, as compared to variation within each region.

In this chapter we describe the physical and biological setting in Bristol Bay, the methods used to quantify energy flux to pelagic birds, and the results of these calculations. Next we describe the physical and biological setting on Georges Bank, the methods used to quantify energy flux in this region, and the results of these calculations. We discuss some of the limits to our estimates of spatial variation in energy flux, and we discuss the implications of our results.

Bristol Bay (southeastern Bering Sea)

Physical environment

The continental shelf of the southeastern Bering Sea is a broad flat plain with a minimum width of 500 km (Fig. 11.1). Notable topographic features include the Pribilof Islands, submarine banks southeast of Cape Newenham, and a shelf break at 170 m. Important hydrographic features include water mass boundaries around the Pribilof Islands (Kinder *et al.*, 1983), along the shelf break (Kinder & Coachman, 1978), along the 50 m isobath (Schumacher *et al.*, 1979), and near the 100 m isobath (Coachman & Charnell, 1979). The front along the shelf break marks the rapid transition from warm, saline oceanic water to colder, less salty shelf water. Interactions between these two water masses, including offshore movement of fresh water and onshore movement of salts, occurs in the outer domain (Fig. 11.2). During the summer the water column in the outer domain consists of a surface layer mixed by the wind and heated by the sun (Fig. 11.2). Between the well-mixed upper and lower layers is an

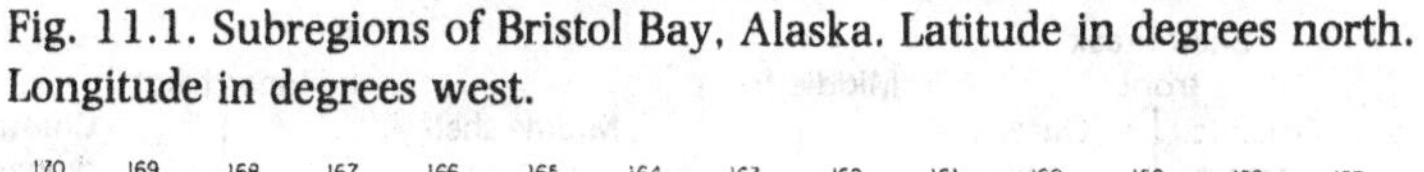

Fig. 11.1. Subregions of Bristol Bay, Alaska. Latitude in degrees north. Longitude in degrees west.

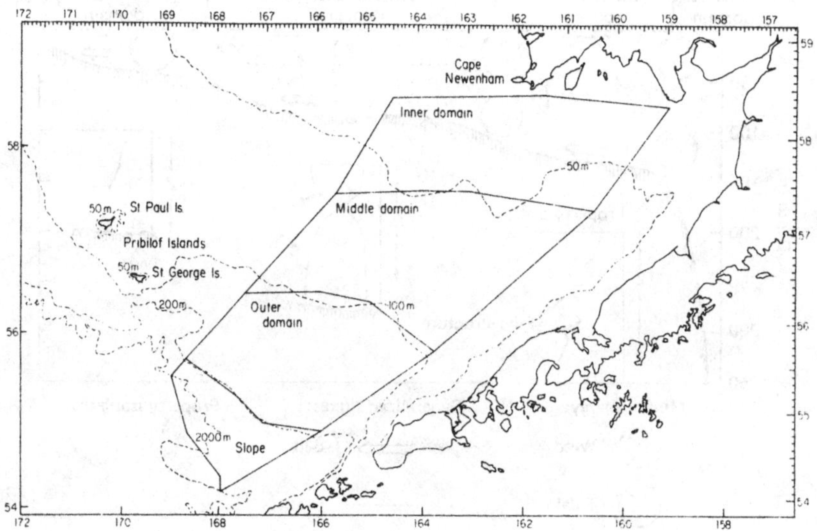

intermediate layer characterized by fine-scale interleaving of oceanic and shelf waters (Coachman & Charnell, 1979).

The broad middle front (Fig. 11.2) marks the transition from a three-layer system on the outer shelf to a two-layer system across the middle domain. During the summer the surface layer is mixed by the wind and heated by the sun. The bottom layer is isolated from interactions with other water masses and it remains relatively cold throughout the summer.

The inner front (Fig. 11.2) is associated with the 50 m isobath and it is the boundary between stratified and vertically homogeneous water masses. The inner front marks the rapid lateral transition in the balance of tidal mixing, which prevails in shallower water inside the front, and buoyant inputs, which prevail in deeper water seaward of the front (Schumacher *et al.*, 1979). Tidal fronts of similar structure and dynamics are found around the Pribilof Islands (Kinder *et al.*, 1983).

Tidal oscillation is the major source of kinetic energy of the water on the shelf (Kinder & Schumacher, 1981). Lateral advection is relatively

Fig. 11.2. Schematic diagram of summer mixing regions in Bristol Bay. Redrawn from Coachman (1982). T = Temperature; S = Salinity; Z = depth.

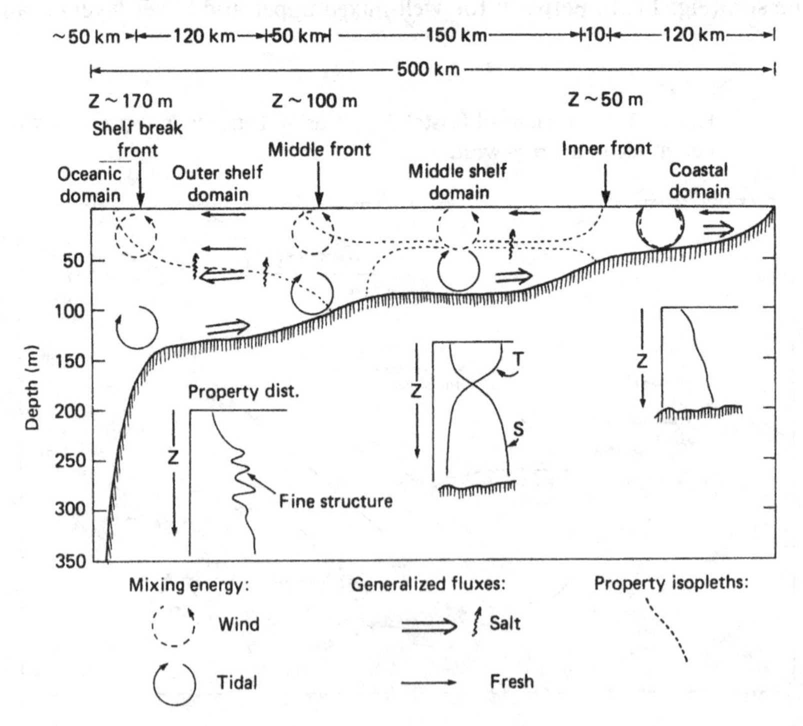

weak and hence net transport of nutrients and plankton across the shelf is small. Currents are strongest near the inner front where speeds are 1–6 cm s^{-1} parallel to the front. Mean current speeds are less than 1 cm s^{-1} in the middle domain, and across-isobath speeds of 1–5 cm s^{-1} occur in the outer domain. The middle front is, as a consequence, often a region of convergence. Lateral flux of salts, including nutrients, occurs in the outer domain but not in the middle domain (Coachman, 1982).

Ice cover is a seasonal feature. Maximum coverage of the shelf and maximum ice thickness (2 m) are reached in March. Stratification of the water column begins in April and continues through October when lower sun angles, decreased day length, and increased frequency of storms result in breakdown of the thermocline.

Biological environment

Seasonal stratification of the water column results in an intense, short-lived bloom of plankton over the shelf in late April and early May (Hattori & Goering, 1981). In the outer domain, much of this bloom is captured by relatively large calanoid copepods, which rise from deep water along the shelf margin in the spring (Cooney, 1981). Grazing by these copepods is restricted to the outer domain. Smaller copepods that overwinter as eggs in the middle and inner domains do not capture the spring bloom, which sinks to the bottom (Iverson *et al.*, 1979). The result is a highly productive benthic community in the middle domain (Haflinger, 1981).

Structure of the pelagic bird community in the Bering Sea was described by Shuntov (1974), who noted the predominance of Northern Fulmars *Fulmarus glacialis* along the outer shelf in the summer. Shuntov also noted that during the summer large feeding aggregations of moulting shearwaters (primarily Short-tailed Shearwaters *Puffinus tenuirostris*) occurred along the landward edge of the 'cold pool' (i.e. the inner front, see also Takenouti & Ohtani, 1974). Irving, McRoy & Burns (1970) found that murres (*Uria* spp.) and gulls (*Larus* spp. and Ivory Gull *Pagophila eburnea*) were abundant along the ice edge in the winter.

Distribution of breeding colonies was catalogued by Sowls, Hatch & Lensink (1978). The largest breeding colonies in Bristol Bay occur on St George Island in the Pribilofs, and at Cape Newenham (Fig. 11.1). Numerically dominant species at the Pribilofs include Northern Fulmars, Thick-billed Murres *Uria lomvia*, Red-legged Kittiwakes *Rissa brevirostris*, and Least Auklets *Aethia pusilla*. Major breeding species in coastal colonies include Black-legged Kittiwakes *Rissa tridactyla*, Glaucous-winged Gulls *Larus glaucescens*, and Common Murres *Uria aalge*. The major prey taken at the Pribilof Islands was the Walleye Pollack *Theragra chalcogramma* (Hunt *et al.*, 1981*a*). Major prey in coastal colonies included sandlance *Ammmodytes* sp., Capelin *Mallotus villosus*, and squid (Sanger, 1983).

Methods for Bristol Bay

Energy flux to pelagic bird populations was estimated from two rate equations, one describing daily individual energy requirement as a function of body mass, and a second describing change in bird abundance as a function of date. Standard metabolic rate (SMR) was calculated from the allometric formula of Lasiewski & Dawson (1967): SMR (kcal day^{-1}) = $78.3\ M^{0.723}$ (kg). The daily intake of active birds was estimated using a conversion of 2.8 SMR (Kooyman *et al.*, 1982) and an assimilation efficiency of 75% (Cooper, 1978). Converting to kilojoules, daily intake (I) was: I (kJ day^{-1}) = $1216\ M^{0.723}$ (kg). Average mass of numerically important species was determined from birds collected in central Bristol Bay in 1981.

Seasonal change in abundance of individual species was estimated from the equation for the normal curve (Preston, 1966; Schneider *et al.*, 1986). Annual occupancy (O) was the integral of the normal curve: O (bird-days $\text{km}^{-2}\ \text{yr}^{-1}$) = $H\ T\ \sqrt{2\pi}$, where H is the maximum annual density (birds km^{-2}) and T is the standard deviation (in days) around the mean date, measured in days from 1 January.

Mean date and standard deviation were computed from counts made at sea in 1980 and 1981. Counts were made in a 90° sector from the beam forward, out to 300 m, while underway between hydrographic stations. The number of birds of each species seen during 10 min was divided by the area scanned during the interval. The standardized counts were then averaged in each of the four sub-regions (Fig. 11.1) on each cruise. Cruise averages were used to determine the maximum density (H), mean date of bird occupancy (M), and standard deviation around the mean date (T) of each species. Mean dates and standard deviations were computed using the mid-point of each cruise (days from 1 January) as the variate, and then using the average standardized count as a weighting factor. This procedure is equivalent to that used to compute a mean and a variance from data grouped into a frequency distribution. These statistics were used to compute occupancy of each species in each of four sub-regions (slope, outer, middle, and inner shelf). Annual energy flux to each species was the product of occupancy (O) and intake (I). Annual energy flux in each of the four sub-regions was the sum of these products. More detailed description of the methods can be found in Hunt *et al.* (1981*b*), Schneider & Hunt (1982), and Schneider *et al.* (1986).

Model estimates of flux to all species were computed for each day of the year in each of the four sub-regions. Model estimates were compared graphically to estimates of flux to all species during each cruise. To obtain a cruise estimate, the daily intake (I) was multiplied by the average standardized count of a species in one sub-region during the cruise. Annual flux to each species was computed by rectangular approximation, as follows. Average energy flux to each species during each cruise was multiplied by half of the time since the last cruise, plus half of the time until the next cruise. These products were summed over two annual cycles – 1 March 1980 to 1 March 1981 and 1 September 1980 to 1 September 1981. Annual estimates were then summed over species for comparison to model estimates, which were computed over the same two annual periods.

Model for Bristol Bay

Model estimates of energy flux in the inner domain were in good agreement with all but two cruise estimates (Fig. 11.3). The model of seasonal occupancy captured the major feature of energy flux to birds in the inner domain – brief periods of high flux. The three seasonal peaks in the inner domain were due to murres (April 1980) and *Puffinus* shearwaters (August 1980 and July 1981).

Fig. 11.3. Seasonal variation in energy flux to pelagic birds in four sub-regions of Bristol Bay. Horizontal lines show energy flux during individual cruises. Curves show model estimates of energy flux.

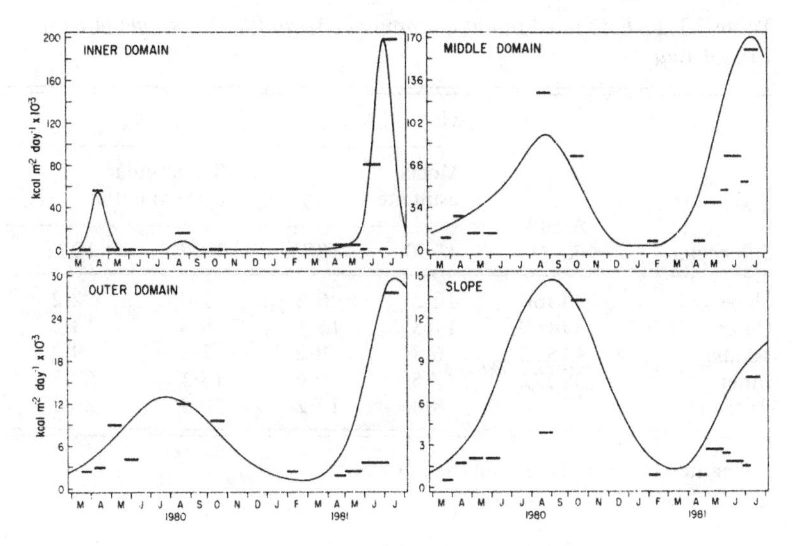

Agreement between model and cruise estimates was less satisfactory in the middle domain (Fig. 11.3). The occupancy model is relatively insensitive to low values of flux near the date of peak flux, but is highly sensitive to low values far from the date of peak abundance. The sensitivity of the model to outlying values resulted in over-estimation of energy flux in the middle domain in March and April of 1981 (Fig. 11.3). The model did capture the main features of energy flux to birds in the middle domain – energy flux occurred over a longer period than in the inner domain, and peak flux was lower. Energy flux in the middle domain was lower in 1980 than in 1981 (Table 11.1). During 1980 the major contributors to energy flux in this domain were murres (27% of the model estimate), shearwaters (23%), Northern Fulmars (24%) and *Larus* gulls (13%). During 1981 the major contributors to the model estimate were shearwaters (65%), and Northern Fulmars (17%).

Agreement between model and cruise estimates was more satisfactory in the outer than in the middle domain (Fig. 11.3). During 1980 the major contributors to the model estimates were Northern Fulmars (27%), shearwaters (23%), and *Larus* gulls (23%). During 1981 the major contributors to the model estimates were Northern Fulmars (60%), *Larus* gulls (13%), and shearwaters (13%).

Agreement between model and cruise estimates was least satisfactory in the slope region (Fig. 11.3). The model was sensitive to high values of flux in October 1980. The model was more sensitive to low values in the spring, 5–6 months before October, than to intermediate values in August, 2 months before October. The major contributors to model estimates of

Table 11.1. *Estimated minimum annual energy flux to pelagic birds in Bristol Bay*

		Flux (kJ m^{-2} yr^{-1})			
		Model estimate		Rectangular approximation	
Sub-region	Area (km^2)	1980	1981	1980	1981
Slope	14360	12·2	9·6	7·0	2·2
Outer	34460	11·3	15·1	9·8	13·3
Middle	44510	6·3	9·2	7·3	9·2
Inner	39372	7·5	30·6	15·1	62·5
Average[a]	—	8·8	17·2	10·2	25·3

[a] Weighted by area of each sub-region.

energy flux beyond the shelf break were *Larus* gulls (50% in 1980, 41% in 1981) and Northern Fulmars (26% in 1980, 29% in 1981).

Cruise estimates tended to fall below model estimates and this suggests that the model over-estimated annual flux. This could not be tested directly but it is at least possible to compare model estimates with estimates made by rectangular approximation. The model estimates were lower than the rectangular approximations in the inner domain (Table 11.1). Model estimates were higher than rectangular approximations in the slope and outer domains. These differences reflect the sensitivities of the two methods of integration – sensitivity to gaps between cruises in the case of rectangular approximation, and sensitivity to peak values and outlying values in the case of the occupancy model. The model estimates were, on the average, conservative relative to the rectangular approximations (Table 11.1).

Georges Bank (northwest Atlantic)

Physical environment

Georges Bank is a shallow bank roughly 300 km long and 150 km wide (Fig. 11.4). Notable topographic features include a shelf break at about 150 m depth, the Northeast Channel (220 m deep), the Great South Channel (70 m deep), and pronounced bathymetric irregularities on the crest of the bank (less than 40 m depth). Minimum depth on the bank is 3 m. Important hydrographic features include water mass boundaries

Fig. 11.4. Sub-regions of Georges Bank, northwest Atlantic Ocean. s = southern edge; e = eastern peak; n = northern flank; sh = shoals; w = western side. NC = Northeast Channel; GSC = Great South Channel; GM = Gulf of Maine; MAB = Middle Atlantic Bight.

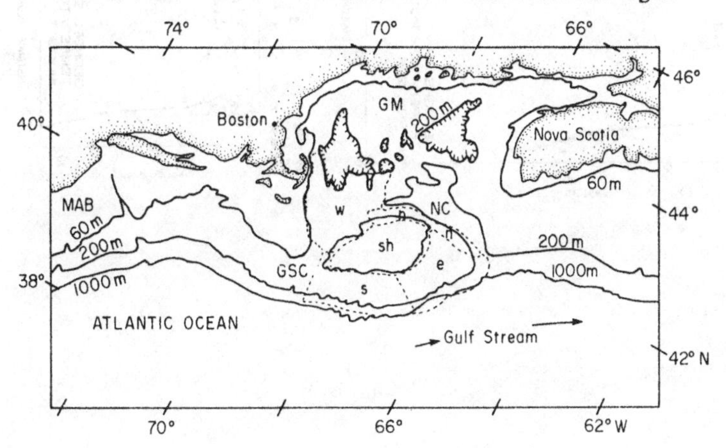

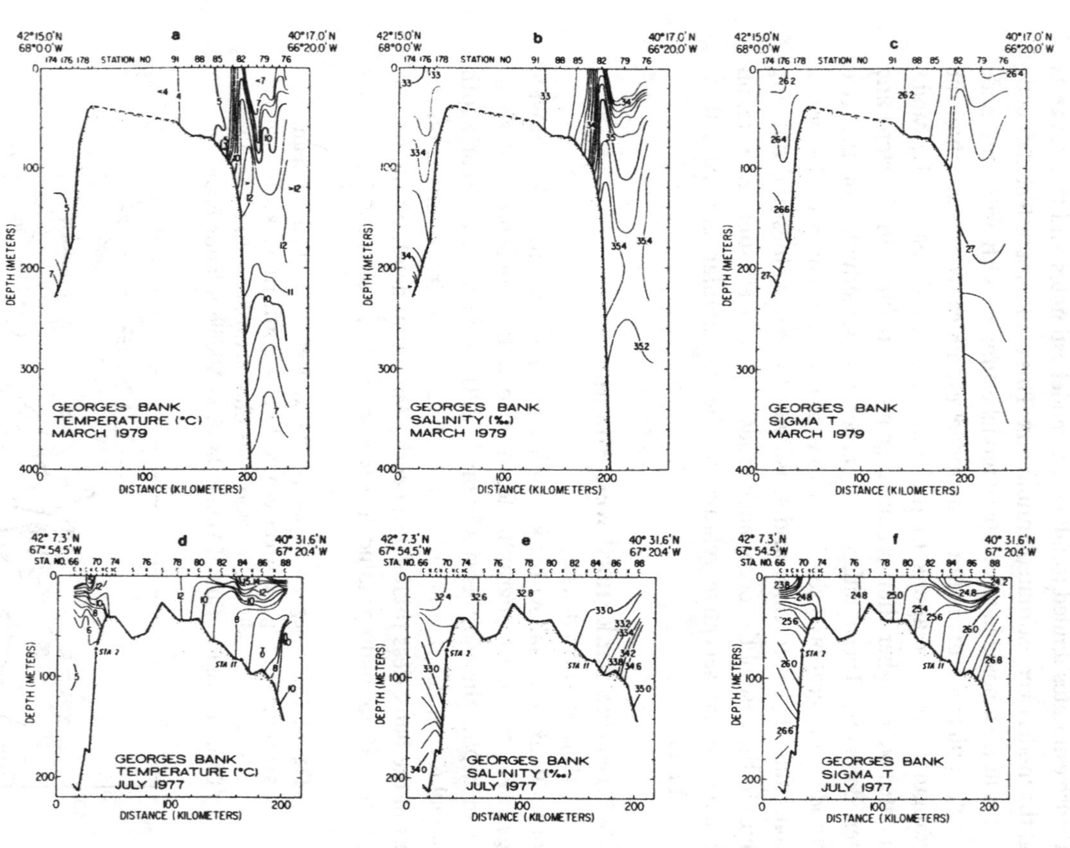

Fig. 11.5. Salinity, temperature, and density (sigma t) profiles across Georges Bank during spring and summer. Redrawn from Allen *et al.* (1983).

along the southern edge, along the northern flank, and along the 60 m isobath (Butman *et al.*, 1982).

The shelf water/slope front along the southern edge is similar in structure and continuous with the front at the shelf break in the middle Atlantic Bight (Mooers, Garvine & Martin, 1979). This front, which marks the boundary between shelf water and saltier slope water, intersects the bottom at approximately 80 m (Fig. 11.5). A second, much weaker saline front along the northeast flank separates Georges Bank water from less-saline Gulf of Maine water (Fig. 11.5).

During the summer a thermal front forms at approximately the 60 m isobath. As in Bristol Bay, a cold pool of water becomes isolated at the bottom between the 60 and 100 m isobaths (Fig. 11.5). A seasonal thermocline, at about 30 m, separates the cold pool from the layer above, which is mixed by the wind and heated by the sun.

Currents on Georges Bank are dominated by diurnal and semidiurnal tides (Allen *et al.*, 1983). Net advection across the shelf is relatively weak. Net circulation consists of a slow clockwise flow around the Bank, with current speeds that are typically 5–10 cm s^{-1} (Butman *et al.*, 1982). This circulation pattern includes strong northeastward flow along the northern flank, southerly flow on the northeast peak, broad southwestward flow along the southern edge, and northward flow on the eastern side of the Great South Channel. High current velocities and a strong thermal front are associated with the 50 m isobath along the northeast flank (Magnell *et al.*, 1980). This jet carries new water onto the Bank during the summer (Hopkins & Garfield, 1982).

Biological environment

Maximum daily primary production on Georges Bank occurs in April and October (O'Reilly, Evans-Zetlin & Busch, in press). Minimum daily production occurs during the winter (December through February). Annual primary production ranged from 455 g C m^{-2} yr^{-1} on the shallow parts of the bank, to 265 g C m^{-2} yr^{-1} in deeper water along the southern edge (O'Reilly *et al.*, in press).

Riley (1947) found that maximum zooplankton biomass occurred in August, and Clark, Pierce & Bumpus (1943) described the association of specific zooplankton assemblages with Georges Bank water. Recently Blake *et al.* (1983) found that maximum benthic biomass occurred on the central part of the Bank, between the 60 and 100 m isobaths, a pattern similar to that of the benthos in the Bering Sea (Haflinger, 1981).

The structure of the pelagic bird community on Georges Bank was described by Powers & Brown (in press). Breeding birds do not forage on the Bank. The Great Shearwater *Puffinus gravis*, a transequatorial migrant, was the most abundant bird on the Bank in the summer and fall. Other numerically important species during the summer were Sooty Shearwaters *Puffinus griseus* and Wilson's Storm Petrels *Oceanites oceanites*. Numerically important species during the winter and spring included Northern Fulmars, Herring Gulls *Larus argentatus*, and Great Black-backed Gulls *L. marinus*. Red Phalaropes *Phalaropus fulicarius* were abundant in deeper waters of the southern edge during their northward spring migration.

Methods for Georges Bank

Energy flux to seabird populations was estimated from the rate equation describing daily energy requirement as a function of body mass (see above) and from the average standardized density of birds in five sub-regions of Georges Bank (Fig. 11.4) in four seasons – spring (March–May), summer (June–August), fall (September–November), and winter (December–February). Sub-regions were defined on the basis of bathymetry, hydrography, and fishing activity (Powers & Brown, in press). Counts were made in a 300 m zone on one side of the ship while underway at speeds greater than 4 knots. The number of birds seen in the zone in 10 min was divided by the area scanned during this time. Standardized counts were averaged within each of the five sub-regions and each of the four seasons. Annual occupancy of a sub-region by a species was the product of average density and length of season (90 days), summed over four seasons. Annual energy flux was the product of occupany and intake. Products were summed across species within each of the sub-regions to obtain annual flux in that sub-region. The area of each sub-region was used as a weighting factor to obtain total flux, averaged over the five sub-regions.

Results for Georges Bank

Total energy flux to birds, averaged over all five sub-regions, was 9 kJ m^{-2} yr^{-1} (Table 11.2). Great Shearwaters accounted for 40% of this flux, with less flux to Northern Fulmars (17%), Herring Gulls (12%), and Great Black-backed Gulls (12%). This flux was highly seasonal. Flux to Great Shearwaters occurred primarily in June and July, with less flux in October. The highest flux to Northern Fulmars occurred in January through March, while flux to Great Black-backed Gulls remained high from September through April.

Maximum energy transfer to Great Shearwaters occurred in the northern entrance of the Great South Channel, along the northern flank, and the

Bank along the 67° meridian (Powers, 1983). This distribution represents the eastern, northern and western periphery of the mixed water mass maintained by tidal energy on the Bank throughout the summer (Riley, 1942). Flux to Northern Fulmars occurred in slightly deeper water. Maximum concentrations of Fulmars were observed along the Great South Channel, along the northern flank, and along the northeast peak of the Bank (Powers, 1983). Maximum concentrations of *Larus* gulls were observed west of the Great South Channel and across the northern flank eastward to the Northeast Channel. Fulmars, gulls and shearwaters were all recorded in association with fishing vessels, but differences in distribution of these species indicate that fishing activity was not the sole determinant of spatial variation in energy flux in these species (Powers & Brown, in press). In general, high energy flux to birds occurred in areas of strong topographic relief and strong hydrographic gradients, especially on the northern edge of the Bank from the Great South Channel eastward along the periphery of the shoals.

Discussion

In our comparison of two boreal shelf ecosystems we found that energy flux to birds averaged 9 kJ m^{-2} yr^{-1} over Georges Bank, and 13 kJ m^{-2} yr^{-1} in Bristol Bay, averaged over 2 years. This difference is small compared with the difference in flux to birds found in structurally different ecosystems such as the open ocean in the eastern Pacific (0.4 kJ m^{-2} yr^{-1}: Sanger, 1972) and coastal upwellings (30 kJ m^{-2} yr^{-1}: Schaeffer, 1970; Wiens & Scott, 1975). The difference between Georges Bank and Bristol Bay was also small compared with the difference in flux within the two regions. Average annual flux in Bristol Bay ranged from 7.8 kJ m^{-2} yr^{-1}

Table 11.2. *Estimated average annual energy flux to pelagic birds on Georges Bank*

Sub-region	Area (km^2)	Flux (kJ m^{-2} yr^{-1})
Shoal	12605	6·5
West	13624	8·5
North	3011	19·9
East peak	12046	14·3
South	11181	3·6
Average[a]		9·0

[a] Weighted by area.

in the middle domain to 19.1 kJ m^{-2} yr^{-1} in the inner domain. Average annual flux to birds on Georges Bank ranged from 3.6 kJ m^{-2} yr^{-1} in stratified water of the southern sector to 19.9 kJ m^{-2} yr^{-1} along the northern edge of the bank. Our calculations indicate that spatial variation in energy flux to birds was greater within than between these two boreal shelf ecosystems. Seasonal counts of pelagic birds in another boreal shelf system, such as the North Sea or the Grand Banks of Newfoundland, are needed to test whether energy flux to birds is similar at this scale in structurally similar ecosystems.

Energy flux to birds in Bristol Bay varied with depth and hydrographic regime. Higher rates of energy flux were observed in shallower water inside the inner front than in stratified water seaward of this front. Energy flux to pelagic birds was lower in stratified water of the middle domain than in stratified water of the outer shelf domain. This pattern was observed in 1975–79 data from Bristol Bay (Schneider & Hunt, 1982). The patterns were confirmed by data from 1980 and 1981 (Schneider *et al.*, 1986). Flux in the outer domain was due primarily to surface-foraging birds, especially Northern Fulmars in the summer and fall, and *Larus* gulls in the fall and winter. Flux in shallower water, especially inside the inner front, was due primarily to subsurface-foraging birds, murres and *Puffinus* shearwaters.

Energy flux on Georges Bank was also structured by hydrographic regime, but with a different spatial pattern. Flux was relatively low in completely mixed water inside the 60 m isobath. Flux was relatively high along the northern and eastern edge of the Bank, especially along topographic features such as the periphery of the shoals and the northern entrance to the Great South Channel. Along the periphery of the bank, flux was due primarily to fulmars and gulls during the winter. Great Shearwaters were the most important avian consumers on Georges Bank and this species was generally found in shallower water than Fulmars or *Larus* gulls.

Energy flux to birds was probably more localized than indicated by our analysis. Midsummer transects in Bristol Bay showed strong localization of shearwaters just inside the inner front, and strong localization of Fulmars near the shelf break (Schneider *et al.*, 1986). Transects across Georges Bank show localization of birds near the edge of the Bank, especially along the northern flank and to a lesser degree along the shelf break (Powers, 1983). These finer-scale analyses suggest that flux to birds is localized in regions of strong topographic and hydrographic gradients (Brown, 1980).

A better understanding of local variation in energy flux will be needed to examine the interaction of seabirds with fisheries. Average energy flux to birds and fisheries from a shared resource may underestimate the degree of interaction of seabirds with fisheries. Average energy flux to birds and fisheries from a shared prey may underestimate the degree of interaction among highly aggregated consumers, especially if prey are also aggregated. If consumers (both seabirds and fisheries) require a resource that is aggregated above some economic threshold, and if spatial and temporal overlap is high, then negative interaction among consumers may occur well before any major reduction in prey stock. A high degree of spatial and temporal overlap, and potentially significant negative interaction, seems inevitable in the case of consumers in which the costs of locomotion dictate reliance on spatially predictable resources. Examples of consumers that are likely to depend on spatially predictable resources include diving birds (Schneider *et al.*, 1986) and factory ships. Management models that do not include spatial structure are likely to underestimate the impact of one aggregated consumer on another.

Until recently research on pelagic birds has focused on patterns in distribution and standing stock, rather than on processes such as energy flux. As our calculations show, rather uneven progress has been made toward satisfactory description of seabird dynamics, including energy flux. To date we have a general model of individual intake, a preliminary model of seasonal abundance, and no functional description of spatial variation in seabird biomass. Average daily intake can be described as an allometric function of body mass and there is considerable empirical support for this generalization. However, there is room for refinement in the application of these models to free-living birds (Walsberg, 1983), including seabirds (Wiens, 1984). The occupancy model used for the Bering Sea calculations (Schneider *et al.*, 1986) was the first attempt to model seasonal change in seabird numbers away from colonies. The model derives some external justification from the prevalence of circannual rhythms in birds (Gwinner, 1977), including breeding seabirds (Marshall & Serventy, 1959). To date little is known about the degree of endogenous control of seasonal movement of pelagic birds away from colonies. The occupancy model assumes equal accumulation and departure rates of birds from an area. The data from the outer and middle domains suggest that seasonal accumulation occurs more rapidly than seasonal dispersal from these areas (Fig. 11.3). Better seasonal models will be needed to quantify this asymmetry, or to evaluate biologically important details such as individual variation in residence time in an area.

No attempt was made in this chapter to go beyond a simple box model and develop a functional description of spatial variation in feeding activity. A satisfactory description of spatial variation in foraging activity is an especially difficult problem. The use of multiple regression to identify important spatial variables has not been notably successful (Abrams & Griffiths, 1981; Kinder *et al.*, 1983). Central place foraging theory has been applied to birds tied to breeding colonies (Ford *et al.*, 1982), but the theory has not been extended to mobile predators foraging on mobile prey.

It seems likely that the dynamics of seabird predation will turn out to be a function of several processes, including localized aggregation ('fast' processes), seasonal movement ('intermediate'-rate processes), and population change ('slow' processes). The spatial scale of these processes is likely to differ and this will need to be included in any functional description of energy flux to pelagic birds. Indeed it is possible that pelagic birds, because they are the most visible of all marine predators, may provide important insights into the operation of scale-dependent interaction of prey and predator populations in marine ecosystems.

Acknowledgements We are indebted to the many people who helped us count birds in Bristol Bay and on Georges Bank, often under difficult conditions. Data collection and analysis was supported by the following agencies: Environmental Protection Agency, National Marine Fisheries Service, National Science Foundation, Office of Marine Pollution Assessment. The manuscript was prepared during a stay by the first author in the Department of Zoology, University of Rhode Island, Kingston, RI 02881, USA.

References

Abrams, R.W., & Griffiths, A.M. (1981). Ecological structure of the pelagic seabird community in the Benguela Current region. *Mar. Ecol.-Prog. Ser.*, **5**, 269–77.

Allen, J.S., Beardsley, R.C., Blanton, J.O., Boicourt, W.C., Butman, B., Coachman, L.K., Huyer, A., Kinder, T.H., Royer, T.C., Schumacher, J.D., Smith, R.L., Sturges, W. & Winant, C.D. (1983). Physical oceanography of continental shelves. *Rev. Geophys. Space Phys.*, **21**, 1149–81.

Blake, J.A., Grassle, J.F., Maciolek-Blake, N., Neff, J.M. & Sanders, H.L. (1983). *The Georges Bank benthic infauna monitoring program.* Duxbury, Massachusetts: Battelle New England Marine Research Laboratory.

Brown, R.G.B. (1980). Seabirds as marine animals. In *Behavior of marine animals*, vol. 4, ed. J. Burger, B.L. Olla & H.E. Winn, pp. 1–19. New York: Plenum Press.

Butman, B., Beardsley, R.C., Magnell, B., Frye, D., Vermesch, J.A., Schlitz, R., Limeburger, R., Wright, W.R. & Noble, M.A. (1982). Recent observations of the mean circulation on Georges Bank. *J. Phys. Oceanogr.*, **12**, 569–91.

Clark, G.L., Pierce, E.L. & Bumpus, D.F. (1943). The distribution and reproduction of *Sagitta elegans* on Georges Bank in relation to hydrographic conditions. *Biol. Bull.*, **85**, 201–26.

Coachman, L.K. (1982). Flow convergence over a broad continental shelf. *Cont. Shelf Res.*, **1**, 1–14.

Coachman, L.K. & Charnell, R.K. (1979). On lateral water mass interaction – a case study, Bristol Bay, Alaska. *J. Phys. Oceanogr.*, **9**, 278–97.

Cooney, R.T. (1981). Bering Sea zooplankton and micronekton communities with emphasis on annual production. In *The eastern Bering Sea shelf: oceanography and resources*, vol. 2, ed. D.W. Hood & J.A. Calder, pp. 947–74. Rockville, Maryland: U.S. Department of Commerce.

Cooper, J. (1978). Energetic requirements for growth and maintenance of the Cape Gannet (Aves: Sulidae). *Zoologica Afr.*, **13**, 307–17.

Ford, R.G., Wiens, J.A., Heinemann, D. & Hunt, G.L. (1982). Modelling the sensitivity of colonially breeding marine birds to oil spills: guillemot and kittiwake populations on the Pribilof Islands, Bering Sea. *J. Appl. Ecol.*, **19**, 1–31.

Furness, R.W. (1978). Energy requirements of seabird communities: a bioenergetics model. *J. Anim. Ecol.*, **47**, 39–53.

Furness, R.W. (1982). Competition between fisheries and seabird communities. *Adv. Mar. Biol.*, **20**, 225–307.

Furness, R.W. & Cooper, J. (1982). Interactions between breeding seabirds and pelagic fish populations in the southern Benguela region. *Mar. Ecol.-Prog. Ser.*, **8**, 243–50.

Gwinner, E. (1977). Circannual rhythms in bird migration. *Ann. Rev. Ecol. Syst.*, **8**, 381–405.

Haflinger, K. (1981). A survey of benthic infaunal communities of the southeastern Bering Sea shelf. In *The eastern Bering Sea shelf: oceanography and resources*, vol. 2, ed. D.W. Hood & J.A. Calder, pp. 1091–104. Rockville, Maryland: U.S. Department of Commerce.

Hattori, A. & Goering, J.J. (1981). Nutrient distribution and dynamics in the eastern Bering Sea. In *The eastern Bering Sea shelf: oceanography and resources*, vol. 1, ed. D.W. Hood & J.A. Calder, pp. 975–92. Rockville, Maryland: U.S. Department of Commerce.

Hopkins, T.S. & Garfield, N., III. (1982). Physical origins of Georges Bank water. *J. Mar. Res.*, **39**, 465–500.

Hunt, G.L., Jr, Burgeson, B., & Sanger, G.L. (1981*a*). Feeding ecology of seabirds of the eastern Bering Sea. In *The eastern Bering Sea shelf: oceanography and resources*, vol. 2, ed. D.W. Hood & J.A. Calder, pp. 629–48. Rockville, Maryland: U.S. Department of Commerce.

Hunt, G.L., Jr, Gould, P.G., Forsell, D.S., & Petersen, H. (1981*b*). Pelagic distribution of marine birds of the eastern Bering Sea. In *The eastern Bering Sea shelf: oceanography and resources*, vol. 2, ed. D.W. Hood & J.A. Calder, pp. 689–715. Rockville, Maryland: U.S. Department of Commerce.

Irving, L., McRoy, C.P. & Burns, J.J. (1970). Birds observed during a cruise in the ice-covered Bering Sea in March, 1968. *Condor*, **72**, 110–12.

Iverson, R.L., Coachman, L.K., Cooney, R.T., English, T.S., Goering, J.J., Hunt, G.L., Jr, MacCauley, M.C., McRoy, C.P., Reeburgh, W.S. & Whitledge, T.E. (1979). Ecological significance of fronts in the southeastern Bering Sea. In *Ecological processes in coastal and marine systems*, ed. R.L. Livingston, pp. 437–65. New York: Plenum Press.

Kinder, T.H. & Coachman, L.K. (1978). The front overlying the continental slope of the eastern Bering Sea. *J. Geophys. Res.*, **83**, 4551–9.

Kinder, T.H., Hunt, G.L. Jr, Schneider, D.C. & Schumacher, J.D. (1983). Correlations between seabirds and oceanic fronts around the Pribilof Islands, Alaska. *Estuar. Cst Shelf Sci.*, **16**, 309–19.

Kinder, T.H. & Schumacher, J.D. (1981). Circulation over the continental shelf of the southeastern Bering Sea. In *The eastern Bering Sea shelf: oceanography and resources*, vol. 1, ed. D.W. Hood & J.A. Calder, pp. 53–78. Rockville, Maryland: U.S. Department of Commerce.

Kooyman, G.L., Davis, R.W., Croxall, J.P. & Costa, D.P. (1982). Diving depths and energy requirements of King Penguins. *Science*, **217**, 726–7.

Lasiewski, R.C. & Dawson, W.R. (1967). A re-examination of the relation between standard metabolic rate and body weight in birds. *Condor*, **69**, 13–23.

Magnell, B.A., Spiegel, S.L., Scarlet, R.I. & Andrews, J.B. (1980). The relationship of tidal and low-frequency currents on the north slope of Georges Bank. *J. Phys. Oceanogr.*, **10**, 1200–12.

Marshall, A.J. & Serventy, D.L. (1959). The experimental demonstration of an internal rhythm of reproduction in a transequatorial migrant, the Short-tailed Shearwater *Puffinus tenuirostris*. *Nature Lond.*, **184**, 1704–5.

Mooers, C.N.K., Garvine, R.W. & Martin, W.W. (1979). Summertime synoptic variability of the middle Atlantic Shelf water/slope water front. *J. Geophys. Res.*, **84**, 4837–54.

O'Reilly, J.E., Evans-Zetlin, C. & Busch, D.A. (In press). Primary production on Georges Bank. In *Georges Bank*, ed. R. Backus. Cambridge, Massachusetts: MIT Press.

Powers, K.D. (1982). A comparison of two methods of counting birds at sea. *J. Field Ornithol.*, **53**, 209–22.

Powers, K.D. (1983). *Pelagic distributions of marine birds off the northeastern United States*. NOAA Technical memorandum NMFS-F/NEC-27, 201 pp. Washington, D.C.: U.S. Dept. Commerce.

Powers, K.D. & Backus, E.H.(In press). Energy transfer to birds. In *Georges Bank*, ed. R. Backus. Cambridge, Massachussetts: MIT Press.

Powers, K.D. & Brown, R.G.B. (In press). Seabirds. In *Georges Bank*, ed. R. Backus, Cambridge, Massachussetts: MIT Press.

Preston, F.W. (1966). The mathematical representation of migration. *Ecology*, **47**, 375–92.

Riley, G.A. (1942). The relationship of vertical turbulence and spring diatom flowerings. *J. Mar. Res.*, **5**, 67–87.

Riley, G.A. (1947). A theoretical analysis of the zooplankton population of Georges Bank. *J. Mar. Res.*, **6**, 104–13.

Sanger, G.L. (1972). Preliminary standing stock and biomass estimates of seabirds in the subarctic Pacific Ocean. In *Biological oceanography of the northern North Pacific Ocean*, ed. A.Y. Takenouti, pp. 589–611. Tokyo: Idemitsu Shoten.

Sanger, G.L. (1983). Diets and food web relationships of seabirds in the Gulf of Alaska and adjacent marine regions. In *Environmental assessment of the Alaskan continental shelf. Final reports of the principal investigators*, 130 pp. Juneau, Alaska: NOAA, Office of Marine Pollution Assessment.

Schaeffer, M.B. (1970). Men, birds, and anchovies in the Peru Current – dynamic interactions. *Trans. Am. Fish. Soc.*, **99**, 461–7.

Schneider, D.C. & Hunt, G.L. (1982). Carbon flux to seabirds in waters with different mixing regimes in the southeastern Bering Sea. *Mar. Biol.*, **67**, 337–44.

Schneider, D.C., Hunt, G.L., Jr & Harrison, N.M. (1986). Mass and energy transfer to seabirds in the southeastern Bering Sea. *Cont. Shelf Res.* **5**, 5, 241–57.

Schumacher, J.D., Kinder, T.H., Pashinski, D.J. & Charnell, R.L. (1979). A structural front over the continental shelf of the eastern Bering Sea. *J. Phys. Oceanogr.*, **9**, 79–87.

Shuntov, V.P. (1974). *Seabirds and the biological structure of the ocean.* Vladivostok: Far East Book Publishers (In Russian). English translation Washington, D.C.: U.S. Department of Commerce, Nat. Tech. Inf. Service NTIS-TT-74-55032.

Sissenwine, M.P. (1984). Why do fish populations vary? In *Workshop on exploitation of marine communities*, ed. R.M. May, pp. 59–94. Berlin: Springer-Verlag.

Sowls, A.L., Hatch, S.A. & Lensink, C.L. (1978). *Catalog of Alaskan seabird colonies.* Washington, D.C.: U.S. Fish and Wildlife Service FWS/OBS 78/78.

Takenouti, A.Y. & Ohtani, L. (1974). Currents and water masses in the Bering Sea: a review of Japanese work. In *Oceanography of the Bering Sea with emphasis on renewable resources*, ed. D.W. Hood & E. Kelley, pp. 39–75. Fairbanks, Alaska: University of Alaska Institute of Marine Science.

Walsberg, G.L. (1983). Avian ecological energetics. In *Avian biology*, vol. 7, ed. D.S. Farner, J.R. King & K.C. Parkes, pp. 166–220. New York: Academic Press.

Wiens, J.A. (1984). Modelling the energy requirements of seabird populations. In *Seabird energetics*, ed. G.C. Whittow & H. Rahn, pp. 255–84. New York: Plenum Press.

Wiens, J.A. & Scott, J.M. (1975). Model estimation of energy flow in Oregon coastal bird populations. *Condor*, **77**, 439–52.

12

Trophic relationships and food requirements of California seabirds: updating models of trophic impact

KENNETH T. BRIGGS
Center for Marine Studies, University of California, Santa Cruz, California 95064, USA
and

ELLEN W. CHU
Contributing Editor, *Bioscience*, American Institute of Biological Sciences, 7334 Champagne Point Road, Kirkland, Washington 98034, USA

Introduction

The California Current system comprises some of the world's most productive marine waters. Seasonal nutrient upwellings along the Pacific coasts of the USA and Baja California, Mexico, promote primary production rates from 130 g to more than 300 g carbon m^{-2} yr^{-1} – considerably higher than in adjacent waters of the central North Pacific (Garrison, 1976; Smith & Eppley, 1982). As in major upwelling regions off Peru and Africa, food chains are relatively short, and pelagic stocks of macroplankton, schooling fish, and squid support large populations of seabirds and marine mammals, as well as important human fisheries.

From an oceanographic and fisheries standpoint this region is perhaps the best studied in the world, but the role of seabirds and other predators is only now emerging. In this chapter we apply new approaches to a familiar bioenergetics model to assess the impact of California coastal seabirds on marine prey populations. The model is based on comprehensive population studies done from mid-1975 through early 1978 in southern California, and 1980 through 1982 in central and northern California (Briggs *et al.*, 1981*a*,*b*, 1984; Chu, 1984), as part of our work focusing on all seabird and marine mammal populations in habitats targeted for oil and gas leasing (Fig. 12.1). Information simultaneously gathered by ships, airplanes, and NOAA weather satellites provides the hydrographic context for interpreting our results.

Fig. 12.1. Map of coastal California showing important geographical features and (shaded) limits of area studied.

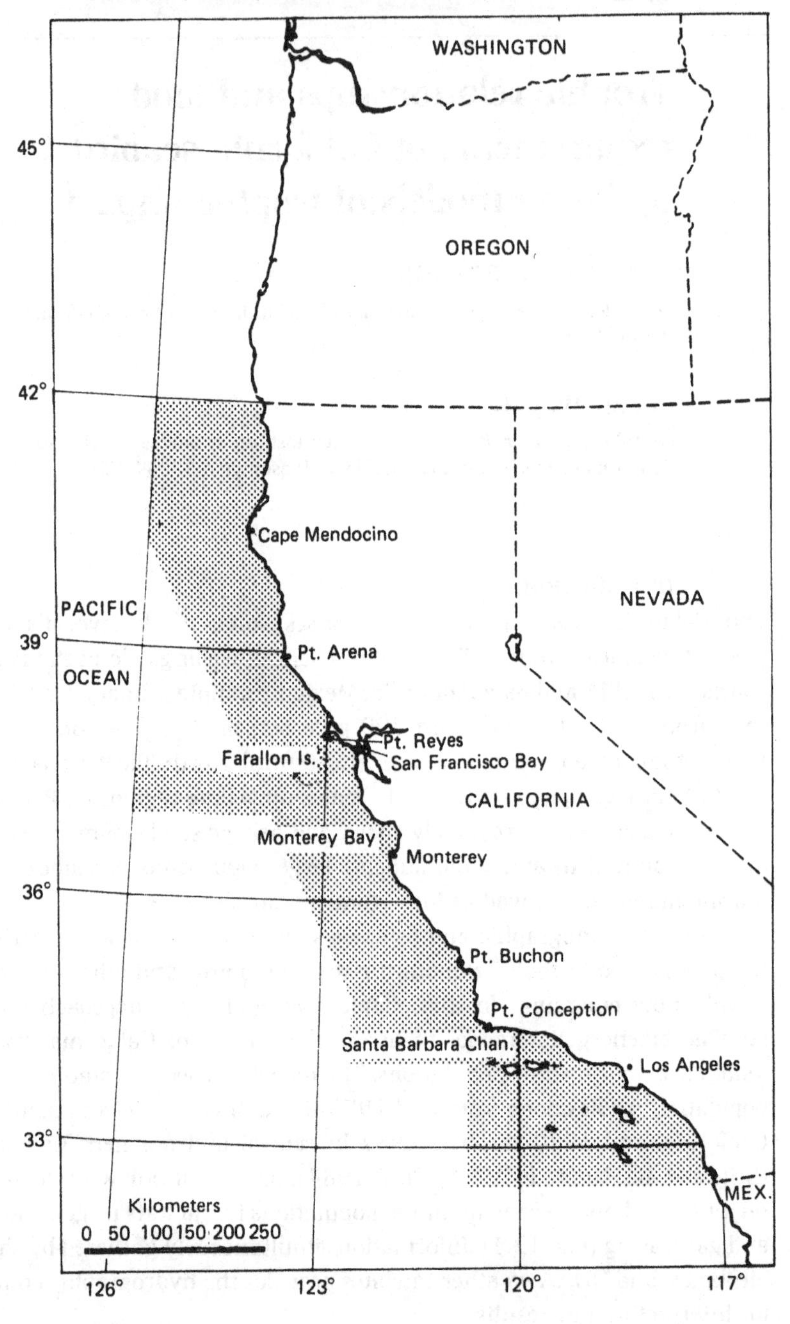

We surveyed seabirds at sea biweekly or monthly by ship and aircraft, estimating bird density as described by Briggs *et al.* (1981*b*, 1984). These surveys produced a series of 'snapshots' of bird populations in the span of a few days. But, because they were temporally coarse-grained, the surveys do not resolve rates of population turnover among migrants, and thus total numbers of birds that might be present throughout the year. Beaches, colonies, and roosts were censused monthly.

In addition, we have been looking at seabird diets in Monterey Bay since 1974; in southern California we studied foods during 1976 and 1977 (Briggs, unpublished specimen notes; Briggs *et al.*, 1981*a*; Hunt *et al.*, 1981).

Seabirds and feeding guilds

Most of California's more than 100 seabird species (Jones, Garrett & Small, 1981) belong to the Procellariiformes, Pelecaniformes, and Charadriiformes. In all seasons, non-breeding species outnumber local breeders. The most abundant summer visitors (e.g. shearwaters *Puffinus*) come from the southern hemisphere; many winter visitors migrate from Alaska and the Pacific northwest.

The 21 nesting species number about 0.85 million birds. Most are Common Murres *Uria aalge* (61%) and Cassin's Auklets *Ptychoramphus aleuticus* (12%) on colonies in the Gulf of the Farallones and north of Cape Mendocino (Sowls *et al.*, 1980). Gulls Laridae (6%), cormorants Phalacrocoracidae (10%), and storm petrels Hydrobatidae (3%) comprise most of the remainder of the nesting total. Subtropical species (Brown Pelican *Pelecanus occidentalis*, Black Storm Petrel *Oceanodroma melania*, Elegant Tern *Sterna elegans*, Xantus' Murrelet *Endomychura hypoleuca*), which represent only a small fraction of the California nesting total, nest on the islands and mainland of southern California and adjacent Baja California, Mexico. These and other warm-water species usually disperse to central California in late summer and fall, when water temperatures are highest (Ainley, 1976).

We estimate that the number of seabirds off California varies from 1.8 million in midsummer to 7.0 million in fall. Biomass densities of seabirds in central and northern California average 20 kg km^{-2} during fall and winter but only 8–12 kg km^{-2} in summer; biomass off southern California is lower, peaking in spring and fall at 12–18 kg km^{-2} and falling to 3–6 kg km^{-2} in late summer and midwinter (K.T. Briggs, unpublished data). Spring and fall peaks reflect the passage of millions of migrating shearwaters, phalaropes *Phalaropus*, and gulls.

Table 12.1. *Synopsis of anatomical and ecological features correlating with dietary differences within feeding 'guilds' of California seabirds. Table lists only abundant species for which local diet studies exist. 'Guilds' are arranged in descending order of specialization*

Feeding guilds							
Predominant food type	Where food captured	Species	Seasons of maximum occurrence	Bill size	Geographic/hydrographic concentrations	Predominant diet items	References
Neuston	Surface	Phalarope sp.	Spring, fall	Small	Convergences bordering upwellings	Insects, fish eggs, euphausiids, copepods	Briggs *et al.*, 1984
		Ashy Storm Petrel *Oceanodroma homochroa*	All	Small	Outer Coastal Upwelling Zone (*CUZ*)	(fish eggs)[a], crustaceans	Ainley *et al.*, 1974
		Leach's Storm Petrel *O. leucorhoa*	All	Small	California Current (CC) seaward of *CUZ*	(fish eggs)[a], crustaceans	Ainley *et al.*, 1974
Plankton	Subsurface	(Sooty Shearwater)[b]	Spring, summer	Medium	11–14 °C neritic waters	Large euphausiids	Chu, 1984
		Xantus' Murrelet *Endomychura hypoleuca*	Spring, summer	Small	14–16 °C, stratified waters of Southern California Eddy	Fish larvae: anchovy, rockfish, saury	Hunt *et al.*, 1981
		Cassin's Auklet *Ptychoramphus aleuticus*	All	Small	10–14 °C, *CUZ*	Fish larvae, euphausiids, amphipods	Hunt *et al.*, 1981; Manuwal, 1974
		(Common Murre)[b]	All	Medium	*CUZ*	Large euphausiids	Ainley & Sanger, 1979
Invertebrates	Benthos	Scoter sp.	Fall, winter spring	Medium	Coastal, shallow	Mollusks, nereids, herring eggs	Vermeer, 1981
Epipelagic fish	Surface only (aerial plunge)	Brown Pelican *Pelecanus occidentalis*	Summer, fall	Large	13–17 °C, translucent, neritic waters	Anchovy	Anderson *et al.*, 1980
		Arctic Tern *Sterna paradisaea*	Fall	Medium	CC seaward of *CUZ*	Epipelagic fish	Briggs, unpublished data
		Coastal tern spp.	Summer	Medium–small	Coastal, shallow	Epipelagic fish	Briggs, unpublished data
Fish/squid	Piracy	Pomarine Jaeger *Stercorarius pomarinus*	Fall–spring	Medium	All	Epipelagic fish	Bent, 1921
Fish/squid	Subsurface only	Arctic Loon *Gavia arctica*	Fall, spring	Large	Coastal shallow	Midwater & demersal fish/squid	Baltz & Morejohn, 1977
		Western Grebe *Aechmophorus occidentalis*	Winter	Medium	Coastal, shallow sandy substrate	Midwater & demersal fish/squid	Palmer, 1962
		Pelagic Cormorant *Phalacrocorax pelagicus*	All	Medium	Neritic, rocky substrates	Solitary demersal fish	Ainley *et al.*, 1981

		Brandt's Cormorant *P. penicillatus*	All	Large	Neritic, above rocky & sandy substrates	Midwater & demersal fish & squid	Ainley *et al.*, 1981
		Common Murre *Uria aalge*	All	Medium	*CUZ*	Midwater & demersal fish & squid	Baltz & Morejohn, 1977
		Rhinoceros Auklet *Cerorhinca monocerata*	Winter	Small	*CC* & Davidson Current waters (*DC*)	Small midwater fish & squid	Sealy, 1973
		Puffin sp.	Winter, spring	Medium	*CC*	Small midwinter fish	Briggs, unpublished data
		Pigeon Guillemot *Cepphus columba*	Spring, summer	Medium	9–14 °C neritic waters	Small, midwater & demersal fish	Follett & Ainley, 1976
Fish/squid/ carrion	Surface only (dipping)	Black-footed Albatross *Diomedea nigripes*	Spring, summer	Large	Seaward of *CUZ*	Neustonic squid & offal	Sanger, 1974
		Northern Fulmar *Fulmarus glacialis*	Winter	Medium	Seaward of *CUZ*	Neustonic squid & offal	Sanger, 1983; Briggs, unpublished records
		Western Gull *Larus occidentalis*	All	Medium	Neritic waters	Surface fish, squid, invertebrates, offal, refuse, carrion	Briggs, 1977; Hunt *et al.*, 1981
		Herring Gull *L. argentatus*	Winter	Medium	Neritic waters	Surface fish, squid offal, refuse, carrion, invertebrates	Briggs, unpublished observations
		California Gull *L. californica*	Fall–spring	Medium	Neritic waters	Surface fish, squid, offal, refuse, carrion, invertebrates	Briggs, unpublished observations
		Heerman's Gull *L. heermanni*	Summer, fall	Medium	Coastal	Surface fish, especially rockfish	Quinlivan, personal communication
		Black-legged Kittiwake *Rissa tridactyla*	Winter	Medium	*CC & DC*	Surface fish, squid	Baltz & Morejohn, 1977
		Bonaparte's Gull *Larus philadelphia*	Fall	Small	Coastal	Surface fish, insects crustaceans	Baltz & Morejohn, 1977
	Surface & midwater	Sooty Shearwater *Puffinus griseus*	Spring, summer	Medium	11–14 °C neritic waters	Epipelagic & midwater fish, spawning shallow-demersal squid	Chu, 1984
		Pink-footed Shearwater *P. creatopus*	Summer	Medium	11–14 °C neritic waters	Surface & midwater fish & squid	Briggs *et al.*, 1981*a*

[a] Originally reported as fish eye lenses, but sampling of plankton and planktivorous birds in the same area suggests otherwise.
[b] Although predominantly fish-eaters, murres and shearwaters are so numerous that they significantly affect plankton.

During late 1977 and late 1982, biomass dropped appreciably compared with the same season during previous years; the drop coincided with pronounced environmental warming.

The most abundant seabirds fall into several overlapping feeding guilds based on prey type and prey capture method (Table 12.1). Most are surface-seizing and diving birds that eat primarily fish and squid; specialists – in either foraging method or prey type – are few (see also Ainley & Sanger, 1979). Nonetheless, several factors reduce co-occurrence and potential competition. The small-billed planktivores (petrels, phalaropes, murrelet, and auklet) differ in foraging method and habitat. The neuston-feeders remain primarily within convergences bordering upwelling waters (phalaropes) or just seaward in stratified, translucent California Current waters (storm petrels; Briggs *et al.*, 1984). Cassin's Auklets forage in the water column within upwellings; Xantus' Murrelets, in contrast, are found outside intense upwellings, primarily in the warmer, stratified waters of southern California – a habit that mirrors the distribution of spawning Northern Anchovies *Engraulis mordax* (Briggs *et al.*, 1981*a*; Lasker, Peleaz & Laurs, 1981).

Medium-sized fish- and squid-eaters overlap considerably more in space and time (Table 12.1). Gull, puffin, and murre populations are largest from fall through winter; shearwaters, on the other hand, predominate from midspring through late summer. It remains to be seen whether diet, geographic, or behavioral differences separate these species ecologically on the wintering grounds.

A model of trophic impact

To estimate the role of California's seabirds in energy exchange and their effect on selected prey populations, we used a familiar allometric model, combining existing information on seabird diets and metabolic requirements with the new population figures. The model estimates daily energy requirements (DEE) for each bird species and each bird from the allometric equations of Kendeigh, Dolnik & Gavrilov (1977), and multiplies DEE by density of each species at sea, calculated from survey data. Densities were adjusted upward by a fraction corresponding to shoreline population size during the same period (0 for some highly pelagic species, and up to 0.67 for some nearshore species during nesting season). This enabled us to estimate the food requirements of whole populations, at sea as well as on or near land, by assuming that birds on or near land fed at the same locations as those recorded at sea.

We applied this model according to methods outlined by Wiens & Scott (1975), with some modification. DEE was interpolated linearly for environmental temperatures between 0° C and 30° C according to the equations:

$$\text{DEE} = 4.142\ W^{0.5444}\ (0°\text{ C})$$

$$\text{DEE} = 1.068\ W^{0.6637}\ (30°\text{ C})$$

where W is mass in g and DEE is in kcal day^{-1} (Kendeigh *et al.*, 1977). Body weights (masses) for each species came from the literature or our collections. Mean sea surface temperature (SST) was obtained routinely during our ship and airplane surveys, and additional SST data came from images produced by the NOAA 6 and 7 weather satellites. For temperature values, we entered the monthly mean SSTs at locations where each species was sighted, a procedure that realistically approximates the thermal environment of foraging marine birds. We assumed that foraging and other activity increased DEE by 40%; assimilation efficiency was assumed to be 80% (Wiens & Scott, 1975; Furness, 1978). Again following Wiens & Scott, we estimated the costs of egg production and feeding young from clutch size, hatching rate, growth rates (either field measurements or estimates after the methods of Ricklefs (1967)), chick growth period, and fledging success (see Appendix 12.1).

To estimate impact geographically, we multiplied bird densities at each location by factors for daily energy requirement (in g C m^{-2} day^{-1}), % fish and squid, % zoo- and ichthyoplankton, % benthic mollusks and other invertebrates, and % other foods. Conversions of kcal to g carbon and wet weights were done for each prey type using values tabulated by Wiens & Scott (1975) and Schneider & Hunt (1982). By extrapolating mean population densities from survey tracks to the entire coastal zone (162 500 km^2 in central and northern California, 60 000 km^2 in southern California; eastern two-thirds of shaded area in Fig. 12.1), we estimated total daily consumption.

Model results

Geographic and temporal patterns of consumption

Seabird biomass in California's neritic zone far exceeds that offshore. During 1980–82, for example, average monthly biomass over the central and northern California shelf (0–199 m deep) averaged 230% of the biomass over the continental slope (200–1999 m) and 650% of that beyond the slope. Our model estimates of carbon flux to seabirds followed a similar pattern: shelf waters, particularly where the shelf is broad, supported much higher consumption per unit area than waters farther offshore.

Year-round centers of estimated high carbon flux were located in shallow water from Cape Mendocino to the California–Oregon border, from Monterey Bay to Point Reyes, and from Point Buchon southeastward for 300 km to Cortés Bank (see Fig. 12.1 for place names). In these areas local carbon flux to birds averaged about 3 mg C m^{-2} day^{-1}, representing perhaps 10% of the carbon available to predators from schooling fish and squid (assuming 10% transfer efficiency for each trophic step and Garrison's (1976) estimates of primary production in Monterey Bay) (Green, 1978). Maximum monthly fluxes within Santa Barbara Channel, Monterey Bay, the Gulf of the Farallones, and north of Cape Mendocino were one to two orders of magnitude higher than coastwide averages. In some seasons many other locations showed high values (e.g. the aprons surrounding each of the southern California Channel Islands).

Fig. 12.2 shows four trends in daily carbon consumption among California seabirds. First, consumption tends to be low in January or February and in June or July, depending on relative contributions of rapidly metabolizing small birds *versus* slowly metabolizing large birds; unfortunately our time series were too short to determine the ranges of year-to-year variation. Secondly, at each point in the annual cycle, estimated average daily consumption is about 60% to 600% higher in central and northern California than in southern California, a pattern stemming from the relatively large populations of shearwaters, phalaropes,

Fig. 12.2. Average daily carbon flux to seabirds off central and northern California (upper) and southern California (lower).

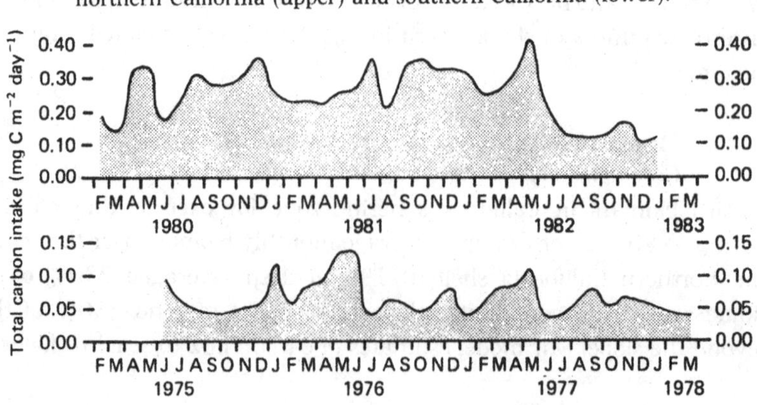

gulls, and alcids north of Point Conception. Thirdly, large populations of single species produce distinct peaks in total consumption. At times, the consumption by Sooty Shearwaters *Puffinus griseus*, phalaropes, or murres in their preferred habitats exceeds the combined consumption of all other co-occurring species (see below and Figs 12.2 and 12.4). Fourthly, concurrent with reduced seabird biomass and warm ocean temperatures, estimated carbon flux declined by up to 60% in fall–winter 1977–78 and 1982–83.

Energy uptake by predominant species

The feeding activities of a small number of species with large seasonal populations dominate total energy uptake by California seabirds. Figs 12.3 and 12.4 show energy flux to each of these species for southern California and central and northern California, respectively.

Sooty Shearwaters dominate the carbon flux pattern in both areas during late spring through about midsummer. Reaching maximum instantaneous populations of 2.7 million birds in May, June, or July, this species consumes an average of up to 0.25 mg C m^{-2} day^{-1} in the north, and about half that off southern California. Assuming a prey composition of 90% fish and squid and 10% zooplankton (primarily euphausiids), we estimate annual consumption to be about 36 000 tonnes wet weight. Sooty Shearwaters consume primarily juvenile rockfish *Sebastes* spp., squid, and euphausiids in spring and early summer, shifting to anchovies

Fig. 12.3. Average daily carbon flux to selected southern California seabirds during 1975–78.

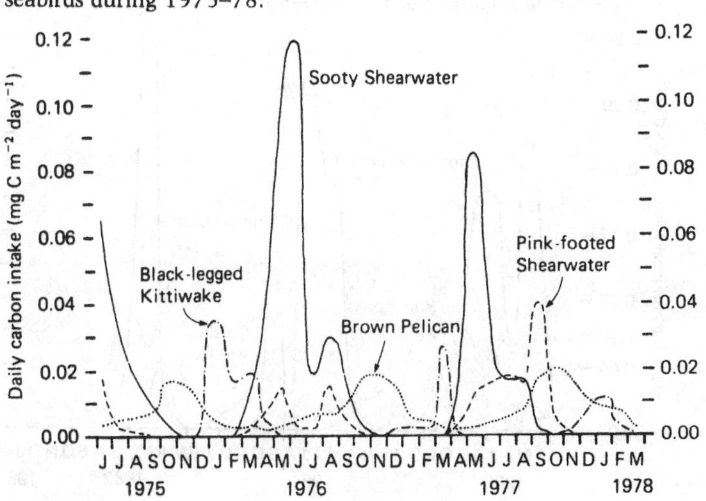

in late summer, when dense schools are available off central California. By doing so, shearwaters take advantage of the oily, high-energy anchovies to put on fat before their long return migration to nesting islands in the southern hemisphere (Chu, 1984). Wiens & Scott (1975) estimated that Sooty Shearwaters off Oregon consumed nearly 31 000 tonnes, with a distinct peak in August. Peak daily rates (0.2–0.3 mg C m^{-2} day^{-1}) in that study were similar to those in ours. Guzman, Myres & Wahl (unpublished) have suggested that distinct waves of migrant Sooty Shearwaters from South America arrive in spring off California. This might have the effect of producing peaks and troughs in total food intake on a scale of 1–2 weeks. Our surveys were made too infrequently (monthly) to resolve variation that might occur on the suggested time scale.

Brown Pelicans nest in small numbers in southern California. The local population increases in late summer as tens of thousands of post-nesting visitors move in from Mexican colonies. Anderson *et al.* (1980; and personal communication) have shown that pelicans feed on more than 30 species of epipelagic fish near colonies in the Gulf of California, but take almost 90% anchovies when nesting in California. The appearance of large pelican populations in central California in fall coincides with a general decline in upwelling frequency and with anchovy migration northward from southern California spawning areas (Briggs *et al.*, 1983*b*). As Ainley (1977) points out, the less turbid waters that occur then facilitate prey location and capture by visually orientated, plunge-diving

Fig. 12.4. Average daily carbon flux to selected central and northern California seabirds during 1980–83.

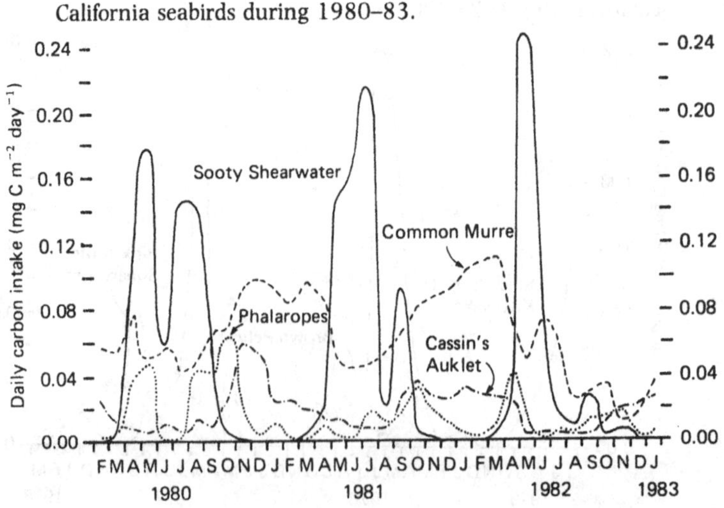

pelicans, boobies, and terns. Maximum carbon flux to pelicans occurs in early fall in southern California, reaching an average of 0.02 mg C m^{-2} day^{-1}. In central and northern California, pelican populations number as high as about 23 000 individuals in early fall, representing an estimated 22–28% of the state total (Briggs *et al.*, 1983*b*), and consume up to about 0.006 mg C m^{-2} day^{-1}. In total, pelicans thus consume about 2300 tonnes of fish per year.

Common Murre populations are large in summer and even larger in winter, when hundreds of thousands of birds from colonies in the Pacific Northwest join local breeders (Smail, Ainley & Strong, 1972; Briggs, *et al.*, 1983*a*). Reaching a maximum average intake of above 0.1 mg C m^{-2} day^{-1}, murres consume about the same total amount of energy in a year (19 mg C m^{-2} yr^{-1}) as do Sooty Shearwaters (17 mg C m^{-2} yr^{-1}). Moreover, the murre population nesting in California is growing rapidly, increasing by an average of more than 8% per year since 1960 (Ainley & Lewis, 1974; Sowls *et al.*, 1980). Like Cassin's Auklets, another species with winter peaks in energy use, the murre population off California declined greatly in fall 1982 compared with previous years, resulting in much lower estimated energy consumption. It is not yet clear whether murres dispersed northward from California in response to ocean warming associated with the major El Niño event occurring at that time (McGowan, 1983; Lynn, 1983), or whether post-breeding mortality of adults and young increased in 1982 as a result of low food availability or prevalence of storms from October onward.

The model results thus suggest temporal separation of peak impact by some potential competitors, but our time series are too short to substantiate this. Phalaropes and auklets generally showed peak consumption at different times in 1980 and 1982, as did murres and shearwaters in 1980 and 1981, and Black-legged Kittiwakes *Rissa tridactyla* and other piscivores in 1976 and 1977. Sooty and Pink-footed *Puffinus creatopus* Shearwaters, which appear to have their greatest impact at much the same time, often are segregated according to water temperature. The Sooty reaches maximum abundance in the cool waters (11–14° C) off central California and in the upwellings characteristic of the northwestern portion of southern California. The Pink-footed reaches peak abundance in warmer waters (15–20° C) in the southern and eastern sectors of the Southern California Bight (Ainley, 1976; Briggs *et al.*, 1981*a*).

Table 12.2. *Major ecological characteristics of California coastal and offshore waters relating to patterns of seabird distribution and feeding*

	Northern 42°00′N–38°30′N	Central 38°30′N–34°30′N	Southern 34°30′N–32°30′N
Shoreline and shelf	Fairly linear; shelf 10 km broad in south, 40 km in north	Fairly linear but with major embayments; shelf width variable	Curving to southeast; sea floor a series of basins and ranges topped by islands, some shelf waters 175 km offshore
Dominant hydrographic conditions			
Spring–summer	Predominant, heavy northwest wind, strong upwelling, shallow mixed layer, weak stratification of thermocline, high turbulence. Cool, less saline California Current water found offshore of cold, more saline upwellings. Transport generally southward	Northwest winds leading to strong upwelling south to Point Sur (36 °N); other conditions similar to northern California but warmer stratified waters of California Curent often approach the coast in eddies south of San Francisco. Transport generally southward	Cyclonic Southern California Eddy (SCE) predominates; water cool in west, warm in east with 5 to 8 Celsius degrees surface gradient. Upwelling intense only in northwest. Stratified waters in east with considerable subtropical influence. Upwelling lessens after May–June
Fall–winter	Prevalence of strong south wind leading to coastal convergence, deep mixed layer, weak stratification. Transport northward over shelf. River run-off important near shore	Upwelling infrequent with (weaker) south winds interspersed with calms or north winds. Flow northward over shelf, southward offshore. Run-off not as important	Upwelling infrequent; winds are the calmest in California Current; stratification strong. SCE predominates with northward flow inshore. Run-off unimportant
Breeding seabirds	Large colonies (10^4–10^5 birds), dominated by murres	Large colony complex in Gulf of the Farallones (10^5 birds) dominated by murres, auklets, cormorants, gulls	Colonies relatively small 10^2–10^4 birds) dominated by auklets, cormorants, gulls, murrelets
Visiting seabirds	Very numerous, mostly Alaskan breeders	Very numerous; Alaskan, Mexican, and austral breeders intermix	Numerous, mostly austral breeders in spring, subtropical breeders in fall, Alaskan breeders in winter
Important fish resources	Rockfish, flatfish, salmon, hake	Flatfish, rockfish, hake, squid, anchovy, tuna, salmon	Anchovy predominant; squid, saury, mackerel, rockfish, tunas, billfish
Years studied	1980–1983	1980–1983	1975–1978
Study area size	66900 km^2	95600 km^2	60000 km^2

Hydrographic context

The California Current – a wide (to 1000 km), slow (20–40 cm s^{-1}) system originating offshore at about 45° N and flowing towards the equator (Hickey, 1979) – dominates coastal circulation (Table 12.2). In spring through late summer, northwest winds promote strong upwelling over the continental shelf and slope, particularly downwind of promontories – Cape Mendocino, Point Arena, Point Reyes, Point Sur, and Point Conception. Nutrient regeneration at local upwelling centers is high, but, on a large scale, southward transport of nutrients in the California Current itself seems to affect year-to-year variations in productivity more strongly than upwelling alone (Bernal & McGowan, 1981; Chelton, 1981). In years of diminished transport (such as may occur during El Niño events), productivity drops in waters off California (Smith & Eppley, 1982).

Upwelling is greatest and water temperatures are lowest in late spring; in late summer and fall, upwelling decreases in frequency and strength and temperatures are highest. Several major pelagic fish stocks – hake *Merluccius* spp., anchovy, and rockfish – spawn primarily in late winter–early spring before surface waters advect offshore during the peak upwelling season (Parrish, Nelson & Bakun, 1981; Bakun & Parrish, 1982). These drifting fish eggs and fry remain near the coast in winter because of coastal convergence driven by south winds; the larger, mobile juveniles can exploit coastal food sources supplemented by upwelling later in the spring.

To begin exploring factors affecting seasonal and geographic patterns of seabird consumption, we have computer-mapped model results for September 1981, February 1982, and July 1982, and compared them with NOAA satellite images of sea surface temperatures (Figs 12.5–12.7); seabird data used in the model and the NOAA images were taken simultaneously, giving us representative seasonal snapshots of actual events.

During the September survey an intense upwelling event occurred from Point Arena to Point Reyes (Fig. 12.5, left). A plume of cool, nutrient-rich water extended from Point Reyes southward for more than 75 km, while a smaller upwelling mass originating near the coast was visible northwest of Monterey Bay. Warm, relatively stratified water in an anticyclonic eddy west of Point Arena entrained upwelled water at the shelf break and forced it offshore near 38° N, 125° W (Briggs *et al.*, 1984); similar structures were reported at this location in August 1982, by Mooers & Robinson (1984). Estimated seabird consumption of planktonic crustaceans (euphausiids, copepods, amphipods) and fish eggs and larvae nearly matched the general

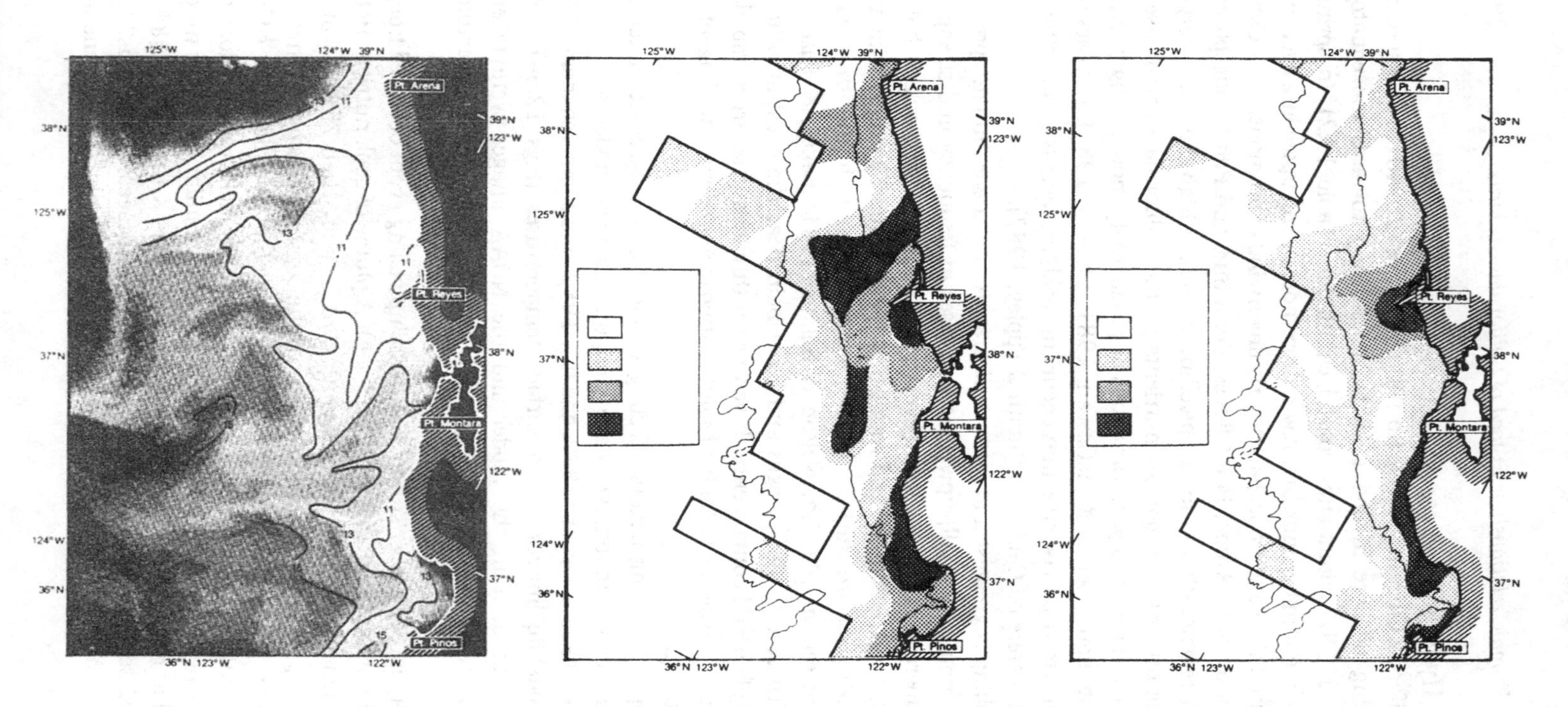

Fig. 12.5. Patterns of seabird trophic impact compared with a satellite image of sea surface temperature off central California during September 1981. Isotherms superimposed on satellite image are from low-altitude infrared measurements. Light grey indicates cool water.

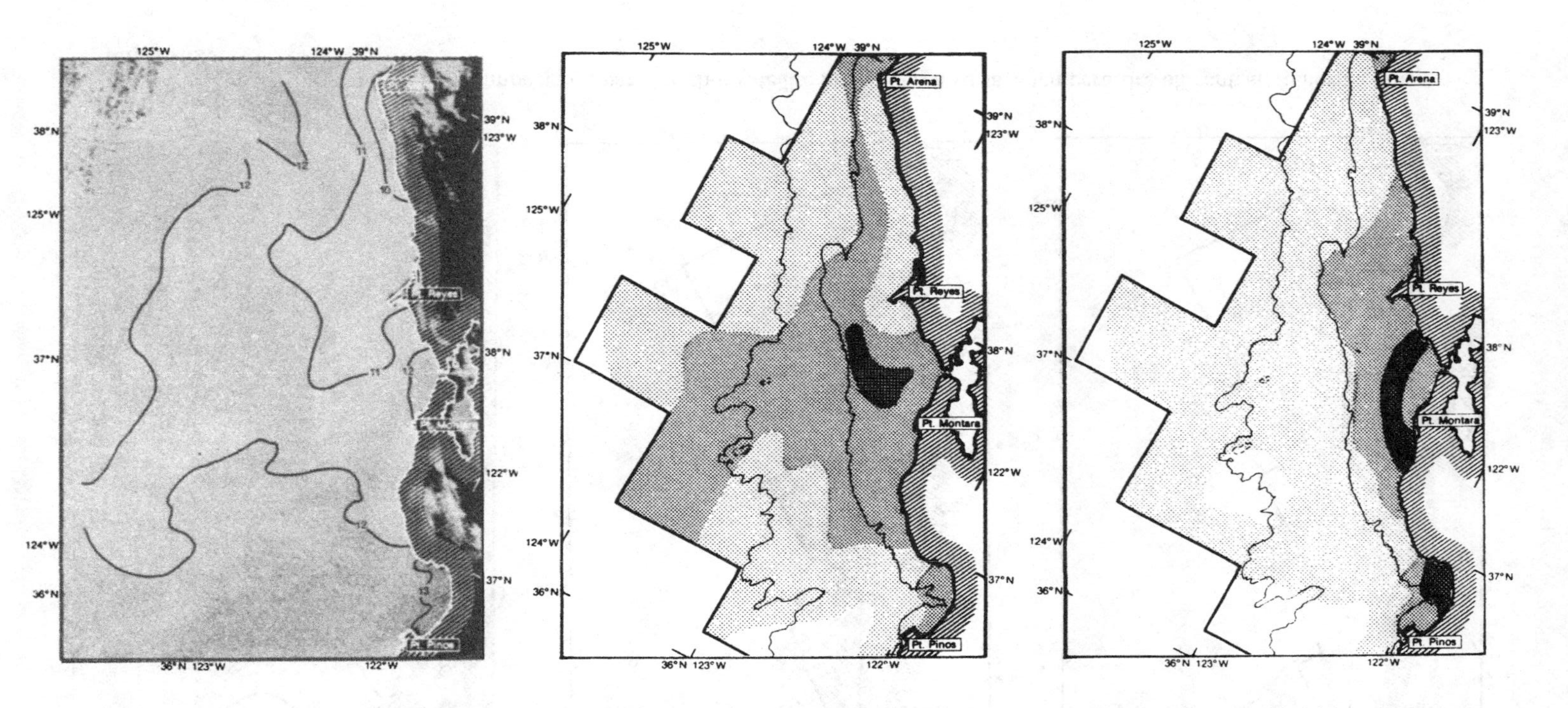

Fig. 12.6. Patterns of seabird trophic impact compared with a satellite image of sea surface temperature off central California during February 1982.

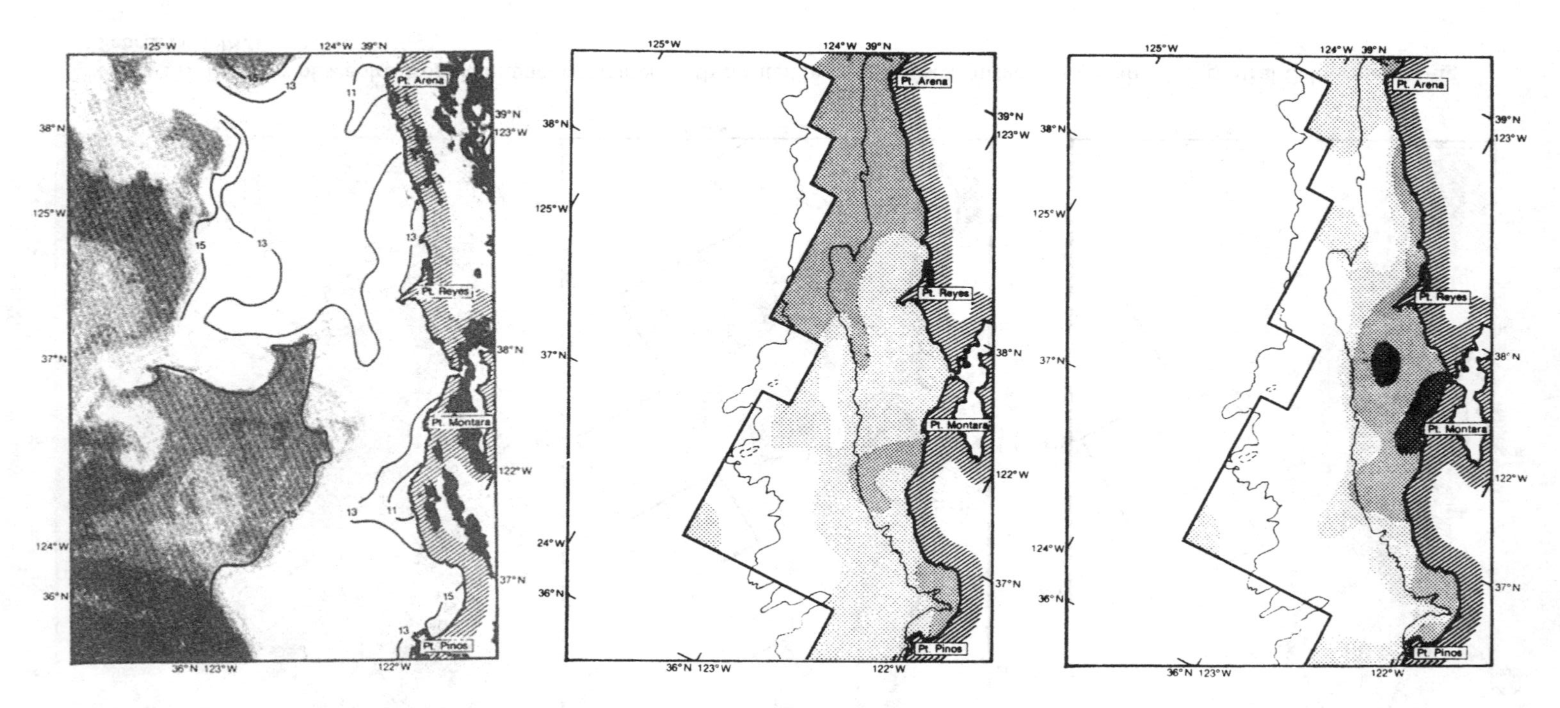

Fig. 12.7. Patterns of seabird trophic impact compared with a satellite image of sea surface temperature off central California during July 1982.

upwelling boundaries (Fig. 12.5, middle). High plankton consumption (> 2.0 mg wet weight m^{-2} day^{-1}), primarily by phalaropes and Cassin's Auklets, appeared at the edges of the Point Reyes upwelling and from Point Montara to Monterey Bay. Consumption of fish and squid (Fig. 12.5, right) by Common Murres, Brown Pelicans, Brandt's Cormorants *Phalacrocorax penicillatus*, and Western Gulls *Larus occidentalis* contributed proportionately much more to overall carbon flux than did planktivory. Fish and squid consumption was highest over the shelf north of Monterey Bay. This suggests that, for fish and squid and their predators as well, optimal combinations of substrate, circulation, and feeding conditions are met downstream from major upwellings in less turbulent waters.

In contrast to the strong thermal gradients of fall, surface waters were more homogeneous in February (Fig. 12.6). A hint of upwelling appeared near the coast at Point Arena, but the general pattern was a broad, gradual warming. Bernal & McGowan (1981) have shown that year-round maximum zooplankton standing stocks occur seaward of the shelf break off central California. Our model results showed plankton consumption throughout the survey area with a band of moderate values over the outer continental slope. The small zone of relatively high planktivory surrounding the Farallon Islands corresponded with sightings of Cassin's Auklets, which may have been making early breeding season visits to the colony.

The July survey took place during a period of upwelling everywhere over the shelf, with strong centers at Point Arena and south of Point Montara, and complex intrusions of warm and cool waters interleaved over the continental slope (Fig. 12.7). Fish and squid consumption – primarily by murres, gulls, cormorants and shearwaters – was mostly confined to the shelf; planktivory was more widespread. Interestingly, in both this period and in February, zones of heavy consumption of fish and squid were associated with the tidal plume emanating from the mouth of San Francisco Bay. It seems likely that feeding conditions here may be enhanced both by nutrient input from terrestrial and estuarine sources associated with the Bay (including the Sacramento–San Joaquin River system draining California's Central Valley) and by physical mechanisms associated with formation of tidal thermohaline fronts (Holligan, 1981). Impact on fish and squid populations was heavy near the Farallones, which still had many cormorant and gull nests; plankton consumption, in contrast, was light because Cassin's Auklets, most of which had probably already fledged young, typically forage at this time at the outer edges of upwellings from the Farallones northward. During some years in July, tremendous shearwater, gull and murre populations overlap to exploit anchovy or squid schools in northern Monterey Bay. At these times calculated carbon flux exceeds 1.5 g C m^{-2} day^{-1} for 1 or 2 weeks, accounting for 50–150% of average local primary production.

These examples point to different patterns of trophic impact among piscivores and planktivores. Birds eating fish and squid concentrate in neritic areas somewhat removed from active upwellings. Here, prey capture may be easier because wind-induced turbulence is less than in upwelling centers (Ainley, 1977; Briggs *et al.*, 1983*a*). Many planktivores feed seaward of the shelf. Phalaropes, for example, feed in surface thermal fronts where upwelling waters meet those of the California Current (Briggs *et al.*, 1984). We suspect that the persistence and seasonal predictability of upwelling features upstream of colonies like the Farallones play a pivotal role in individual and colony reproductive success.

Impact on prey populations

Although seabirds take a wide variety of prey off California (Table 12.1, Appendix 12.1), relatively few prey species are dominant; these same prey species are among the most important constituents of epipelagic crustacean, mollusk, and fish biomass off California: the crustaceans *Calanus pacificus*, *Thysanoessa* spp., and *Euphausia pacifica*; the squid *Loligo opalescens*, *Gonatus* spp., and *Onychoteuthis borealijaponicus*; and anchovy, saury *Cololabis* spp., rockfish, and several flatfish such as *Citharichthys* sp. (Fleminger, 1964; Longhurst, 1967; Brinton, 1976; Horn, 1980; Chu, 1984). These populations sustain an estimated 93% of all seabird consumption, with the heaviest impact on *Euphausia*, *Loligo*, anchovy, and rockfish (60–70% of the annual total consumption).

To estimate the total daily seabird consumption for the state, we have added model results for southern California to those for central and northern California, based on overall yearly similarity of surface temperatures – the near-normal southern year (1975–76) with normal northern years (1980–81), and warm years in the south (1977–78) with warm years in the north (1982–83). This matching procedure masks some important seasonal changes in temperature anomalies but enables us to compare yearly figures with at least some basis for assuming population similarities. Overall, we find that total daily consumption of fish and squid exceeds that of plankton by 200–300% in most months, and the relative contribution of southern California populations is considerably less than that of the larger northern area. Combining the values from all regions yields total of 500–600 tonnes day^{-1} of prey consumed by California seabirds.

Fish and squid consumption declined steeply in the latter half of 1982; we attribute this to the influence of early El Niño conditions. At that time, waters off California were 1–4° C warmer than usual, and the thermocline was quite deep (Lynn, 1983). Phytoplankton stocks were low and concentrated deeper than usual (McGowan, 1983), zooplankton biomass was declining to 90% below normal in spring 1983 (J.McGowan, personal communication), and important seabird prey fish such as anchovy and rockfish were experiencing very low reproductive and growth rates (D. McLain, personal communication).

The total annual consumption, then, amounts to about 44 000 tonnes of plankton and 149 000 tonnes of fish and squid in years of near-normal temperatures, compared with about 28 000 tonnes and 94 000 tonnes, respectively, during warm-water years. Our data are incomplete for southern California in early 1975 but indicate that in that anomalously cool period, when primary production was high (Smith & Eppley, 1982), consumption by birds was somewhat higher than these figures.

Primary production varies regionally, seasonally, and annually off California, from about 130 to perhaps 300 g C m^{-2} yr^{-1} (Garrison, 1976; Smith & Eppley, 1982). Total seabird consumption, which we estimate to be 0.09 g C m^{-2} yr^{-1} in a year of near-normal water temperatures, thus represents about 4–7% of tertiary production, assuming 10% transfer efficiency at each step. Cushing (1975), Smith & Eppley (1982), and others have found transfer efficiencies to range as high as 18–20% for the herbivore–anchovy step in the trophic web. However, efficiencies characteristic of other carnivore–herbivore interactions are poorly known and may be much lower. In a model of energy consumption by seabirds in the Bering Sea, Schneider & Hunt (1982) estimated that seabirds consumed 0.03–0.05% of primary production in summer (3–5% of tertiary reproduction, if tertiary indeed represents 1% of primary production). However, they concluded that because of conservative assumptions, their model probably underestimated actual consumption. The model estimates presented by Wiens & Scott (1975) and Furness & Cooper (1982) are also in the range reported here, after appropriate allowances are made for numbers of species considered by each model.

Although diet and energy requirements have yet to be determined for most marine mammals, we can make a few estimates for comparison. The results of marine mammal population surveys conducted by our colleagues off California during 1975–82 indicate a standing stock averaging 170–500 kg km^{-2} through the year, including migrating baleen whales that are not feeding (M.L. Bonnell, personal communication; T.P. Dohl, personal communication). After excluding the baleen whales, and assuming a daily ration of 5% of body weight, we calculate that these marine mammals might consume about 1800–5500 tonnes of fish and squid per day, or roughly 25–75% of tertiary production each year. Green (1978) estimated total fish plus squid consumption by mammals in the region extending offshore to 370 km from the California–Oregon border to the tip of Baja California, Mexico (690 000 km^2) to be 10 529–14 457 tonnes day^{-1}. Allowing for the size of the region considered here (32% of Green's study area), our gross estimates of marine mammal prey consumption are quite similar. The total commercial and sport fishing catches off California averaged 0.3–0.6 million tonnes yr^{-1} during the 1970s, or about 25–95% of the assumed catch by marine mammals and 200–400% of that by seabirds (references cited in Green, 1978).

California fisheries scientists have identified several commercially exploitable pelagic fish stocks that are under-used for a variety of reasons. Prominent among these are squid, rockfish, and mackerel populations heavily utilized by seabirds and marine mammals. Current fisheries policies stipulate that predator needs be considered when setting fishery regulations. We hope that, by using models such as the one presented here, managers can set catch limits consistent with maintenance and recovery of predator populations.

Acknowledgements We wish to thank the following colleagues who have contributed over the years to collection, analysis, and interpretation of seabird data: Kathleen Dettman, George Hunt, Jr, David Lewis, Mark Pierson, and William Breck Tyler. Discussions of these and related ideas with John Croxall, David Schneider, Gerald Sanger, Daniel Anderson, and R.G.B. Brown have helped us immeasurably. Critical reviews of an early draft by D.W. Anderson and D.G. Ainley helped us to strengthen and clarify our presentation. We thank Cynthia Campbell and Caroline Caldwell for preparation of manuscript drafts, Darwin Briggs for graphics assistance, and David Carlson for computer analyses. Thomas Dohl and Michael Bonnell provided us with synoptic data on marine mammal populations; Laurence Breaker and Ronald Gilliland supplied us with satellite images and aided with their interpretation. This paper is a contribution from the Center for Marine Studies, University of California, Santa Cruz, under financial support from the Pacific Outer Continental Shelf Office of the U.S. Minerals Management Service.

References

Ainley, D.G. (1976). Occurrence of seabirds in the coastal region of California. *Western Birds*, 7, 33–68.

Ainley, D.G. (1977). Feeding methods in seabirds: a comparison of polar and tropical nesting communities in the eastern Pacific Ocean. In *Adaptations within Antarctic ecosystems*, ed G.A. Llano, pp. 669–85. Washington, D.C.: Smithsonian Institution.

Ainley, D.G., Anderson, D.W. & Kelly, P.R. (1981). Feeding ecology of marine cormorants in southwestern North America. *Condor*, **83**, 120–31.

Ainley, D.G. & Lewis, T.J. (1974). The history of Farallon Island marine bird populations, 1854–1972. *Condor*, **76**, 432–66.

Ainley, D.G., Morrell, S. & Lewis, T.J. (1974). Patterns in the life histories of storm petrels on the Farallon Islands. *Living Bird*, **4**, 295–311.

Ainley, D.G. & Sanger, G.A. (1979). Trophic relations of seabirds in the northeastern Pacific Ocean and Bering Sea. In *Conservation of marine birds of northern North America*, ed. J.C. Bartonek & D.N. Nettleship, pp. 95–122. Washington, D.C.: U.S. Fish & Wildlife Service, Wildlife Research Report 11.

Anderson, D.W. & Gress, F. (1983). Status of a northern population of California Brown Pelicans. *Condor*, **85**, 79–88.

Anderson, D.W., Gress, F., Mais, K.F. & Kelly, P.R. (1980). Brown Pelicans as anchovy stock indicators and their relationships to commercial fishing. *Calif. Coop. Ocean. Fish. Invest. Rep.*, **21**, 54–61.

Bakun, A. & Parrish, R.B. (1982). Turbulence, transport, and pelagic fish in the California and Peru Current systems. *Calif. Coop. Ocean. Fish. Invest. Rep.*, **23**, 99–112.

Baltz, D.M. & Morejohn, G.V. (1977). Food habits and niche overlap of seabirds wintering on Monterey Bay. *Auk*, **94**, 526–43.

Bent, A.C. (1921). Life histories of North American gulls and terns. *Bull. U.S. Natl Mus.*, **113**, 1–345.

Bernal, P.A. & McGowan, J.A. (1981). Advection and upwelling in the California Current. In *Coastal upwelling*, ed. F.A. Richards, pp. 381–99. Washington, D.C.: American Geophysical Union.

Briggs, K.T. (1977). Social dominance among young Western Gulls: its importance in survival and dispersal. Unpublished thesis, University of California, Santa Cruz.

Briggs, K.T., Chu, E.W., Lewis, D.B., Tyler, W.B., Pitman, R.L. & Hunt, G.L., Jr (1981*a*). Distribution, numbers, and seasonal status of seabirds of the Southern California Bight. In *Investigators' reports. Summary of marine mammal and seabird surveys of the Southern California Bight area, 1975–1978*, book I, part III, pp. 1–399. Springfield, Virginia: Publication no. PB-81-248-205, U.S. National Technical Information Service.

Briggs, K.T., Dettman, K.F., Lewis, D.B. & Tyler, W.B. (1984). Phalarope feeding in relation to autumn upwelling off California. In *Marine birds: their feeding ecology and commercial fisheries relationships*, ed. D.N. Nettleship, G.A. Sanger & P.F. Springer, pp. 51–62. Ottawa: Canadian Wildlife Service.

Briggs, K.T., Lewis, D.B., Tyler, W.B. & Hunt G.L., Jr (1981*b*). Brown Pelicans in southern California: habitat use and environmental fluctuation. *Condor*, **83**, 1–15.

Briggs, K.T., Tyler, W.B., Lewis, D.B. & Dettman, K.F. (1983*a*). *Seabirds of central and northern California, 1980–1983: status, abundance, and distribution*, pp. 1–246. Springfield, Virginia: Publ. no. 85-183846, U.S. Natl. Tech. Inf. Service.

Briggs, K.T., Tyler, W.B., Lewis, D.B., Kelly, P.R. & Croll, D.A. (1983*b*). Brown Pelicans in central and northern California. *J. Field Ornithol.*, **54**, **353–73.**

Brinton, E. (1976). Population biology of *Euphausia pacifica* off southern California. *Fish. Bull.*, **74**, 733–62.

Chelton, D.B. (1981). Interannual variability of the California Current – physical factors. *Calif. Coop. Ocean. Fish. Invest. Rep.*, **22**, 34–48.

Chu, E.W. (1984). Sooty Shearwaters off California: diet and energy gain. In *Marine birds: their feeding ecology and commercial fisheries relationships*, ed. D.N. Nettleship, G.A. Sanger & P.F. Springer, pp. 64–71. Ottawa: Canadian Wildlife Service.

Crossin, R.S. (1974). The storm-petrels (Hydrobatidae). In *Pelagic studies of seabirds in the central and eastern Pacific Ocean*, ed. W.B. King, pp. 154–205. Washington, D.C.: Smithsonian Institution.

Cushing, D.H. (1975). *Marine ecology and fisheries*. Cambridge: Cambridge University Press.

Fleminger, A. (1964). Distributional atlas of calanoid copepods in the California Current region, Part 1. *Calif. Coop. Fish. Invest. Atlas*, **2**, 1–313.

Follett, W.I. & Ainley, D.G. (1976). Fishes collected by Pigeon Guillemots, *Cepphus columba* Pallas, nesting on southeast Farallon Island, California. *Calif. Fish Game*, **62**, 28–31.

Furness, R.W. (1978). Energy requirements of seabird communities: a bioenergetics model. *J. Anim. Ecol.*, **47**, 39–53.

Furness, R.W. & Cooper, J. (1982). Interactions between breeding seabird and pelagic fish populations in the southern Benguela region. *Mar. Ecol. Progr. Ser.*, **8**, 243–50.

Garrison, D.L. (1976). Contributions of net plankton and nanoplankton to the standing stocks and primary productivity in Monterey Bay, California during the upwelling season. *Fish. Bull.*, **74**, 183–94.

Green, K.A. (1978). *Ecosystem description of the California Current*. Washington, D.C.: Report no. MMC-77/11 of the U.S. Marine Mammal Commission.

Hickey, B.M. (1979). The California Current system – hypotheses and facts. *Progr. Oceanogr.*, **8**, 191–279.

Holligan, P.M. (1981). Biological implications of fronts on the northwest European continental shelf. *Phil. Trans. R. Soc. Lond.*, **A302**, 547–62.

Horn, M.H. (1980). Diversity and ecological roles of noncommercial fishes in California marine habitats. *Calif. Coop. Ocean. Fish. Invest. Rep.*, **21**, 37–47.

Hunt, G.L., Jr, Pitman, R.K., Naughton, M., Winnet, K., Newman, A., Kelly, P.R. & Briggs, K.T. (1981). Distribution, status, reproductive ecology, and foraging habits of breeding seabirds. In *Summary report 1975–1978; marine mammal and seabird survey of the Southern California Bight area*, vol. 3, book 4, pp. 1–399. Springfield, Virginia: Publication no. PB-81-248-205, US National Technical Information Service.

Johnsgard, P.A. (1975). *Waterfowl of North America*. Bloomington, Indiana: Indiana University Press.

Jones, L., Garrett, K. & Small, A. (1981). Checklist of the birds of California. *Western Birds*, **12**, 57–72.

Kendeigh, S.K., Dolnik, V.R. & Gavrilov, V.M. (1977). Avian energetics. In *Granivorous birds in ecosystems*, ed. J. Pinkowski & S.C. Kendeigh, pp. 127–204. Cambridge: Cambridge University Press.

Lasker, R., Peleaz, J. & Laurs, R.M. (1981). The use of satellite infrared imagery for describing ocean processes in relation to spawning of the Northern Anchovy (*Engraulis mordax*). *Remote Sens. Environ.*, **12**, 439–53.

Longhurst, A.R. (1967). Diversity and trophic structure of zooplankton communities in the California Current. *Deep-Sea Res.*, **14**, 393–408.

Lynn, R.J. (1983). The 1982-83 warm episode in the California Current. *Geophys. Res. Lett.*, **10**, 1093–5.

McGowan, J.A. (1983). El Niño and biological production in the California Current. *Trop. Ocean-Atmos. Newsl.*, **21**, 23.

Manuwal, D.A. (1974). The natural history of Cassin's Auklet *Ptychoramphus aleuticus*. *Condor*, **76**, 421–31.

Mooers, C.N.J. & Robinson, A.R. (1984). Turbulent jets and eddies in the California Current and infrared cross-shore transports. *Science*, **223**, 51–3.

Palmer, R.W. (1962). *Handbook of North American birds*. New Haven, Connecticut: Yale University Press.

Parrish, R.H., Nelson, C.S., & Bakun, A. (1981). Transport mechanisms and reproductive success of fishes in the California Current. *Biol. Oceanogr.*, **1**, 175–203.

Ricklefs, R.E. (1967). A graphical method of fitting equations to growth curves. *Ecology*, **48**, 978–83.

Sanger, G.A. (1974). Black-footed Albatross (*Diomedea nigripes*). In *Pelagic studies of seabirds in the central and eastern Pacific Ocean*, ed. W.B. King, pp. 96–128. Washington, D.C.: Smithsonian Institution.

Sanger, G.A. (1983). Diets and food web relationships of seabirds in the Gulf of Alaska and adjacent marine regions. In *Environmental assessment of the Alaskan continental shelf*. Juneau, Alaska: NOAA.

Schneider, D. & Hunt, G.L., Jr (1982). Carbon flux to seabirds in waters with different mixing regimes in the southeastern Bering Sea. *Mar. Biol.*, **67**, 337–44.

Sealy, S.G. (1973). Interspecific feeding assemblages of marine birds off British Columbia. *Auk*, **90**, 796–802.

Smail, J., Ainley, D.G. & Strong, H. (1972). Notes on birds killed in the 1971 San Francisco oil spill. *Calif. Birds*, **3**, 25–32.

Smith, P.E. & Eppley, R.W. (1982). Primary production and the anchovy population in the Southern California Bight: comparison of time-series. *Limnol. Oceanogr.*, **27**, 1–17.

Sowls, A.L., DeGange, A.R., Nelson, J.W. & Lester, G.S. (1980). *Catalog of California seabird colonies*. Washington, D.C.: U.S. Department of Interior, Fish and Wildlife Service, Report FWS/OBS-80/37. 371 pp.

Vermeer, K. (1981). Food and populations of Surf Scoters in British Columbia. *Wildfowl*, **32**, 107–16.

Wiens, J.A. & Scott, J.M. (1975). Model estimation of energy flow in Oregon coastal seabird populations. *Condor*, **77**, 439–52.

Appendix 12.1 *Input values for energetics model for California seabirds. Dash in columns for chick data indicates not nesting in California*

		Diet (%)							
Species	Mass (kg)	Fish & squid	Crustacean & ichthyo-plankton	Mollusk & other benthic inverte-brate	Other	Clutch size	Hatching rate[a]	Fledging rate[a]	Sources[b]
All loons	1.956	100	0	0	0	—	—	—	3, 15
Western Grebe *Aechmophorus occidentalis*	0.866	90	10	0	0	—	—	—	15
Small grebes	0.236	20	60	20	0	—	—	—	7, 15
All albatrosses	2.800	90	0	0	10	—	—	—	15, 16
Northern Fulmar *Fulmarus glacialis*	0.670	85	5	0	10	—	—	—	7, 17
Pink-footed Shearwater *Puffinus creatopus*	0.750	95	5	0	0	—	—	—	7
Buller's Shearwater *P. bulleri*	0.700	90	10	0	0	—	—	—	5
Sooty Shearwater *P. griseus*	[c]	90	10	0	0	—	—	—	8, 13
Black-vented Shearwater *P. opisthomelas*	0.480	90	10	0	0	—	—	—	15
Leach's Storm-Petrel *Oceanodroma leucorhoa*	0.035	25	75	0	0	1	0.8	0.7	9, 10, 15, 19, 21
Fork-tailed Storm-Petrel *O. furcata*	0.035	25	75	0	0	1	0.8	0.7	9, 10, 15, 19
Ashy Storm-Petrel *O. homochroa*	0.035	25	75	0	0	1	0.8	0.7	9, 10, 15, 19
Black Storm-Petrel *O. melania*	0.060	25	75	0	0	1	0.8	0.7	9, 10, 15, 19

Brown Pelican *Pelecanus occidentalis*	3.000	100	0	0	0	3	?	0.6	1, 2, 15
Double-crested Cormorant *Phalacrocorax auritus*	2.100	100	0	0	0	2.3	?	0.8	3, 11, 15, 19
Brandt's Cormorant *P. penicillatus*	2.100	100	0	0	0	2.4	?	0.8	3, 11, 15, 19
Pelagic Cormorant *P. pelagicus*	1.500	100	0	0	0	2.4	?	0.8	3, 11, 15, 19
All scoters	0.980	20[d]	0	70	10	—	—	—	12, 15, 20
All phalaropes	0.050	0	100	0	0	—	—	—	6
Pomarine Jaeger *Stercorarius pomarinus*	0.670	100	0	0	0	—	—	—	5, 15
Black-legged Kittiwake *Rissa tridactyla*	0.420	94	6	0	0	—	—	—	6, 17
Sabine's Gull *Xema sabini*	0.420	70	30	0	0	—	—	—	4
Bonaparte's Gull *Larus philadelphia*	0.177	10	40	0	50	—	—	—	3, 4, 6
Heermann's Gull *L. heermanni*	0.492	100	0	0	0	—	—	—	4, 5
California Gull *L. californica*	0.682	80	10	0	10	—	—	—	3, 4
Herring Gull *L. argentatus*	0.900	80	10	0	10	—	—	—	3, 4
Western Gull *L. occidentalis*	1.000	85	5	0	10	3	0.8	1.8	5, 11, 19
Glaucous-winged Gull *L. glaucescens*	1.050	85	5	0	10	—	—	—	4, 5
Common/Arctic Tern *Sterna hirundo/paradisaea*	0.200	90	10	0	0	—	—	—	4, 5
Forster's Tern *S. forsteri*	0.200	90	10	0	0	3	0.8	1.0	4, 5, 19
Royal Tern	0.300	100	0	0	0	—	—	—	4

Appendix 12.1

Species	Mass (kg)	Diet (%) Fish & squid	Crustacean & ichthyo-plankton	Mollusk & other benthic inverte-brate	Other	Clutch size	Hatching rate[a]	Fledging rate[a]	Sources[b]
S. maxima									
Elegant Tern *S. elegans*	0.300	100	0	0	0	—	—	—	4, 5
Caspian Tern *S. caspia*	0.400	100	0	0	0	3	0.8	1.2	4, 5, 19
Pigeon Guillemot *Cepphus columba*	0.450	90	10	0	0	1.7	0.9	1.3	10, 11, 18
Common Murre *Uria aalge*	0.980	90	10	0	0	1	0.7	0.4	3, 17, 19, 21
Xantus' Murrelet *Endomychura hypoleuca*	0.167	90	10	0	0	1.7	0.4	?	11, 19
Rhinoceros Auklet[e] *Cerorhinca monocerata*	0.600	90	10	0	0	—	—	—	3, 18
Cassin's Auklet *Ptychoramphus aleuticus*	0.164	20	80	0	0	1	0.5	0.4	3, 11, 14, 19
All puffins	0.900	95	5	0	0	1	?	0.5?	17, 19

[a] Rate is number of young per breeding attempt (nest with eggs).

[b] Sources; 1, Anderson & Gress, 1983; 2, Anderson *et al.*, 1980; 3, Baltz & Morejohn, 1977; 4, Bent, 1921; 5, Briggs, unpublished specimens; 6, Briggs *et al.*, 1984; 7, Briggs *et al.*, 1981*a*; 8, Chu, 1984; 9, Crossin, 1974; 10, Follett & Ainley, 1976; 11, Hunt *et al.*, 1981; 12, Johnsgard, 1975; 13, L. Krasnow, personal communication; 14, Manuwal, 1974; 15, Palmer, 1962; 16, Sanger, 1974; 17, Sanger, 1983; 18, Sealy, 1973; 19, Sowls *et al.*, 1980; 20, Vermeer, 1981; 21, Wiens & Scott, 1975.

[c] Variation in weights of Sooty Shearwaters is as follows: May, 0.800 kg; June, 0.850; July, 0.725; August, 0.950; September, 1,050; October–April 0.700 (Chu, 1984).

[d] In the form of herring eggs.

[e] Numbers of Rhinoceros Auklets breeding in California are insignificant compared to numbers of winter visitors.

13

Trophic relationships among tropical seabirds at the Hawaiian Islands

CRAIG S. HARRISON
Environment and Policy Institute, East–West Center, 1777 East West Road, Honolulu, Hawaii 96848, USA
and
MICHAEL P. SEKI
Southwest Fisheries Center Honolulu Laboratory, National Marine Fisheries Service, National Oceanic and Atmospheric Administration, 2570 Dole Street, Honolulu, Hawaii 96822-2396, USA

Introduction

Seabird communities in tropical and subtropical waters are often composed of large numbers of species that have complex trophic relationships. For example, 17 species breed on Laysan Island, Hawaii (Ely & Clapp, 1973) and 18 species breed on Christmas Island (Pacific) (Schreiber & Ashmole, 1970); trophic relationships are probably more complex than those in cold-water communities because of the larger number of food species available.

An important characteristic of tropical and subtropical zones is a relative lack of seasonal change in surface waters. Ryther (1963) and Ashmole (1971) defined these zones to include all areas where sea surface temperatures remain above 23° C all year. More precise definitions are probably impossible. A permanent thermocline is usually present which limits vertical enrichment of the euphotic zone throughout the year. Nutrient depletion in tropical oceans results in low primary productivity. This phenomenon is more pronounced in oceanic areas than near continents because land-based nutrients in neritic zones can increase productivity (Raymont, 1966).

Seabirds have adapted to life in tropical waters in various ways. Whereas cold-water species of a genus frequently have multi-egg clutches, tropical species often lay a single egg, presumably because the relatively impoverished food supplies allow fewer young to be raised successfully (Lack, 1967). Cold-water species tend to breed at predictable times each year to maximize the advantages of a seasonal abundance of food and to minimize the effects of cold weather and storms on vulnerable young. In contrast, tropical species often have ill-defined breeding seasons because they do not face severe winter weather and have fewer predictable feeding opportunities. Non-seasonal breeding occurs with many tropical species and at various locations (Nelson, 1979). Tropical seabirds have adopted foraging strategies that minimize contact with predatory fish such as tunas, sharks, and billfish that eat birds. They plunge rather than dive (Ainley, 1977), which restricts feeding to surface waters, and prey largely on epipelagic fauna (Harrison & Hida, 1980). Tropical oceanic areas do not have seasonal abundances of food that would attract vast numbers of migratory birds. Most migrant seabird species avoid tropical and subtropical areas and fly through them quickly to spend their non-breeding period in polar or subpolar areas (Ainley & Boekelheide, 1984). For example, Sooty and Short-tailed Shearwaters *Puffinus griseus* and *P. tenuirostris*, from the southern hemisphere migrate to Alaska to forage during their non-breeding season (Harrison, 1982), but few stop to feed in the tropical Pacific (King, 1970). These shearwaters are important consumers on migration in the colder water systems of California (see Chapter 12) and Oregon (Wiens & Scott, 1975).

From 1977 to 1983 we studied the ecology of Hawaiian seabirds to discover how they might be influenced by the growing fishery. Research on the biology of important commercial fishes has also helped to clarify the role of Hawaiian seabirds in the marine ecosystem. In this chapter we review the trophic relationships among Hawaiian seabirds and compare these with other tropical and subtropical seabird communities, discuss fishery interactions, and suggest directions for future research.

The Hawaiian Archipelago

We concentrated our work in the Northwestern Hawaiian Islands (NWHI) where 18 seabird species (two albatrosses, two shearwaters, one storm petrel, two petrels, three boobies, one frigatebird, one tropicbird, and six terns) breed. The NWHI are subtropical, extending from latitude 23° N to 28° N and longitude 162° W to 178° W in the North Pacific (Fig. 13.1). This location has several important influences on the seabird community. Many species are at or near the northern limit of their breeding range. The depth of the mixed layer is much shallower than in the main Hawaiian Islands (Hirota *et al.*, 1980), and the NWHI are often in the path of North Pacific winter storms. Apparently the food supply is seasonal. Fish larvae are much more abundant during summer than winter (Hirota *et al.*,

1980). Like the situation around Tahiti (Rougerie & Chabanne, 1983), the migratory tunas are much more abundant in Hawaii during summer (Waldron, 1964). These factors are presumably responsible for the tendency for seabirds to breed during summer, although albatrosses and petrels are important exceptions (Harrison, Hida & Seki, 1983).

Resource partitioning

The diets of Hawaiian seabirds are very complex. Representatives of 56 families, 86 genera, and 74 species of fish were found in the diets of the 18 species that breed in the NWHI (Harrison *et al.*, 1983). Numerous squid families and various groups of crustaceans were also represented in their diets. As previously, we group the seabirds of the NWHI into five guilds based on similarities in composition of diet, size of prey taken, and feeding strategies. These guilds are the albatrosses, pelecaniforms, terns and shearwaters that associate with predatory fishes, nocturnal petrels, and neuston-feeding terns. We acknowledge that tropical seabirds do not fit neatly into these categories and there are some difficulties with the characterization. For example, boobies and frigatebirds occasionally feed in association with tunas; however, they occur only with about 1% of the tuna schools sighted in Hawaiian waters (National Marine Fisheries Service (NMFS), unpublished data). Sooty Terns *Sterna fuscata* and Wedge-tailed Shearwaters *Puffinus pacificus* occasionally feed at night (Gould, 1967). Our information suggests that these feeding strategies do not

Fig. 13.1. The Hawaiian Archipelago.

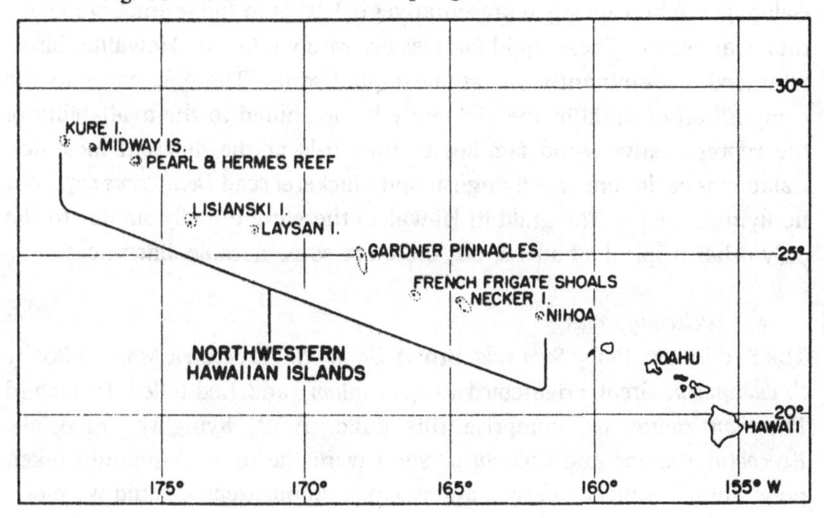

represent modal feeding behaviour in Hawaiian waters, and we have ignored such details in our analysis.

Albatrosses

Black-footed and Laysan Albatrosses *Diomedea nigripes* and *D. immutabilis*, eat primarily squid, flyingfish eggs, and deep-water crustaceans (Table 13.1). Many of the crustaceans and squid that were consumed possess photophores and may be bioluminescent at night. At least eight squid families were identified, but Ommastrephidae was the most common. When adequate keys are developed to identify squid beaks in Hawaii, it will be interesting to determine the differences and similarities between these birds with respect to squid predation. The proportions of prey consumed by each congener was very different. By volume, Laysan Albatrosses consumed twice as much squid, and Black-footed Albatrosses ate 11 times more flyingfish eggs. The differences in diets may result from the apparent tendencies of Laysan Albatrosses to feed at night and of Black-footed Albatrosses to feed during the day. Both species forage in the cool waters north of the NWHI, much farther from the islands than other seabirds in this community. Their feeding locations suggest a tenuous relationship with the tropical marine environment.

The albatross guild in Hawaii can be compared with the Waved Albatross *D. irrorata* in the Galapagos Islands, the only other tropical albatross. Like Hawaiian birds, Waved Albatrosses eat primarily squid, fish, and deep-water crustaceans (Harris, 1973). The squid families consumed in Hawaii, however, were different from those consumed in the Galapagos, where mostly representatives of Histioteuthidae and Octopoteuthidae are eaten. These squid families are rarely taken by Hawaiian birds, who feed predominantly on ommastrephid squid. The differences in the composition of squid in the diets may be attributed to the availability of the representative squid families to the birds at the different localities. Galapagos birds consume flyingfish and mackerel scad *Decapterus* spp., but no flyingfish eggs. The guild in Hawaii is therefore broadly similar to the only other tropical albatross, but there are some notable differences.

Pelecaniformes

The Red-footed Booby *Sula sula*, Brown Booby *S. leucogaster*, Masked Booby *S. dactylatra*, Great Frigatebird *Fregata minor*, and Red-tailed Tropicbird *Phaethon rubricauda*, comprise this guild. Adult flyingfish (especially *Exocoetus volitans* and *Cypselurus* spp.) were the most commonly taken prey, but adult mackerel scad and juvenile ommastrephid squid were also very important components of the diet (Table 13.1). Members of this guild have diets with different percentage compositions of major prey items and consume prey of different sizes. Masked Boobies took the largest prey; the mean prey length was almost twice that of the other species. Masked Boobies frequently took fish larger than 20 cm. The other birds consumed prey primarily in the 8–15 cm range, although Brown Boobies took many 5 cm juvenile goatfish (Mullidae). Red-footed Boobies, an offshore species, ate much more squid than Brown Boobies, an inshore species. Flyingfish were especially common (> 60% by volume) in the diet of Great Frigatebirds. Frigatebirds have a structural inability to take flight from the ocean's surface and therefore do not enter the water. Certain prey were taken exclusively by a single species at one location when presumably such prey was abundant. For example, virtually all Pacific Saury *Cololabis saira* were taken by Red-footed Boobies during the winter months at the northern end of the archipelago.

Table 13.1. *Average percent volumes for major prey (families or groups that constitute 2% or more of the volume of a guild) items of Northwestern Hawaiian Islands birds*

Prey	Albatrosses	Pelecaniforms	Terns and shearwaters associated with predatory fishes	Nocturnal petrels	Neuston-feeding terns
Carangidae	17.8	9.8	—	—	—
Clupeidae	—	—	2.2	—	2.7
Coryphaenidae	—	2.9	—	—	2.0
Exocoetidae	25.1	44.3	10.2	—	10.2
Hemiramphidae	—	4.2	—	—	—
Mullidae	—	3.6	20.8	—	8.9
Myctophidae	—	—	—	10.2	—
Ostraciontidae	—	—	—	—	21.1
Scombridae	—	2.5	—	—	—
Sternoptychidae	—	—	—	10.8	—
Synodontidae	—	—	4.9	—	7.4
Other fishes	4.2	8.8	19.1	24.4	24.1
Squid	48.2	13.2	30.2	23.5	2.7
Crustaceans	6.6	—	—	5.6	10.6
Marine insects	—	—	—	—	9.7
Unidentified remains	11.0	—	—	—	12.9

This guild may be compared with seabird communities at Ascension Island (Stonehouse, 1962; Dorward, 1963), Christmas Island (Ashmole & Ashmole, 1967; Schreiber & Hensley, 1976), the Seychelles Islands (Diamond, 1974, 1984), Rose Atoll (Harrison, Hida & Seki, 1984), and the Galapagos Islands (Harris, 1969). It consumes primarily flyingfish and ommastrephid squid throughout its range, but inshore feeding Brown Boobies have much greater diet diversity (Diamond, 1984). In Hawaii, this guild is distinguished by the prominence of juvenile goatfish and adult mackerel scad in the diet.

Terns and shearwaters associated with predatory fish

This guild includes many of the most common tropical seabirds: the Sooty Tern, Brown Noddy *Anous stolidus*, Black Noddy *A. tenuirostris*, White Tern *Gygis alba*, Wedge-tailed Shearwater, and Christmas Shearwater *Puffinus nativitatis*. These species feed largely in association with predatory fishes, especially tunas (Murphy & Ikehara, 1955; Waldron, 1964; Ashmole & Ashmole, 1967; Erdman, 1967). Recent observations in the eastern tropical Pacific indicate little feeding in the absence of fish schools (R.L. Pitman, personal communication).

Juvenile forms of goatfish, ommastrephid squid (especially *Symplectoteuthis* spp.), mackerel scad, and flyingfish were the main prey consumed by this guild (Table 13.1). Black Noddies and to a lesser extent White Terns feed inshore in association with the jacks *Caranx* spp. and nearshore tunas (e.g. *Euthynnus affinis*). Consequently, inshore-feeding terns eat more herrings Clupeidae and juvenile lizardfish Synodontidae and somewhat fewer mackerel scads and squid than the remaining birds in this guild. The offshore-feeding Sooty Terns, Brown Noddies, Wedge-tailed Shearwaters, and Christmas Shearwaters associate with Skipjack Tuna *Katsuwonus pelamis*, Yellowfin Tuna *Thunnus albacares*, and Dolphinfish *Coryphaena hippurus* (Table 13.4). Squid account for 30–50% of the dietary volumes of the offshore birds. The mean prey length of shearwaters, the heaviest species in this guild, is almost twice that of the lightest tern. For many of the common prey taxa, all species in the guild took similar size classes, usually 3–8 cm. This guild takes prey much smaller than that consumed by pelecaniforms. Feeding techniques account for some resource partitioning. Shearwaters can pursue prey underwater, whereas terns feed at the surface and rarely wet their feathers. Feeding locations may account for many of the differences among the diets of these species, but detailed information does not exist. Seasonal variation was apparent in the diet of each species in this guild.

The Hawaiian guild consumes higher proportions of fish than squid compared with guilds of Ascension Island, Christmas Island (Ashmole & Ashmole, 1967), and the Seychelles (Diamond, 1984). Hawaiian birds generally eat fewer flyingfish, substituting goatfish and mackerel scad. The Seychelles are the only other location where the latter two families are important components of the diet.

Nocturnal petrels

The Bonin Petrel *Pterodroma hypoleuca*, Bulwer's Petrel *Bulweria bulwerii*, and Sooty Storm Petrel *Oceanodroma tristrami*, apparently feed extensively at night. Each species feeds offshore in surface waters, usually alone but occasionally in association with other birds. This guild feeds primarily on squid, hatchetfish Sternoptychidae and lanternfish Myctophidae (Table 13.1). Most of their prey possess photophores and occur in surface waters only under reduced light conditions (Harrison *et al.*, 1983). Food samples from this guild are usually in very poor condition, making identification to genus or species difficult. Timing of breeding may be an important means to avoid competition for prey resources. Bulwer's Petrels breed during summer whereas Bonin Petrels and Sooty Storm Petrels breed during winter. Bulwer's and Bonin Petrels take larger prey than the storm petrels. They also consume more fish than squid, whereas the storm petrels take fish and squid in equal volumes. Improved identification technique for midwater fish may enable a better understanding of resource partitioning among the species in this guild.

Few comparisons with other tropical areas are possible because little work on these or similar species is published. Great-winged Petrel *Pterodroma macroptera* in New Zealand consumed similar midwater fish but different families of squid (Imber, 1973). Phoenix Petrels *P. alba* at Christmas Island ate 78% squid and 14% fish (Ashmole & Ashmole, 1967). Bonin Petrels are unusual among *Pterodroma* because they eat primarily fish rather than squid.

Neuston-feeding terns

The Gray-backed Tern *Sterna lunata* and Blue-gray Noddy *Procelsterna cerulea* have somewhat similar diets and, unlike other terns in Hawaii, rarely feed in association with predatory fishes. They feed inshore and take small prey during the breeding season. Their diets are remarkable because of the very small proportion of squid (Table 13.1). Their primary preys were marine insects (*Halobates sericeus*), crustaceans, and juvenile forms of Cowfish *Lactoria fornasini*, flyingfish, goatfish, and lizardfish. Their diets differed widely in the proportions of several taxa. Gray-backed Terns feed far more on Cowfish, whereas Blue-gray Noddies feed far more on marine insects (Cheng & Harrison, 1983) and lizardfish. Some competition is avoided because of slightly different breeding seasons: Blue-gray Noddies feed most of their young in March–May and Gray-backed Terns feed their young in May–July. In addition, Gray-backed Terns feed throughout the NWHI, but Blue-gray Noddies are restricted to the southern islands because of an absence of suitable breeding habitat elsewhere. Gray-backed Terns are larger birds and take somewhat larger prey than the noddies.

Neuston-feeding terns are little studied elsewhere. Blue-gray Noddies on Christmas Island have similar diets to Hawaiian birds in regard to the consumption of *Halobates* and minute crustaceans. However, Hawaiian birds ate goatfish and lizardfish and Christmas Island birds ate tunas, blennies, and snake mackerels (Ashmole & Ashmole, 1967).

Consumption of major prey items

It is very difficult to measure directly the amount of food that seabirds consume (Furness, 1982). Consequently, we have developed a model to estimate the energy and food requirements of the NWHI seabird community.

Table 13.2. *Recent population estimate of seabird breeding pairs in the Hawaiia*

	Main Hawaiian Islands	Kaula	Nihoa	Necker	French Frigate Shoals	Gardner Pinnacles
Diomedea nigripes	0	25–50	40–60	200–250	2000–2250	0
D. immutabilis	10	30–70	1–5	800–900	900–1000	10–15
Pterodroma phaeopygia	400–600	0	0	0	0	0
Pt. hypoleuca	0	0	0	0	30–50	0
Bulweria bulwerii	500–1000	20–50	75000–100000	250–500	200–500	10–15
Puffinus pacificus	40000–60000	1500–2500	30000–40000	1500–2500	1500–1750	50–100
P. nativitatis	40–60	75–125	200–250	0	15–20	0
P. puffinus newelli	4000–6000	0	0	0	0	0
Oceanodroma castro	+	0	0	0	0	0
O. tristami	0	0	2000–3000	+	+	0
Phaethon rubricauda	200–400	250–400	250–300	100–150	550–600	20–23
P. lepturus	500–3000	0	0	0	0	0
Sula dactylatra	3–4	200–400	250–300	250–300	500–600	125–150
S. leucogaster	75–100	300–500	150–200	20–25	40–60	5–10
S. sula	900–1200	250–350	1500–2000	650–750	550–600	0
Fregata minor	5–10	250–350	3500–4500	700–900	350–375	0
Sterna lunata	10	500–600	9000–12000	3500–4500	750–1000	1500–2500
S. fuscata	60000–90000	35000–50000	10000–25000	12500–25000	60000–78000	250–500
Procelsterna cerulea	0	?	2000–2500	1000–1500	+	+
Anous stolidus	15000–20000	15000–25000	25000–35000	10000–15000	5000–7500	1000–1500
A. tenuirostris	300–800	20–30	1000–5000	300–500	750–850	200–300
Gygis alba	50–100	30–40	1000–5000	100–300	500–750	150–250

The model was largely developed by Pettit, Whittow & Ellis (1984) and was expanded to encompass the entire NWHI seabird community. Estimates will be refined in the future as improved date become available.

Model methods and input

The model includes a component for adult maintenance energy. An estimate of the energy required for daily existence for each of the 18 species in the NWHI community was multiplied by the total number of days each year that each species is associated with the breeding islands. This amount was then multiplied by the total number of birds on each island. The estimate of daily existence energy includes the energy for foraging based upon gliding flight. It is particularly difficult to estimate the costs of foraging flight, yet we recognize it to be an extremely important component of the model (Furness, 1982). More than 5 million seabirds breed in Hawaii. The populations of each species and island are presented in Table 13.2. Non-breeding populations are included in this model and have been estimated as a percentage of the known breeding population, ranging from 31% to 70% (Fefer *et al.*, 1984). Tropical seabirds moult throughout the year, and this model assumed that daily existence energy requirements reflect the daily cost of moulting.

Archipelago (*from Harrison et al., 1984, where full references are given*)

Laysan	Lisianski	Pearl & Hermes Reef	Midway	Kure	
14000–21000	2800–3800	8000–11000	6500–7500	700–1300	*Diomedea nigripes*
105000–132000	23000–30000	9000–12000	150000–200000	3000–4000	*D. immutabilis*
0	0	0	0	0	*Pterodroma phaeophgia*
50000–75000	150000–250000	400–600	2500–5000	400–600	*Pt. hypoleuca*
1000–2000	50–100	< 10	0	0	*Bulweria bulwerii*
125000–175000	10000–30000	5000–10000	500–1000	900–1100	*Puffinus pacificus*
1500–2000	400–600	< 10	25–50	20–30	*P. nativitatis*
0	0	0	0	0	*P. puffinus newelli*
0	0	0	0	0	*Oceanodroma castro*
500–2500	?	300–500	0	?	*O. tristrami*
1500–2500	900–1300	10–60	4000–5000	1000–1300	*Phaethon rubricauda*
0	0	0	1	0	*P. lepturus*
400–425	300–350	140–160	5–10	65–75	*Sula dactylatra*
34	15–25	50–60	0	50–60	*S. leucogaster*
700–800	350–450	40–60	450–500	400–450	*S. sula*
2000–2500	750–850	300–400	60–75	200–250	*Fregata minor*
5000–10000	15000–20000	650–750	100–200	30–50	*Sterna lunata*
375000–500000	400000–600000	35000–45000	30000–45000	8000–12000	*S. fuscata*
0	0	0	0	0	*Procelsterna cerulea*
10000–15000	7500–15000	1700–2000	500–1000	700–800	*Anous stolidus*
1500–2500	500–1000	75–125	2000–6000	0	*A. tenuirostris*
1000–2000	50–100	10–20	5000–7500	5–10	*Gygis alba*

Added to the adult maintenance energy in the model is a component that estimates the costs of producing an egg. This was determined using bomb calorimetry and estimations of the efficiency of egg production. We assumed that birds laid a single egg and that each breeding pair produced one egg per year. The cost of producing an egg was then multiplied by an estimate of the number of eggs laid each year for each species. The amount of energy expended in raising a chick was estimated using growth equations and chick weight data. This amount was then multiplied by the number of chicks raised, using the highest observed reproductive success for that species in the NWHI (Pettit *et al.*, 1984).

The food requirement for each species was determined by use of volumetric estimates of the percentage of the diet for each prey species for each seabird (Harrison *et al.*, 1983). These estimates were converted into estimates of food consumption by using bomb calorimetric data for common prey items. The estimates of food consumed were calculated individually for each species and later pooled into feeding guilds.

Model results and discussion

Seabirds in the NWHI consume 410 000 tonnes of fish, squid, crustaceans, and other food sources each year (Table 13.3). Squid is the largest component, comprising over 54% of the total consumption. Fish accounts for 35% of the prey consumption of this community, the most important families being Exocoetidae (7%), Mullidae (5%), and Carangidae (primarily mackerel scad) (4%). Most of the squid were Ommastrephidae, especially *Symplectoteuthis oualaniensis*, but the large amount of unidentified squid in the albatross diets (Harrison *et al.*, 1983) makes this conclusion tentative.

Two feeding guilds account for almost all of the prey consumed. Albatrosses take 264 000 tonnes (64%) and birds that feed in association with predatory fishes consume 117 000 tonnes (29%). For individual species, Laysan Albatrosses account for 60% of the consumption. Laysan Albatrosses, Black-footed Albatrosses, Bonin Petrels, Wedge-tailed Shearwaters, and Sooty Terns together account for 94% of the prey consumed. Each of these birds consumes a large proportion of squid. The impact of neuston-feeding terns on the marine ecosystem is trivial, as biological intuition might suggest, but the minor role of pelecaniforms (2%) is unexpected.

It is interesting to compare the estimates of prey consumed by the NWHI seabird community with the present and projected fishery landings in the Hawaiian Archipelago. Although fish landings fluctuate each year and may be under-reported by as much as a factor of two, 6100 tonnes were reported as landed in 1978. The most optimistic estimates for increased fishery landings are 34 000–53 000 tonnes (Swerdloff, 1980). Hawaiian seabirds take 67 times the current reported landings of commercial fish in Hawaii, and more than seven times the highest projected fishery yields. Clearly seabirds are a very important component of the marine ecosystem in Hawaii, as they take more than a trivial quantity of the production of lower trophic levels. For example, an ecosystem model applied at French Frigate Shoals estimates that seabirds consume 42% of the annual production of small surface pelagic fishes and squid (Polovina, 1984). Several potential errors complicate this estimate of total food consumption. The largest is the population estimate of non-breeding seabirds. This population is very difficult to estimate accurately (Fefer *et al.*, 1984) and accounts for almost two-thirds of the consumption estimated by this model. In addition, there is considerable imprecision in the breeding

Table 13.3. *Annual consumption (thousand tonnes) of major prey (families or groups that constitute 2% or more of the volume of the diet of a guild)*

Prey	Albatrosses	Pelecaniforms	Terns and shearwaters associated with predatory fishes	Nocturnal petrels	Neuston feeding terns	Total
Carangidae	—	1.1	15.6	—	—	16.7
Clupeidae	—	—	0.2	—	0.1	0.3
Coryphaenidae	—	0.2	0.1	—	0.1	0.4
Exocoetidae	14.8	5.1	8.9	0.1	0.1	29.0
Hemiramphidae	—	0.4	0.1	—	—	0.5
Mullidae	—	0.1	20.2	0.6	0.1	21.0
Myctophidae	—	—	1.6	4.5	—	6.1
Ostraciontidae	—	—	—	—	0.5	0.5
Scombridae	—	0.2	—	—	—	0.2
Sternoptychidae	—	—	—	2.6	—	2.6
Synodontidae	—	—	0.8	0.3	—	1.1
Other fishes	6.8	0.8	16.0	0.9	0.2	24.7
Squid	167	1.3	51.4	3.4	0.1	223.2
Crustaceans	22.2	—	0.4	1.2	0.1	23.9
Unidentified remains	28.6	—	1.1	3.7	—	33.4
Other	25	0.1	1.0	0.3	—	26.4
Total	264.4	9.3	117.4	17.6	1.3	410.0

population estimates. Many species are very difficult to census. Nesting in the NWHI is often protracted over several months, and the colonies are remote and difficult to visit. We have inadequate knowledge of the budgets of adult foraging activites and do not know their feeding locations. Most estimates of feeding distances from colonies are speculation (Harrison & Stoneburner, 1981). It is important to learn the extent of foraging so that an area can be defined to estimate fish production.

These estimates of the amounts of food taken by seabirds are the first for a subtropical or tropical community. Seabirds play as important a role in tropical waters as they do in temperate areas. Wiens & Scott (1975) estimated that seabirds consume as much as 22% of the annual production of pelagic fish in Oregon. Schaefer (1970) estimated that seabirds in Peru consume 2.8 million tonnes of anchovies each year (one-third of the commercial catch) and account for 20% of the fish mortality (Furness & Cooper, 1982). Seabirds at one colony in Shetland, Scotland, consume about 29% of the fish production within a 45 km radius of the colony. Seabirds in the Benguela region take about 30% of the mean annual catch in the adjacent fishing grounds and about 17% of the annual fish production (Furness & Cooper, 1982). Primary productivity in Hawaii is about 36 g C m^{-2} (Hirota *et al.*, 1980), or 12–36% of the productivity of continental shelf or upwelling areas (Ryther, 1963). To the extent that fish production can be correlated with primary productivity, Hawaiian seabirds may consume a greater percentage than those in other areas.

Fishery interactions

Do seabirds in Hawaii compete with commercial fishermen? Because much of the NWHI remains unexploited, it is difficult to compare this community with areas that have established fisheries. Tropical ecosystems are complicated by the migratory nature of tuna stocks and other prey, including squid. It is difficult to define a discrete ecosystem, and annual fish production estimates could be misleading. The juvenile forms of fish and squid that comprise much of the food consumed by seabirds are below the trophic levels that would directly affect commercial fisheries. Prey of the seabirds may also be forage for commercial fish. Commercial fisheries for large species might result in increased forage for seabirds, as has occurred in the North Sea (Furness, 1982). Fishery–seabird competition is most severe when humans and birds compete for identical species in identical size classes (Idyll, 1973). The situation in tropical waters is greatly complicated by the dependency of many birds on tunas to drive their prey to the surface.

Bird flocks and predatory fish

The occurrence of terns and shearwaters in flocks provides invaluable assistance to tuna fishermen. Gould (1971) estimated that a trained observer could see and recognize a large feeding flock at a distance of at least 8 km and stated that some fishermen use the distinctive feeding patterns of birds as clues to determine the species of tuna feeding below the flock. If the flocking birds remain active low over the water, they are believed to be feeding over surface-feeding Skipjack Tuna. If the flock activity alternates between low and high altitudes, the fish beneath are presumed to be the deeper-foraging Yellowfin Tuna.

Between 1953 and 1969, the Southwest Fisheries Center Honolulu Laboratory, NMFS, NOAA (formally Pacific Oceanic Fishery Investigations (POFI) and the Bureau of Commercial Fisheries (BCF)) conducted tuna pole-and-line research cruises around the Hawaiian Islands aboard the research vessel *Charles H. Gilbert*. On these cruises, observations of bird flocks and fish schools were recorded using the scouting method (described and evaluated by Royce & Otsu (1955)) by which an experienced fisherman logs all observations of bird flock (as well as solitary individuals) and fish schools. During the *Gilbert* cruises (with at least 30 flock observations) 2138 bird flocks were investigated and 681 (32%) of the fish schools under these flocks were identified (Table 13.4). Tunas constituted 88% of the identified schools, and 86% of these tuna schools were composed of Skipjack Tuna. Because actual fishing verified the composition of schools, the high proportion of unidentified schools (68%) is not unusual. On these pole-and-line cruises, live baitfish were chummed to attract and hold the school close to the vessel. Often, fish will not respond to chumming. Royce

Table 13.4. *Number and percent of fish species associated with 681 bird flocks investigated on pole-and-line cruises in the Hawaiian Islands, 1953–69. Unidentified schools were excluded from this analysis*

Species	Vernacular name	Number of schools	Percentage of schools
Fish			
Katsuwonus pelamis	Skipjack Tuna	514	75.4
Thunnus albacares	Yellowfin Tuna	25	3.7
Euthynnus affinis	Kawakawa	25	3.7
	Mixed tunas	37	5.4
Coryphaena hippurus	Dolphinfish	68	10.0
Mammals			
Delphinidae	Porpoises	12	1.8

& Otsu (1955) and Yuen (1959) reported that fish were caught from only 43 and 47% of the schools chummed, respectively. The high rate of occurrence of Skipjack Tuna under flocks is not surprising. Murphy & Ikehara (1955) found Skipjack Tuna not only the most commonly sighted tuna species in Hawaiian waters, but also the single dominant species in oceanic areas. John J. Naughton (unpublished data) investigated bird flock–fish school interactions in the oceanic waters of the central Pacific and found 73% of identified fish schools to be Skipjack Tuna. Yellowfin Tuna accounted for only 15% of the schools. Naughton suggested that barring various degrees of difficulty in identifying different species, Skipjack Tuna would make up an equally high percentage of the unidentified schools.

Bird flock–fish school interaction is particularly important in Hawaiian commercial fisheries, especially in the Skipjack Tuna pole-and-line fishery. Thus tuna species contribute more than 80%, and Skipjack Tuna more than 65%, of the total catch of ten top-ranked species, which together comprise over 92% of the commercial landing reported to the State of Hawaii (Table 13.5).

Most of the major fisheries in Hawaii have been concentrated in the main Hawaiian Islands. There is, however, increasing interest in extending the various fisheries into the NWHI, where the Hawaiian Islands National Wildlife Refuge protects the breeding grounds of about 5.4 million seabirds

Table 13.5. *Commercial fish landings of the top 10 species for the State of Hawaii, 1961–79*

	Vernacular name	Total landings (MT)	19-year average (MT)
Katsuwonus pelamis	Skipjack Tuna	77 394.6	4073.4
Thunnus albacares	Yellowfin Tuna	8257.9	434.6
T. obesus	Bigeye Tuna	5322.7	280.1
Selar crumenophthalmus	Bigeye Scad	5251.7[a]	208.2[a]
Decapterus macarellus	Mackerel Scad	2322.7	122.2
Tetrapterus audax	Striped Marlin	2272.4	119.6
Makaira nigricans	Blue Marlin	1592.8	83.8
Pristipomoides filamentosus	Pink Snapper	1001.0	68.2
Coryphaena hippurus	Dolphinfish	960.6	50.6
Caranx sp.[b]	Jacks	678.2	35.7

[a] Includes the landings for juvenile *Selar crumenophthalmus*. [b] Includes the catches of 11 species of the deep-bodied carangids although *Caranx ignobilis* comprises most of the catch

(Harrison *et al.*, 1983). At present, the fisheries development in the NWHI concerns predominantly benthic stocks, such as lobsters, shrimps and bottom fishes. Although surface trollers have fished for Albacore *Thunnus alalunga*, in waters north of Midway, little effort has focused on other pelagic species. It seems likely that efforts to exploit other pelagic stocks will extend to the waters of the NWHI.

Tern–shearwater feeding guild

Sooty Terns and Wedge-tailed Shearwaters are the most abundant of the 18 species that breed in the NWHI (Harrison *et al.*, 1983). Sooty Terns are by far the most numerous, comprising 48% of the total population; Wedge-tailed Shearwaters comprise 17%. Among the many species of tropical seabirds, terns and shearwaters are also the most frequently associated with schools of tuna and other commercially valuable fishes (King, 1967). Sooty Terns were found 75% of the time and shearwaters 38% within bird flocks on the *Gilbert* pole-and-line cruises. Naughton (NMFS, unpublished data) found that the species composition of bird flocks is most diverse close to islands and that a substantial reduction in the number of species occurs with distance from land.

Because of the strong relationship between the feeding activities of birds and tunas, it is important to determine how much of a dietary overlap exists. Ashmole & Ashmole (1967) examined the food of Yellowfin Tuna captured in surface waters near Christmas Island. They found major differences between the diets of the fish and birds: fish consumed fewer squid and a wider variety of invertebrates. The fish consumed by the tunas also differed significantly from those eaten by birds. They concluded, however, that a large proportion of the items taken by the birds are made available at the surface only by the feeding activities of the tunas. They attributed the difference in the diets mainly to the types of tuna prey that do not come to the surface even when pursued, whereas other organisms such as flyingfish are subject to predation by the birds while escaping the tunas.

Waldron & King (1963) examined 707 stomach samples from Skipjack Tuna in Hawaiian waters, including the Line and Phoenix Islands, and found material representing 11 invertebrate orders and 42 fish families. Fish occurred in 67% of the stomachs and, based on their frequency of occurrence, the highest ranking families were Gempylidae, Scombridae, Mullidae, Chaetodontidae, and Holocentridae in order of importance. Carangids (especially mackerel scad) were important only in the Skipjack Tuna sampled in Hawaii. Crustaceans and squid occurred in 36% and

35% of the stomachs, respectively. Most of the identified squid were ommastrephids. By volume, fish composed 75%, squid 20%, and crustaceans 4% of the Skipjack Tuna food. Scombridae, Carangidae, Mullidae, Nomeidae, and Molidae were the most important fish families. Yuen (1959) found similar results in his study of 573 Hawaiian Skipjack Tuna stomachs. Members of Carangidae (primarily mackerel scad) were by far the most important item in terms of both volume and frequency of occurrence. As in Waldron & King's study, Yuen found nomeids, molids, and scombrids among the highest-ranked fish families. Squid, stomatopods, and decapod crustaceans were also significant contributors to the diet.

By contrast, our study demonstrated that for Sooty Terns, squid (mostly ommastrephids) were the highest-ranked prey items, constituting 53% of the sample volumes (Harrison *et al.*, 1983). Fish accounted for 46% of the volumes; Mullidae, Exocoetidae, Gempylidae, Carangidae (mostly mackerel scad), Nomeidae, and Holocentridae were the highest ranked. Wedge-tailed Shearwaters ate 66% fish, 28% squid, and 1% crustaceans, by volume. Prey items ranked high were the fish species in Mullidae and carangidae (again, mostly mackerel scad) and the squids in Ommastrephidae. Other common prey fish were the monacanthids and exocoetids.

Based on these fish and bird feeding studies, there appears to be a considerable degree of overlap in the diet of Sooty Terns, Wedge-tailed Shearwaters, and Skipjack Tuna. Among the highest-ranking prey items, squid (Ommastrephidae), goatfish (Mullidae), and mackerel scad were very important in all the diets. Other fish belonging to Nomeidae, Gempylidae, and Holocentridae were also commonly taken by both birds and tunas. The higher degree of overlap between Skipjack Tuna and the birds as compared with Yellowfin Tuna and the birds is not surprising. Skipjack Tuna and seabirds are generally surface feeders, and most fish schools found beneath feeding bird flocks are Skipjack Tuna (Table 13.4). Yellowfin Tuna tend to forage at greater depth.

Skipjack Tuna and the seabirds also appear to prey upon organisms of similar sizes. Based on the various Skipjack Tuna feeding studies in the Pacific (Alverson 1963; Waldron & King, 1963; Nakamura, 1965), Blackburn & Laurs (1972) found that the prey of Skipjack Tuna fell mainly in the 10–100 mm size range. In comparison, Harrison *et al.* (1983) found the standard length of Sooty Tern prey items (n = 326) to range from 1 to 120mm (mean 48 mm s.d. 19). All of the common prey items, however, were 20–70 mm long. The prey of Wedge-tailed Shearwaters (n = 212) ranged from 4 to 145 mm in standard length (mean 57 mm s.d. 25).

To a lesser degree, seabirds also feed in association with schools of Dolphinfish. Unlike Skipjack Tuna, which traverse the water column vertically, Dolphinfish normally remain in the surface waters. This behavior is reflected in the composition of Dolphinfish forage items. In the Pacific, flyingfish and other surface-dwelling species were the dominant prey taken (Tester & Nakamura, 1957; Kojima, 1961; Rothschild, 1964). The size composition of the principal Dolphinfish food organisms is also similar to that taken by the Sooty Terns and Wedge-tailed Shearwaters. Kojima (1961) reported that although the size range of prey was wide, from about 1 to 34 cm, most of the prey were between 2–4 and 10–15 cm.

Future research needs and conclusions

Unlike subpolar and temperate communities, tropical seabird communities have rarely been studied. Research on seabirds is usually restricted to those ecosystems that are faced with major developments such as offshore oil and gas activities or large commercial fisheries. Our recent studies in Hawaii complement the work by Ashmole & Ashmole (1967) at Christmas Island and represent a refinement of many of their ideas concerning diets and trophic relationships among tropical seabirds. However, much work remains for a better understanding of the role of seabirds in tropical and subtropical ecosystems.

Our model that estimates the consumption of prey by the NWHI seabird community would be improved with more precise input parameters. Better population estimates, especially for non-breeding birds, would refine our estimates of prey consumption. In addition, estimates of consumption rates of seabirds could be improved by studies utilizing captive birds in zoos, marine parks, or oceanaria. Most of our current estimates are largely based on theoretical considerations rather than direct measurements.

A major challenge in the study of seabirds in all ecosystems is to improve our understanding of foraging ranges and locations. Speculations concerning foraging locations are too often accepted as fact despite a virtual absence of data. Tropical seabirds range widely in warm waters and, like polar and subpolar species, live many years before breeding. Birds observed feeding at sea are not necessarily **breeding** birds, nor should they be assumed to be associated with the nearest colony. How far away from colonies birds feed may be inferred from incubation bouts and chicks' feeding frequencies (Harrison & Hida, 1980). However, such indices are very imprecise, and developments in radiotelemetry are necessary to determine feeding locations. In tropical waters, such technological progress would also be of great interest to fishery managers and fishermen. The movements of birds such as Sooty Terns could be used to locate tuna schools efficiently.

Our understanding of the role of marine birds in tropical ecosystems is most limited by our understanding of tropical marine ecology and fisheries. Improvements in our ability to identify juvenile and larval fish and squid in tropical waters will result in refinements in our analysis of dietary specialization among tropical species. In our work in Hawaii (Harrison *et al.*, 1983), we encountered many unidentifiable flyingfish, goatfish, lizardfish and squid, representing many species. For example, 10 goatfish and 9 flyingfish occur in Hawaiian waters (Tinker, 1978). Although there is undoubtedy some theoretical interest in examining feeding overlap using Morisita's Overlap Index (Diamond, 1984), dietary differences among tropical seabird species are masked by our inability consistently to identify prey beyond the family level. Conclusions based on such analyses repeat the unremarkable proposition that tropical seabirds eat much flyingfish and squid. This is roughly equivalent to concluding that Serengeti herbivores have similar diets because they eat various grasses. Because identification of prey is particularly difficult in tropical ecosystems, it is especially important to utilize a multidisciplinary approach to research and to include fishery biologists. Too often dietary studies of seabirds utilize untrained laboratory personnel whose opinions on the identity of half-digested prey are suspect. Improved estimates of available prey must be developed by fishery biologists if projections concerning the impacts of tropical fisheries on seabird communities are to be improved. This is a prerequisite if marine birds are to be used to monitor changes with time in epipelagic marine fauna (Ashmole & Ashmole, 1968). The remarkable Pacific-wide changes during the 1982–83 El Niño event underscore the interest in such studies.

The interaction between seabirds and predatory fish needs careful study. Only about half of the feeding bird flocks observed during the *Gilbert* cruises were confirmed to be associated with predatory fish. Are some seabirds strictly dependent on fish and marine mammals to drive prey to the surface, or are they merely opportunistic? Additional comparisons between the diets of seabirds and predatory fish are needed to clarify differences and similarities in size classes of prey species.

Lastly, the role of tropical seabird colonies in nutrient cycling deserves careful study. The consumption of prey is not a one-way transfer of energy. As much as 30% of the energy ingested by birds is voided as waste (Wiens & Scott, 1975). Birds enrich the waters surrounding their colonies with calories and nutrients (Hutchinson, 1950: 373; Tuck, 1960; Lindeboom, 1984). Ashmole & Ashmole (1967) hypothesized that seabird guano is an important source of nutrients for coral reef communities adjacent to bird colonies. This notion has recently been rejected in certain Canadian waters (Bédard, Therriault & Bérubé, 1980). Careful nearshore measurements of nutrients and estimates on quantities of guano from NWHI seabird colonies should render a realistic assessment of nutrient cycling as it pertains to tropical seabirds.

Acknowledgements The senior author gratefully acknowledges support from the National Marine Fisheries Service during the preparation of this chapter. Richard S. Shomura provided valuable insights, discussion, and direction. John J. Naughton allowed extensive use of his unpublished reports and data.

References

Ainley, D.G. (1977). Feeding methods in seabirds: a comparison of polar and tropical nesting communities in the eastern Pacific Ocean. In *Adaptations within Antarctic ecosystems*, ed. G.A. Llano, pp. 669–85. Washington, D.C.: Smithsonian Institution.

Ainley, D.G. & Boekelheide, R.J. (1984). An ecological comparison of oceanic seabird communities of the South Pacific Ocean. In *Studies in avian biology, vol. 8. Tropical seabird biology*, ed. R.W. Schreiber, pp. 2–23.

Alverson, F.G. (1963). The food of Yellowfin and Skipjack Tunas in the eastern tropical Pacific Ocean. *Inter-Am. Trop. Tuna Comm., Bull.*, 7, 293–396.

Ashmole, M.J. & Ashmole, N.P. (1968). The use of food samples from sea birds in the study of seasonal variation in the surface fauna of tropical oceanic areas. *Pac. Sci.*, **22**, 1–10.

Ashmole, N.P. (1971). Sea bird ecology and the marine environment. In *Avian biology*, vol. 1, ed. D.S. Farner & J.R. King, pp. 223–87. New York: Academic Press.

Ashmole, N.P. & Ashmole, M.J. (1967). Comparative feeding ecology of sea birds on a tropical oceanic island. *Peabody Mus. Nat. Hist., Yale Univ. Bull.*, **24**, 131 pp.

Bédard, J., Therriault, J.C. & Bérubé, J. (1980). Assessment of the importance of nutrient cycling by seabirds in the St. Lawrence Estuary. *Can. J. Fish. Aquat. Sci.*, **37**, 583–8.

Blackburn, M. & Laurs, R.M. (1972). Distribution of forage of Skipjack Tuna (*Euthynnus pelamis*) in the eastern tropical Pacific. *U.S. Dep. Commer., NOAA Tech. Rep.*, NMFS SSRF-649, 16 pp.

Cheng, L. & Harrison, C.S. (1983). Seabird predation on the Sea-skater *Halobates sericeus* (Heteroptera: Gerridae). *Mar. Biol.*, **72**, 303–9.

Diamond, A.W. (1974). Biology and behaviour of frigatebirds *Fregata* spp. on Aldabra Atoll. *Ibis*, **117**, 302–23.

Diamond, A.W. (1984). Feeding overlap in some tropical and temperate seabird communities. In *Tropical seabird biology*, ed. R.W. Schreiber. *Studies in Avian Biology*, **8**, 24–46.

Dorward, D.F. (1963). Comparative biology of the White Booby and the Brown Booby *Sula* spp. at Ascension. *Ibis*, **103b**, 174–220.

Ely, C.A. & Clapp, R.B. (1973). The natural history of Laysan Island, Northwestern Hawaiian Islands. *Atoll Res. Bull.*, **171**, 1–361.

Erdman, D.S. (1967). Sea birds in relation to game fish schools of Puerto Rico and the Virgin Islands. *Carib. J. Sci.*, **7**, 79–85.

Fefer, S.I., Harrison, C.S., Naughton, M.B. & Shallenberger, R.J. (1984). Synopsis of results of recent seabird research conducted in the Northwestern Hawaiian Islands. In *Proceedings of the symposium on resource investigations in the northwestern Hawaiian Islands*, May 25–27, 1983, University of Hawaii, Honolulu, Hawaii, ed. R.W.Grigg & K.Y. Tanoue. Sea Grant.

Furness, R.W. (1982). Competition between fisheries and seabird communities. *Adv. Mar. Biol.*, **20**, 225–307.

Furness, R.W. & Cooper, J. (1982). Interactions between breeding seabird and pelagic fish populations in the southern Benguela region. *Mar. Ecol. Prog. Ser.*, **8**, 243–50.

Gould, P.J. (1967). Nocturnal feeding of *Sterna fuscata* and *Puffinus pacificus*. *Condor*, **69**, 529.

Gould, P.J. (1971). Interactions of seabirds over the open ocean. Ph.D. dissertation, University of Arizona, Tucson.

Harris, M.P. (1969). Breeding seasons of seabirds in the Galapagos Islands. *J. Zool., Lond.*, **159**, 145–65.

Harris, M.P. (1973). The biology of the Waved Albatross *Diomedea irrorata* of Hood Island, Galapagos. *Ibis*, **115**, 483–510.

Harrison, C.S. (1982). Spring distribution of marine birds in the Gulf of Alaska. *Condor*, **84**, 245–54.

Harrison, C.S. & Hida, T.S. (1980). The status of seabird research in the Northwestern Hawaiian Islands. In *Proceedings of the symposium on status of resource investigations in the northwestern Hawaiian Islands*, April 24–25, 1980, University of Hawaii, Honolulu, Hawaii, ed. R.W.Grigg & R.T. Pfund, pp. 17–31. Sea Grant.

Harrison, C.S., Hida, T.S. & Seki, M.P. (1983). Hawaiian seabird feeding ecology. *Wildl. Monogr.*, **85**, 71 pp.

Harrison, C.S., Hida, T.S. & Seki, M.P. (1984). The diet of the Brown Booby (*Sula leucogaster*) and the Masked Booby (*Sula dactylatra*) on Rose Atoll, Samoa. *Ibis*, **126**, 588–90.

Harrison, C.S., Naughton, M.B. & Fefer, S.I. (1984). The status and conservation of seabirds in the Hawaiian Archipelago and Johnston Atoll. In *Status and conservation of the world's seabirds*, ed. J.P. Croxall, P.G.H. Evans & R.W. Schreiber, pp. 513–26. Cambridge: ICBP.

Harrison, C.S. & Stoneburner, D.L. (1981). Radiotelemetry of the Brown Noddy in Hawaii. *J. Wildl. Mgmt.*, **45**, 1021–5.

Hirota, J., Taguchi, S., Shuman, R.F. & Jahn, A.E. (1980). Distributions of plankton stocks, productivity, and potential fishery yield in Hawaiian waters. In *Proceedings of the symposium on status of resource investigations in the northwestern Hawaiian Islands*, April 24–25, 1980, University of Hawaii, Honolulu, Hawaii, ed. R.W. Grigg & R.T. Pfund, pp. 191–203. Sea Grant.

Hutchinson, G.E. (1950). The biogeochemistry of vertebrate excretion. *Bull. Am. Mus. Nat. Hist.*, **96**, 1–554.

Idyll, C.P. (1973). The anchovy crisis. *Sci. Am.*, **228**, 22–9.

Imber, M.J. (1973). The food of Grey-faced Petrels (*Pterodroma macroptera gouldi*

(Hutton)), with special reference to diurnal vertical migration of their prey. *J. Anim. Ecol.*, **42**, 645–62.

King, W.B. (1967). *Seabirds of the tropical Pacific Ocean*. Washington, D.C.: Smithsonian Institution.

King, W.B. (1970). The trade wind zone oceanography pilot study. Part VII: Observations of seabirds, March 1964 to June 1965. *U.S. Fish Wildl. Serv., Spec. Sci. Rep. Fish.*, **586**, 136 pp.

Kojima, S. (1961). Studies of dolphin fishing conditions in the western Sea of Japan – III. On the stomach contents of dolphin. *Bull. Jpn. Soc. Sci. Fish.*, **27**, 625–9. (In Japanese.) (Translation by W.G. Van Campen, Bur. Commer. Fish. Biol. Lab., Honolulu, Hawaii, 1963, 8 pp.)

Lack, D. (1967). Interrelationships in breeding adaptations as shown by marine birds. In *Proceedings of the 14th International Ornithological Congress*, Oxford, 24–30 July 1966, ed. D.W. Snow, pp. 3–42. Oxford: Blackwell Scientific Publications.

Lindeboom, H.J. (1984). The nitrogen pathway in a penguin rookery. *Ecology*, **65**, 269–77.

Murphy, G.I. & Ikehara, I.I. (1955). A summary of sightings of fish schools and bird flocks and of trolling in the central Pacific. *U.S. Fish Wildl. Serv., Spec. Sci. Rep. Fish.*, **154**, 19 pp.

Nakamura, E.L. (1965). Food and feeding habits of Skipjack Tuna (*Katsuwonus pelamis*) from the Marquesas and Tuamotu Islands. *Trans. Am. Fish. Soc.*, **94**, 236–42.

Nelson, B. (1979). *Seabirds: their biology and ecology*. New York: A & W Publishers, Inc.

Pettit, T.N., Whittow, G.C. & Ellis, H.J. (1984). Preliminary model of food and energetic requirements of seabirds at French Frigate Shoals, Hawaii. In *Proceedings of the symposium on resource investigations in the northwestern Hawaiian Islands*, May 25–27, 1983, University of Hawaii, Honolulu, Hawaii, ed. R.W. Grigg & K.Y. Tanoue. Sea Grant.

Polovina, J.J. (1984). Model of a coral reef ecosystem, part I: the ECOPATH model and its application to French Frigate Shoals. *Coral Reefs* **3**, 1–11.

Raymont, J.E.G. (1966). The production of marine plankton. *Adv. Ecol. Res.*, **3**, 117–205.

Rothschild, B.J. (1964). Observations on dolphins (*Coryphaena* spp.) in the central Pacific Ocean. *Copeia*, 1964, 445–7.

Rougerie, F. & Chabanne, J. (1983). Relationship between tuna and salinity in Tahitian coastal waters. *Trop. Ocean-Atmos. Newsl.*, **17**, 12–13.

Royce, W.F. & Otsu, T. (1955). Observation of skipjack scools in Hawaiian waters, 1953. *U.S. Fish Wildl. Serv., Spec. Sci. Rep. Fish.*, **147**, 31 pp.

Ryther, J.H. (1963). Photosynthesis and fish production in the sea. *Science*, **166**, 72–6.

Schaefer, M.B. (1970). Men, birds and anchovies in the Peru Current – dynamic interactions. *Trans. Am. Fish. Soc.*, **99**, 461–7.

Schreiber, R.W. & Ashmole, N.P. (1970). Sea-bird breeding seasons on Christmas Island, Pacific Ocean. *Ibis*, **112**, 363–94.

Schreiber, R.W. & Hensley, D.A. (1976). The diets of *Sula dactylatra*, *Sula sula*, and *Fregata minor* on Christmas Island, Pacific Ocean. *Pac. Sci.*, **30**, 241–8.

Stonehouse, B. (1962). Ascension Island and the British Ornithologists' Union centenary expedition 1957–59. *Ibis*, **103b**, 107–23.

Swerdloff, S.N. (1980). The Hawaii fisheries development plan. In *Proceedings of the symposium on status of resource investigations in the northwestern Hawaiian Islands*, April 24–25, 1980, University of Hawaii, Honolulu, Hawaii, ed. R.W. Grigg & R.T. Pfund, pp. 309–22. Sea Grant.

Tester, A.L. & Nakamura, E.L. (1957). Catch rate, size, sex and food of tunas and other pelagic fishes taken by trolling off Oahu, Hawaii, 1951–55. *U.S. Fish Wildl. Serv., Spec. Sci. Rep. Fish.*, **250**, 25 pp.

Tinker, S.W. (1978). *Fishes of Hawaii*. Honolulu: Hawaiian Service, Inc.

Tuck, L.M. (1960). *The murres: their distribution, population and biology: a study of the genus* Uria. Ottawa: Queen's Printer.

Waldron, K.D. (1964). Fish schools and bird flocks in the central Pacific Ocean, 1950–61. *U.S. Fish Wildl. Serv., Spec. Sci. Rep. Fish.*, **464**, 20 pp.

Waldron, K.D. & King, J.E. (1963). Food and Skipjack in the central Pacific. In *Proceedings of the world scientific meeting on the biology of tunas and related species*, 2–14 July 1962, ed. H. Rosa, Jr. *FAO Fish. Rep.*, **6**(3), 1431–57.

Wiens, J.A. & Scott, J.M. (1975). Model estimation of energy flow in Oregon coastal seabird populations. *Condor*, **77**, 439–52.

Yuen, H.S.H. (1959). Variability of skipjack response to live bait. *U.S. Fish Wildl. Serv., Fish. Bull.*, **60**, 147–60.

14

Historical variations in food consumption by breeding seabirds of the Humboldt and Benguela upwelling regions

DAVID CAMERON DUFFY and W. ROY SIEGFRIED
Percy FitzPatrick Institute of African Ornithology, University of Cape Town, Rondebosch 7700, South Africa

Introduction

Estimates of food consumption by seabirds are necessary for the management of many commercially important fish species and the ecosystems in which they live. Cohort or Virtual Population Analysis (VPA) (Ricker, 1975), widely used to estimate the sizes of fish stocks, is dependent on estimates of mortality caused by natural predators such as seabirds. The results of VPA are sensitive to even relatively small alterations in the estimates of natural mortality (e.g. Armstrong *et al.*, 1983).

For most marine predators, such as marine mammals, squid, or predatory fish, estimates of populations and their food consumptions are either unavailable or are accurate to only one or two orders of magnitude. Seabirds are one of the few groups for which, despite the often formidable problems of censusing populations and assessing diets, reasonably accurate estimates of food consumption can be provided. These estimates can be used to 'calibrate' or rank vaguer estimates for other predators in relation to seabirds. Fluctuations in avian consumption levels can also signal changes in the total natural mortality of fish, which may have major implications for VPA estimates, although the possibility of such changes appears to have received relatively little attention (but see Schaefer, 1967, 1970; MacCall, 1983).

Estimates of the amounts of food consumed by seabirds are also essential where multi-species management is necessary, either because of legal or treaty requirements as in the California Current (Green, 1978) and the Southern Ocean (Mitchell & Sandbrook, 1980), or because a suite of commercially important fish species may alternate in dominance within a marine ecosystem (Ahlstrom, 1966; Daan, 1980; Newman & Crawford, 1980; Skud,1982) so that the ecosystem can only be managed by understanding the competitive milieu, including predators, of the fish species (May *et al.*, 1979). This appears to be the situation in both the Benguela upwelling ecosystem off southwestern Africa (Shannon, Crawford & Duffy, 1984) and the Humboldt or Peruvian upwelling ecosystem off the west coast of South America (Serra, 1983; Zuta, Tsukayami & Villanueva, 1983).

Finally, seabirds, in addition to being potentially useful as monitors of fish stocks, are in themselves important economic resources as guano producers in Peru, South Africa and Namibia, and as tourist attractions in Galapagos and many places in Europe and North America. For economic and conservation reasons, it is desirable to known how much food is necessary to maintain these populations.

As a result of these and other needs, numerous studies have estimated consumption by seabirds in different marine environments, with varying degrees of apparent precision and accuracy (e.g. Vogt, 1942; Hutchinson, 1950; Davies, 1958; Schaefer, 1967, 1970; Sanger, 1972; Wiens & Scott, 1975; Furness, 1978; Croxall & Prince, 1982, 1984; Furness & Cooper, 1982; Schneider & Hunt, 1982; see also Chapter 11). Most of these studies have provided single values for food consumption by seabirds. Both the Benguela and Humboldt ecosystems are characterised by long- and short-term spatial and temporal variability (e.g. Hutchinson, 1950; DeVries & Schrader, 1981; Hendey, 1981; Shannon *et al.*, 1984). Given the natural variability of seabird populations (reviewed by Drury, 1979) and of marine ecosystems (e.g. Smith, 1978; Walsh, 1978), as it may affect seabirds (Myres, 1979), single estimates for consumption are likely to have only transitory value or to even be dangerous, if they imply existing or potential competition with a commercial fishery. Inflated consumption figures have been used by politicians and commercial fishermen in Peru (Nelson, 1978), South Africa (Jarvis, 1971), Namibia (Green, 1950) and elsewhere to support attempts to cull large numbers of seabirds in the belief that this would increase fish stocks available for commercial exploitation.

In this chapter we use guano production to estimate the food consumption of seabirds in the Southern Benguela and the Humboldt or Peruvian coastal current ecosystems at different stages of the commercial exploitation of their fish stocks. Peru and the west coast of South Africa have been chosen because relatively little is known about seabird population sizes off Chile or fluctuations in the abundance of seabirds breeding off

Namibia. We compare the extent and variability of consumption between years and between decades. Finally, for the Peruvian ecosystem, we examine the possibility that the importance of seabird predation may be in terms of its timing rather than total consumption.

Methods

We have confined our analysis to the major breeding populations of seabirds of each ecosystem: the Guanay Cormorant *Phalacrocorax bougainvillii*, Peruvian Booby *Sula variegata*, and Peruvian Brown Pelican *Pelecanus thagus* in the Humboldt ecosystem; and the Jackass Penguin *Spheniscus demersus*, Cape Gannet *Sula* (*Morus*) *capensis*, and Cape Cormorant *Phalacrocorax capensis* in the Benguela ecosystem.

Abundance of birds

We have little idea of the abundance of migratory seabirds in the Humboldt ecosystem (cf. Murphy, 1936; Brown, 1981; Duffy, 1981) or of seasonal variation in the abundance of migrants in the Benguela ecosystem (Summerhayes, Hofmeyr & Roux, 1974; Abrams & Griffiths, 1981). Hence migratory species which do not breed in our regions are excluded from our calculations.

To examine annual and decadal changes in the abundances of resident breeding species, we used guano production from nesting and roosting islands to estimate bird numbers. Since the early 1900s guano, as a renewable resource, has been collected annually and biennially, which should reflect bird abundance since the last collection (Murphy, 1925; Green, 1950; Hutchinson, 1950; Jordan & Fuentes, 1966*a,b*). Jordan & Fuentes (1966*a,b*) estimated numbers of adult guano birds in Peru between 1909 and 1964 by dividing guano collections by the average amount of guano deposited per adult bird per year (= 15.9 kg; Vogt, 1942). For the period 1960 to 1981, we used estimates of Peruvian guano bird numbers calculated by Tovar (1983).

For the South African seabird population, we calculated numbers of birds from guano production and an estimate made by Jarvis (1971) that each Cape Gannet deposits 8.7 kg guano per year on its breeding island. This was obtained by dividing average guano production (457.6 tonnes) on Malagas Island (33° 03'S, 17° 56'E) during 1950–60 by the estimated maximum population of Cape Gannets on the island (Jarvis, 1971). Although nestlings also contributed to the guano, their numbers are not included in these estimates. Estimates of guano deposition per bird are lower for the slightly larger Benguela species, because there is much more rain, and wind speeds are higher off the South African coast than off Peru and thus there is more run-off of guano into the sea. In our analyses we included all Peruvian guano collections between 1909 and 1980 and all South African collections made at islands between Bird (Penguin) Island, Lambert's Bay (32° 05'S, 18° 17'E), and Dyer Island (34° 14'S, 19° 25'E) before 1975.

Energy requirements of birds

We assumed that all Peruvian guano birds had the energetic requirements of the Guanay Cormorant (mass = 1.8 kg), rather than the lighter Peruvian Booby (mass = 1.3 kg) or the heavier Peruvian Brown Pelican (mass = 6.0 kg), based on data from Duffy (1980). The average mass of a Benguela seabird was assumed to be 2.65 kg, equivalent to that of the Cape Gannet, rather than that of the Jackass Penguin (3.00 kg) or the Cape Cormorant (1.22 kg), based on data in Furness & Cooper (1982). This assumption was necessary, because we have data on the relative abundances of the three species only since 1960 in Peru (Tovar, 1983) and 'complete' data for the entire Benguela region for only three years (1956, 1967, 1978) (Crawford & Shelton, 1981). The 'average' bird masses calculated from these figures do not differ much from those of the Guanay Cormorant (1.80 kg *versus* 1.81 kg), assuming that Guanay Cormorants, Peruvian Boobies and Brown Pelicans account for 83%, 15% and 2%, respectively, of the total Humboldt population (Jordan & Fuentes, 1966*b*), or Cape Gannet (2.65 kg *versus* 2.43 kg), assuming that Jackass Penguins, Cape Gannets and Cape Cormorants account for 57%, 14% and 29%, respectively, of the total population for the Saldanha Bay area in the Benguela region (Furness & Cooper, 1982).

We used the equation of Lasiewski & Dawson (1967) to calculate the SMRs (Standard Metabolic Rates) of the Guanay Cormorant and Cape Gannet. These were multiplied by 2.5 (Cooper, 1978; Kooyman *et al.*, 1982; Nagy, Siegfried & Wilson, 1984) to estimate the energy requirements of a single Guanay Cormorant or Cape Gannet per day and per year, and by 4.186 to convert calories to joules.

Food consumption by birds Estimates of daily annual energy requirements were converted to mass (wet weight) of anchovy *Engraulis* spp. by dividing by a conversion factor of 6.37 kJ g^{-1} (22.35 kJ $g^{-1} \times 0.285$ dry/wet weight, derived from Cooper, 1978) for Cape Anchovy *Engraulis capensis*. We multiplied this by 1.33 to allow for a 75% assimilation efficiency (Dunn, 1975; Cooper, 1978). The birds also eat other species, such as sardines *Sardinops* spp. and horse mackerels *Trachurus* spp. (Murphy,

1925; Rand, 1959, 1960*a*,*b*; Crawford & Shelton, 1981; Cooper, 1984), which have differing energetic values (Nagy *et al.*, 1984).

With regard to fish, the term 'population' in this paper refers to the total sum of interbreeding individuals of a species, whereas the term 'stock' is that proportion of the population that is exploitable by commercial fishing. Fish that are too large, too small, or otherwise unavailable for commercial fishing are considered part of the population but not part of the stock. Population estimates of Anchoveta *Engraulis ringens* stocks were taken from Csirke (1980) and commercial fishing landings from Schaefer (1970) and Idyll (1973). For South African fish stocks, we used estimates from Armstrong *et al.* (1983).

Estimating variability of food consumption We investigated variability of food consumption by seabirds over time by calculating the coefficient of variation (CV) for increasing intervals of years since 1909 in Peru and 1905 in South Africa. Variability should increase slowly and at a relatively constant rate if the population remains stable but should show sharp increases at short time intervals when major changes occur. We tested the theory that decreases in avian predation on young fish in Peru result in increases in recruitment to the commercial fish stock, using data on the strength of annual recruitment of Anchoveta (Csirke, 1980) before and after 1965.

Results

Guano as a measure of seabird populations

Comparisons of direct counts of seabirds, usually by aerial surveys or sketches of occupied areas, with guano yields produced Spearman rank correlations (Siegel, 1956) of 0.66 for Peru ($P < 0.05$; $n = 10$), based on data from Jordan & Fuentes (1966*a*,*b*) and Tovar (1978), and 0.60 for South African islands on which Gape Gannets nest ($P < 0.05$; $n = 14$), using data from figure 2 of Crawford & Shelton (1978). Randall & Ross (1979), however, reported no relation between guano production and area occupied by Cape Gannets during two decades ($n = 3$) at Bird Island (33° 50'S, 26° 17'E) off the south coast of South Africa. On the other hand, Valdivia (1960) showed a significant correlation between Peruvian guano collections and areas occupied by guano birds ($r_S = 0.73$; $P < 0.05$; $n = 16$). On balance, therefore, guano does apparently reflect the abundance of breeding seabirds in both ecosystems, suggesting that change in guano yields can be used to investigate the extent and variability of food consumption by seabirds.

Population trends

In Peru, seabird populations, based on guano production, rose from approximately 4 million birds during 1909–20 to about 8 million in the 1930s, before decreasing sharply in the 1940s (Fig. 14.1). The increase was the result of better protection of the birds and their nesting islands (Jordan & Fuentes, 1966*b*). The decrease in the mid-1940s was the result of a series of El Niño oceanographic events which killed off large numbers of birds by making their main prey, the Anchoveta, scarce or unavailable. The subsequent recovery of the avian predators in the 1950s led to populations of up to 20 million birds, considerably above levels in the 1930s. This was probably a response to an increase in nesting space, following the walling off of coastal headlands with predator-proof concrete fences (Jordan & Fuentes, 1966*b*; Duffy, 1983*b*). The sharp decreases in the late 1950s and during 1965 were again the results of El Niño events. The avian population never recovered fully from the 1965 El Niño, probably because heavy commercial fishing for the Anchoveta reduced the birds' food supply, leading to a large non-breeding population and reduced rates of increase (Tovar, 1978, 1983; Duffy, 1983*a*).

Except for a peak in the 1920s, bird populations resident off the west of South Africa showed no long-term trends until the 1960s when heavy

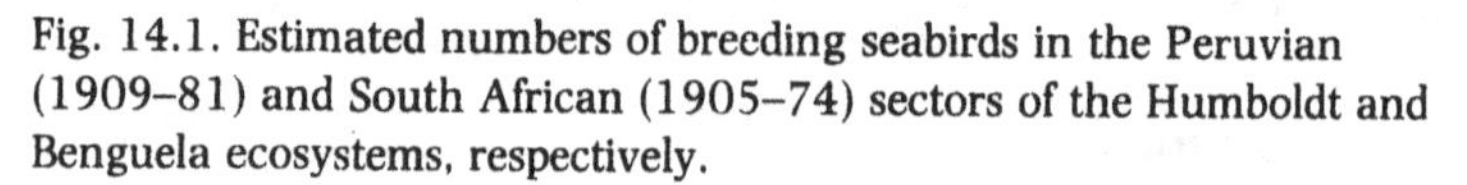

Fig. 14.1. Estimated numbers of breeding seabirds in the Peruvian (1909–81) and South African (1905–74) sectors of the Humboldt and Benguela ecosystems, respectively.

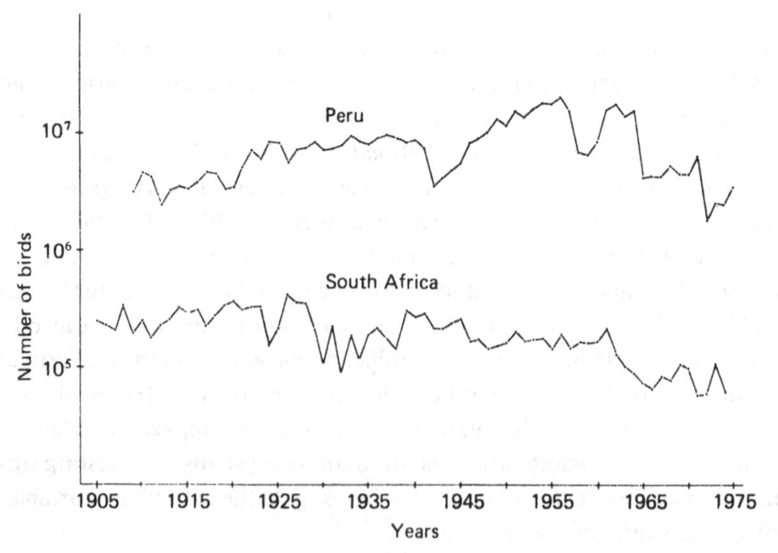

commercial fishing caused a population decrease (Fig. 14.1; Crawford & Shelton, 1981).

Consumption and availability of food Based on an estimated energy expenditure of 1657.7 kJ day^{-1}, a Cape Gannet requires 0.346 kg day^{-1} and 126.0 kg yr^{-1} of Cape Anchovy. A Guanay Cormorant expends 946.0 kJ day^{-1} and requires 0.197 kg day^{-1} and 71.5 kg yr^{-1} of Anchoveta.

In Peru, population estimates for the Anchoveta exist since 1960 (Csirke, 1980). In the first years of heavy fishing (1960–62), the stock was as large as 20.1–25.8 million tonnes. Between 1963 and 1971, the Anchoveta stock fluctuated greatly from 21.6 million tonnes in 1967 to 10.0 million tonnes in 1971. In 1972, the stock collapsed to 1.5 million tonnes and has subsequently remained at about 4–5 million tonnes, supposedly replaced by the Sardina *Sardinops sagax* (Walsh, 1981). If values of 20–25 million tonnes represent the size of the unexploited stock, then consumption of Anchoveta by the seabird population was < 5% in the early 1960s (Fig. 14.2). Even during the peak of 20 million birds in 1956, only 1.73 million tonnes or 6.9–8.6% of the Anchoveta stock was consumed by breeding seabirds. The greatest proportionate consumption by seabirds occurred in 1963–65 when they apparently took up to 11% of the Anchoveta stock during a short decrease in the fish population (Fig. 14.3). Frequent El Niños appear to have kept the avian population at levels

Fig. 14.2. Consumption of food by breeding Peruvian seabirds (1909–80). Solid lines represent annual estimates; broken lines, 3-year means.

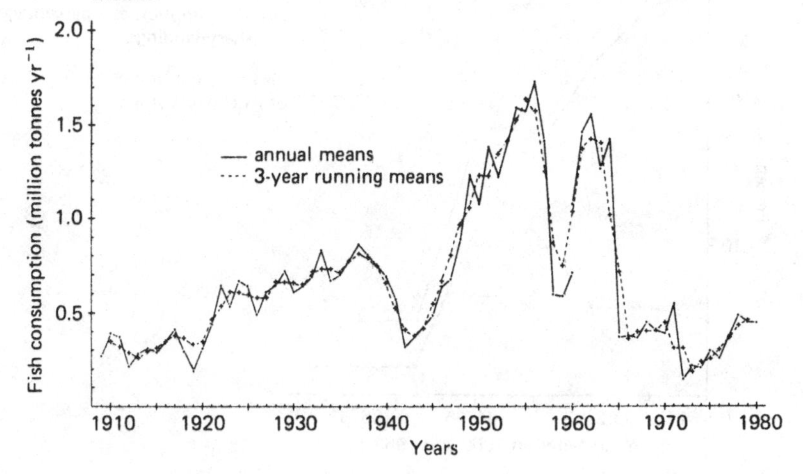

below what is necessary for the consumption of a relatively large proportion of the Anchoveta population.

In South Africa, stock estimates for the Cape Pilchard *Sardinops ocellata* are available since 1950 but Cape Anchovy values are only available since 1963 (Shannon *et al.*, 1984). The Cape Anchovy is believed to have replaced the Cape Pilchard during the 1960s, following commercial over-exploitation of the latter. In the late 1950s, the Cape Pilchard stock was as large as 2 million tonnes but in the mid-1960s it decreased to < 0.5 million tonnes. Cape Anchovy stocks appeared to be increasing rapidly when they were first exploited in the early 1960s, rising from 0.3 million tonnes or less to over 1.0 million tonnes during 1974 and since 1979. Assuming that, before the collapse of the Cape Pilchard stock following commerical over-exploitation, seabirds ate only Cape Pilchards and that this species' pre-exploitation population was about 1 million tonnes, then even the greatest bird consumption (0.050 million tonnes in 1926) accounted for only about 5% of the Cape Pilchard's biomass (Fig. 14.4). Consumption was approximately half this level in most years. If the Cape Anchovy was also common during this period, then the proportionate consumption of Cape Pilchard was even lower.

Fig. 14.3. Consumption of food by breeding Peruvian seabirds, as a percentage of fish stocks and of the commercial fishery landings.

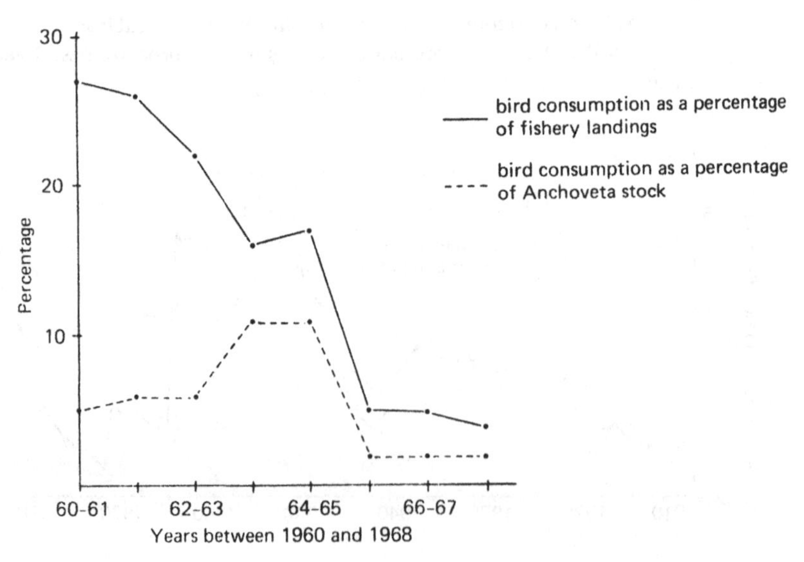

Fig. 14.4. Consumption of food by breeding South African seabirds (1905–74). Solid lines represent annual estimates; broken lines, 3-year running means.

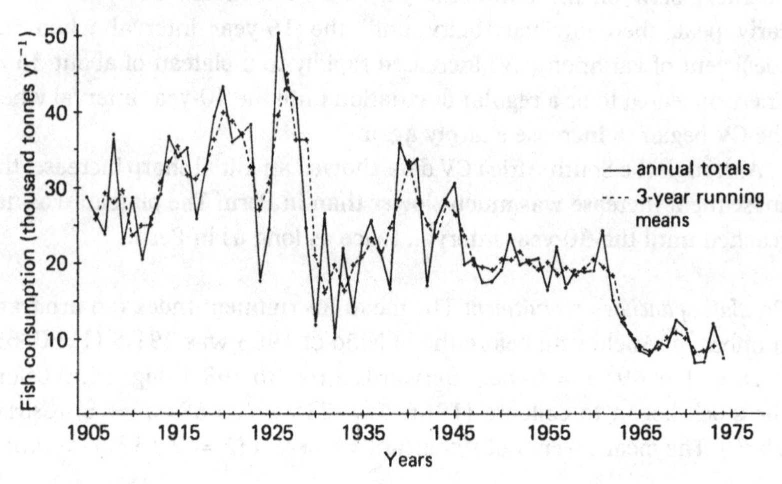

Fig. 14.5. Cumulative coefficients of variation for abundance of breeding populations of seabirds (and their food consumption) as a function of increasing number of years since 1909 in Peru and 1905 in South Africa.

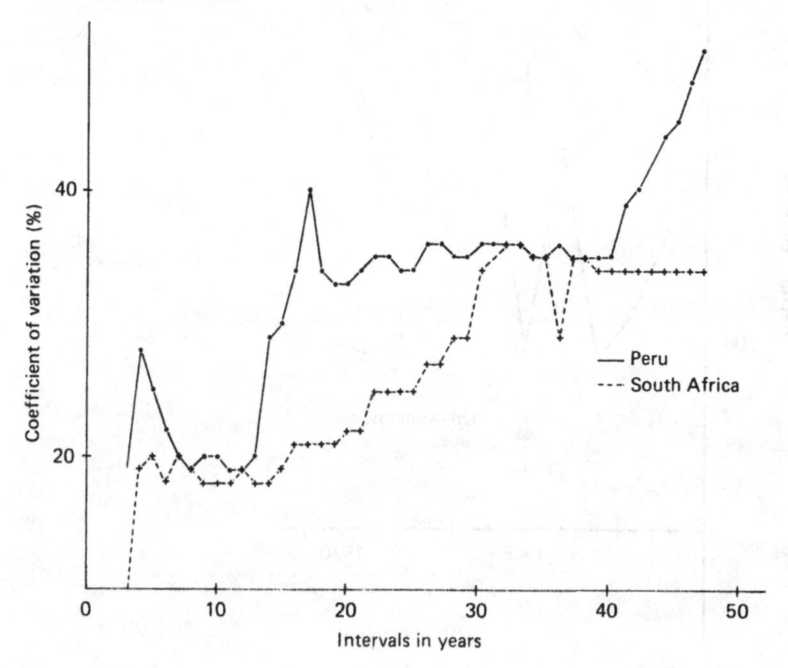

Variability of consumption Variability in consumption data differed considerably between the two areas (Fig. 14.5). Peruvian data showed an early peak then low variability until the 15-year interval when the coefficient of variation (CV) increased rapidly to a plateau of about 35%. There appeared to be a regular fluctuation until the 40-year interval when the CV began to increase sharply again.

Although the South Africa CV data showed an initial sharp increase, the subsequent increase was much slower than in Peru. The plateau was not reached until the 30-year interval, twice as long as in Peru.

Regulating anchovy recruitment The mean Recruitment Index (an arbitrary number) of Anchoveta before the El Niño of 1965 was 291.5 (1960–65) (S.D. = 108.49; n = 6) but afterwards it rose to 398.4 (Fig. 14.6) before the stock began to collapse (1966–70) (S.D. = 75.69; n = 5) (Csirke, 1980). The means were not significantly different ($F = 2.137$; $P > 0.05$),

Fig. 14.6. Recruitment indices and stock size before and after the 1965 collapse of the Peruvian guano bird population (data from Csirke, 1980).

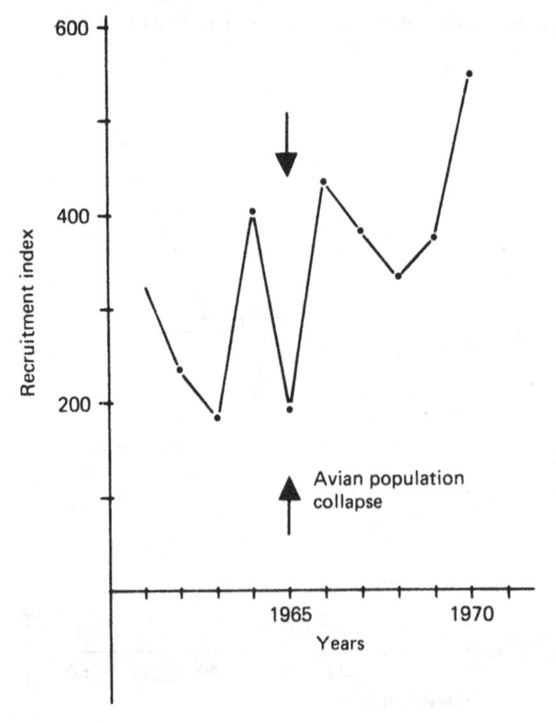

suggesting that recruitment was not increased by the reduction in the population of guano birds. Stock estimates did not increase following 1965 (Csirke, 1980).

Discussion

Guano as a measure of seabird populations

Using information on guano yields to estimate the abundance of seabird populations has major advantages not available from conventional census methods. Long runs of historical data exist, beginning in the early 1900s. Population estimates derived from guano data also provide a summation of bird activity on an island throughout an entire breeding season or year. These data can be used to supplement more exact estimates based on single counts. Both methods have disadvantages. For example, conventional population estimates based on one or more actual counts of birds are subject to a series of potentially serious biases, especially for species with extended or continuous breeding seasons such as occur in both the Benguela and Humboldt avian communities. Even counts made at peak breeding times may seriously under-estimate numbers of breeders if other individuals nest at other times during the year, as do Jackass Penguins and Peruvian Boobies. The accuracy of single counts depends on the time of day and on weather conditions, because these may affect the number of birds absent from their nests (Frost, Siegfried & Cooper, 1976). Ground censuses of individuals or nests are not practical when attempting to count colonies of tens or hundreds of thousands of birds in Peru and South Africa. Replication of counts to obtain confidence limits would be out of the question, even for much smaller colonies. Inter-observer biases would further complicate matters. Aerial photographs have the advantage that a permanent record exists (Nelson, 1978), but counts from photographs are subject to counting bias (Harris & Lloyd, 1977; Shelton *et al.*, 1982).

Counting problems can be avoided by measuring areas used for nesting, on a photograph, and multiplying by a nesting density (e.g. Randall & Ross, 1979). This assumes accuracy in identifying nesting areas (Barrett & Harris, 1965; Shelton *et al.*, 1982) and a relatively constant nesting density. Estimating numbers of nests for species such as the Cape Gannet and Guanay Cormorant may be possible with this method but Jackass Penguin nesting densities are very variable. Even for the Cape Gannet, estimates of nest density vary from 2.25 to 4.54 nests per square metre between islands and from 2.34 to 3.08 nests per square metre on one island (Randall & Ross, 1979). Jackass Penguins also nest in burrows which presents problems in interpreting aerial photography (Frost *et al.*, 1976).

On the other hand, population estimates from guano yields can be biased by three main sets of factors: (1) the estimation of guano deposition per bird per year; (2) the percentage of non-breeders not present on the islands; and (3) economic and environmental factors which may affect the size of the guano collection. Although we have fairly reliable data on the accuracy of guano deposition as an index of the abundance of Peruvian guano birds (see Hutchinson (1950) for a review), similar information is meagre for field conditions in South Africa. More accurate estimates of guano production are needed to refine South African data presented here.

Data on the abundance of seabird populations, as derived from guano production, are probably under-estimates, since pre-breeding immature birds and non-breeding adults may spend most of their time at sea. This problem also affects other census methods. Numbers of pre-breeders can be estimated from calculations of life history tables (e.g. Furness & Cooper, 1982) but these usually involve major assumptions such as population stability, constant reproductive success and fixed longevity. None of these applies to Peruvian seabirds (Nelson, 1978), and they are unlikely to be valid for Benguela species.

Guano harvests can be affected by a number of economic and environmental factors. Government control would tend to buffer the guano industry against free-market economic forces, and provide a continuity which would be less likely under private management. Apparently, guano extraction and management of the birds in both South Africa and Peru have altered little since descriptions by Murphy (1925) and Green (1950). However, in the last decade, in both regions, islands which have been almost deserted by nesting birds or are difficult to supply no longer have watchmen stationed on them. Moreover, the collecting of guano has recently been prohibited on those South African islands where guano is produced primarily by Jackass Penguins. Finally, the guano industry in South Africa has been operated by private companies since 1976, although management and protection of the islands remain the responsibility of the South African government (Sea Birds and Seals Protection Act, 1973). In Peru and South Africa, the collection of guano may not be undertaken during years of poor reproduction by the birds. Hence, guano is allowed to accumulate for more than one year in order to justify its collection. On the other hand, in years of good recruitment and an abundance of birds, it may be impossible to collect all the guano, so the excess is left for the next year. Three-year running means (which we have also used here) should reduce the effects of these factors on population estimates.

Rainfall might affect guano deposits, especially during anomalous conditions such as El Niño in the Pacific (Murphy, 1936; Hutchinson, 1950; Jordan & Fuentes, 1966*a*) and similar, but less severe, events in South African waters (Hutchinson, 1950; Duffy *et al.*, 1984). However, most Peruvian islands receive very little rainfall, even during El Niño years when the coastal desert may be flooded (Vogt, 1940, 1942). Thus, their guano yields are likely to be relatively unaffected by rainfall compared to the loss of guano caused by seabird breeding failures and the abandonment of islands by birds (Murphy, 1925, 1936; Vogt, 1940; Duffy, 1983*a*). In South Africa, where rainfall is more likely to cause run-off of guano from islands, desertions of nests and islands have also been linked to heavy rainfall and El Niño-like events (Hutchinson, 1950; Duffy *et al.*, 1984). During such events, rainfall and nesting failures would both reduce guano yields.

Additional factors, such as relative changes in seabird nesting populations on islands, might affect guano yields. Some species, such as the Guanay Cormorant and the Cape Gannet, produce more or purer guano than others (Vogt, 1942; Rand, 1952). Unfortunately, we lack quantitative information on species composition before the 1950s in Peru (Tovar 1978, 1983). Although changes are known to have occurred (Hutchinson, 1950; Cooper, 1984), we have assumed that these have had minor effects from year to year on guano production, compared to changes caused by El Niño and other environmental factors.

Food consumption

Our calculations of the birds' energy requirements are conservative, compared to other methods (Schneider & Hunt, 1982; Laugksch & Duffy, unpublished). Consumption values twice as great as those obtained here could be produced by selecting the 'correct' metabolic equations and assimilation efficiencies for the birds, and energetic values for fish. We believe, however, that our results are sufficient to show that breeding seabirds have been relatively minor consumers of fish in the two regions. Breeding Peruvian seabirds consumed two orders of magnitude more food than the Benguela birds, even though the areas of the two upwellings considered in this paper only differ by approximately a factor of two (assuming both upwellings to be approximately 100 km wide, the Peruvian coastline is 2250 km, the South African coastline, 900 km). The difference in numbers may, in part, be the result of a relative scarcity of suitable nesting habitat in the Benguela region. Brooke & Crowe (1982) reported 15 islands with a total land area of 389 ha off the 900 km South African coastline between the Orange River (28° 35′S, 16° 27′E) and Cape Agulhas (34° 08′S, 20° 00′E). Only about 7% of this island space was occupied by nesting birds in 1956 (calculated from Rand, 1963), primarily by low-density nesting Jackass Penguins. The total area of Peruvian islands and coastal headlands is approximately 8116 ha (Gonzalez, 1952). The nesting area used by seabirds totalled as much as 332 ha (Valdivia, 1960) but averaged 196 ha (Gonzalez, 1952) along a coastline of approximately 2250 km.

The low levels of food consumption reported in this paper do not necessarily mean that total avian consumption is minor; non-breeding seabirds from the northern and southern hemispheres are abundant in both upwelling ecosystems (Murphy, 1925; Summerhayes *et al.*, 1974) and probably consume large amounts of food, including anchovy and sardines.

Comparison with previous estimates of consumption

In South Africa, Davies (1958) estimated that the Benguela populations of Jackass Penguins, Cape Gannets and Cape Cormorants together consumed 89 000 tonnes yr^{-1} of fish in the early 1950s. This translates to a daily food consumption equal to 31% of the body weight of the Jackass Penguin and Cape Cormorant and 45% for the Cape Gannet. Comparable figures for our study would be 12.6% for the Jackass Penguin, 13.1% for the Cape Gannet and 16.1% for the Cape Cormorant. Davies' values appear to be excessively high. From studies of captive birds, Cooper (1977, 1978) estimated food consumption as 20% of body mass for both adult Cape Gannets and almost fully grown Jackass Penguin chicks. More recently, Furness & Cooper (1982) estimated total fish consumption by seabirds as 16 435 tonnes yr^{-1} in the Saldanha Bay area which is the central part of our Benguela region. Although they used a more elaborate method of calculating energy requirements, their values for daily consumption work out as 13% of body mass for the Jackass Penguin, 19% for the Cape Gannet and 18% for the Cape Cormorant.

Davies' (1958) estimate of 89 000 tonnes would represent a consumption of 30% of the Cape Pilchard stock of about 300 000 tonnes in the early 1950s (Armstrong *et al.*, 1983). Furness & Cooper's (1982) value represents a consumption of 7–8% of the southern Benguela Cape Anchovy stock (204000–227 000 tonnes) in 1977–78 (Armstrong *et al.*, 1983). Furness & Cooper estimated that seabirds consumed 23% of the Cape Anchovy stock in their more restricted, local study area. The stock in this area is part of a larger population which extends from at least the Orange River (28° 35′S, 16° 27′E) to Cape Agulhas (34° 08′S, 20° 00′E) (Duffy & Boyd, 1983). We know that a large portion of this population migrates through the Saldanha Bay area (Crawford, 1981), so that comparisons of local avian consumption with total fish stock estimates are more appropriate in attempts to assess the impact of seabirds on commercially important fish (Bourne, 1983). In other words, avian consumption should be calculated over the entire ranges occupied by fish populations, rather than over limited areas with especially high densities of breeding birds.

Variability of consumption Variability in the abundance of breeding seabirds and their food consumption is clearly much greater in Peru than in South Africa. Two factors may contribute to this: the periodic Peruvian population crashes caused by El Niño, and the expansion of Peruvian nesting space in the 1940s. In South Africa, El Niño-like events lead to nesting failures, rather than mass mortalities of adults (G.D. La Cock, unpublished). In both cases, changes in the variability of estimates of food consumption occurred in pulses rather than continually. Relatively little change occurred over several decades, followed by short periods of high variability. This suggests that estimates of food consumed by the seabirds are likely to remain valid over long periods in ecosystems not subject to commercial fishing. The variability of avian consumption in exploited ecosystems may be much more difficult to predict, because it would change in response to levels of commercial fishing.

Conclusions

Guano yields, despite the problem of calibrating precisely numbers of birds with guano produced, provide a long time-series of the kind of data needed to gain insights into the dynamics of seabird populations in upwelling areas. Our data suggest that breeding seabirds take $< 5\%$ of the populations of their principal fish prey in the Benguela and Humboldt ecosystems. These estimates could be doubled by means of 'judicious' use of selected variables, but the result would still be that these resident breeding seabirds are relatively minor predators in terms of the quantity of fish they remove from their respective ecosystems. Their populations may be limited by El Niño in Peru and by a shortage of nesting space and by milder El Niño-like events off South Africa. Culling breeding seabirds is thus unlikely to improve the landings of commerical fisheries in either area.

Seabird populations in the two ecosystems showed relatively constant degrees of variability over long periods of time. This suggests that even with frequent environmental perturbations, the bird populations, and the communities of which they are a part, did not experience chaotic fluctuations but varied within certain limits. The relative stability of the birds suggests that other predators are also likely to show similar degrees of variability in unexploited systems. Further analysis of guano yields in both regions may provide more information on the dynamics of the avian populations and their ecosystems.

Although the resident breeding seabirds may not consume large quantities of fish in either ecosystem, they merit further study for many reasons. Their depredation, although minor in quantity, may significantly affect the populations of their prey at particularly sensitive stages in their life cycles as proposed by Schaefer (1970), even if his suggestion that seabirds limited recruitment of Anchoveta before 1965 is not justified by our examination of the data. Secondly, seabirds, when judiciously used, may prove highly effective for sampling and monitoring pelagic fish stocks. Seabirds can complement existing sampling methods, reducing their sources of bias (e.g. VPA estimates rely at present entirely on commercial catches). Also, seabirds in the Humboldt and Benguela ecosystems take smaller and younger fish than the commercial fisheries (personal observation). This may allow an early warning of reproductive failures of certain short-lived species, which if not heeded could lead to the extirpation of fish stocks through commercial overfishing (Paulik, 1971). Finally, guano run-off from islands may provide an important source of nutrients for plankton during temporary halts in upwelling (Murphy, 1925; Hutchinson, 1950).

Acknowledgements We thank R. Laugksch, I. Newton, and C. Walter for assistance with calculations and illustrations. A. Berruti, D. Butterworth, D. Cabrera, J. Cooper, M. Plenge, D. Schneider, and V. Shannon provided helpful advice. This paper is based on research supported by the Benguela Ecology Programme of the South African National Committee for Oceanographic Research. Additional assistance was provided by the International Council for Bird Preservation, the National Geographic Society, the United States National Science Foundation (DEB 7716077), the Organization of American States, Princeton University, and the University of Cape Town.

References

Abrams, R.W. & Griffiths, A.M. (1981). Ecological structure of the pelagic seabird community in the Benguela Current region. *Mar. Ecol. Progr. Ser.*, **5**, 269–77.

Ahlstrom, E.H. (1966). Distribution and abundance of sardine and anchovy larvae in the California Current region off California and Baja California, 1951–1964: a summary. *Spec. Scient. Report, U.S. Fish Wildl. Serv.*, **534**, 71 pp.

Armstrong, M.H., Shelton, P.A., Prosch, R.M. & Grant, W.S. (1983). Stock assessment and population dynamics of anchovy and pilchard in ICSEAF Division 1.6 in 1982. *Coll. Sci. Pap. Int. Comm. SE Atl. Fish.*, **10**, 7–25.

Barrett, J.H. & Harris, M.P. (1965). A count of the gannets on Grassholm in 1964. *Br. Birds*, **58**, 201–3.

Berry, H.H. (1975). History of the guano platform on Bird Rock, Walvis Bay, South West Africa. *Bokmakierie*, **27**, 60–4.

Bourne, W.R.P. (1983). Birds, fish and offal in the North Sea. *Mar. Poll. Bull.*, **14**, 294–6.

Brooke, R.K. & Crowe, T.M. (1982). Variation in species richness among the offshore islands of the southwestern Cape. *S. Afr. J. Zool.*, **17**, 49–58.

Brown, R.G.B. (1981). Seabirds in northern Peruvian waters, November-December 1982. *Boln. Inst. Mar. Peru*, volumen extraordinario, 34–42.

Cooper, J. (1977). Energetic requirements of growth of the Jackass Penguin. *Zoologica Afr.*, **12**, 201–13.

Cooper, J. (1978). Energetic requirements for growth and maintenance of the Cape Gannet (Aves: Sulidae). *Zoologica Afr.*, **13**, 307–17.

Cooper, J. (1984). Changes in resource partitioning among four breeding seabirds in the Benguela upwelling system. In *Proceedings of the Fifth Pan-African Ornithological Congress*, ed. J. Ledger, pp. 217–30. Johannesburg.

Crawford, R.J.M. (1981). Distribution, availability and movements of anchovy *Engraulis capensis* off South Africa, 1964–1976. *Fish. Bull. S. Afr.*, **14**, 51–94.

Crawford, R.J.M. & Shelton, R. (1978). Pelagic fish and seabird interrelationships off the coast of South West and South Africa. *Biol. Conserv.*, **14**, 85–109.

Crawford, R.J.M. & Shelton, R. (1981). Population trends for some southern African seabirds related to fish availability. In *Proceedings of the symposium on birds of the sea and shore, 1979*, ed. J. Cooper, pp. 15–41. Cape Town: African Seabird Group.

Croxall, J.P. & Prince, P.A. (1982). A preliminary assessment of the impact of seabirds on marine resources at South Georgia. *CNFRA*, **51**, 501–9.

Croxall, J.P. & Prince, P.A. (1984). The impact of seabirds on marine resources, especially krill, at South Georgia. In *Seabird energetics*, ed. G.C. Whitton & H. Rahn, pp. 285–318. New York: Plenum Press.

Csirke, J. (1980). Recruitment in the Peruvian anchovy and its dependence on the adult population. *Rapp. R.-v. Reun. Cons. Perm. Int. Explor. Mer*, **177**, 307–13.

Daan, N. (1980). A review of replacement of depleted stocks by other species and the mechanism underlying such replacement. *Rapp. P.-v. Reun. Cons. Perm. Int. Explor. Mer*, **177**, 405–21.

Davies, D.H. (1958). The South African pilchard (*Sardinops ocellata*) and maasbanker (*Trachurus trachurus*): the predation of seabirds in the commercial fishery. *Invest. Rep. Div. Fish. Un. S. Afr.*, **31**, 16 pp.

DeVries, T.J. & Schrader, A. (1981). Variation of upwelling/oceanic conditions during the late Pleistocene through Holocene off the central Peruvian coast: a diatom record. *Mar. Micropalaeont.*, **6**, 157–67.

Drury, W.H. (1979). Population dynamics in northern marine birds. In *Conservation of marine birds of northern North America*, ed. J.C. Bartonek & D.N. Nettleship, pp. 123–39. Washington, D.C.: United States Fish and Wildlife Service.

Duffy, D.C. (1980). Patterns of piracy by Peruvian seabirds: a depth hypothesis. *Ibis*, **122**, 521–5.

Duffy, D.C. (1981). Seasonal changes in the seabird fauna of Peru. *Ardea*, **69**, 109–13.

Duffy, D.C. (1983*a*). Environmental uncertainty and commercial fishing: effects on Peruvian guano birds. *Biol. Conserv.*, **26**, 227–38.

Duffy, D.C. (1983*b*). Competition for nesting space among Peruvian guano birds. *Auk*, **100**, 680–8.

Duffy, D.C., Berruti, A., Randall, R.M. & Cooper, J. (1984). The effects of the 1982–1983 warm event on breeding South African seabirds. *S. Afr. J. Sci.*, **80**, 65–9.

Duffy, D.C. & Boyd, A.J. (1983). Anchovy distribution and migration. *S. Afr. J. Sci.*, **79**, 387–8.

Dunn, E.H. (1975). Caloric intake of nestling Double-crested Cormorants. *Auk*, **92**, 53–65.

Frost, P.G.H., Siegfried, W.R. & Cooper, J. (1976). Conservation of the Jackass Penguin (*Spheniscus demersus* (L.)). *Biol. Conserv.*, **9**, 79–99.

Furness, R.W. (1978). Energy requirements of seabird communities: a bioenergetic model. *J. Anim. Ecol.*, **47**, 39–53.

Furness, R.W. & Cooper, J. (1982). Interactions between breeding seabirds and pelagic fish populations in the southern Benguela region. *Mar. Ecol. Progr. Ser.*, **8**, 243–50.

Gonzalez, O. (1952). Mejor utilizacion del guano de islas. *Boln. Cia. Adm. Guano*, **28**, 51–115.

Green, K.A. (1978). *Ecosystem description of the California Current. Final Report.* U.S. Marine Mammal Commission, contract number MM 7AC-026.

Green, L.G. (1950). *At daybreak for the Isles.* Cape Town: H.B. Timmins.

Harris, M.P. & Lloyd, C.S. (1977). Variation in counts of seabirds from photographs. *Br. Birds*, **70**, 200–5.

Hendey, Q.B. (1981). Geological succession at Langebaanweg, Cape Province, and global events of the late Tertiary. *S. Afr. J. Sci.*, **77**, 33–8.

Hutchinson, G.E. (1950). The biogeochemistry of vertebrate excretion. 3. Survey of contemporary knowledge of biogeochemistry. *Bull. Am. Mus. Nat. Hist.*, **96**, 1–554.

Idyll, C.P. (1973). The anchovy crisis. *Sci. American*, **228**, 22–9.

Jarvis, M.J.F. (1971). Interactions between man and the South African Gannet *Sula capensis*. *Ostrich*, **8**, suppl., 497–513.

Jordan, R. & Fuentes, H. (1966*a*). Estudio preliminar sobre las fluctuaciones de las poblaciones de aves guaneras. *Memorias de Premier Seminario Latinoamericano de Oceanografia sobre el Oceano Pacifico Oriental*, pp. 88–94. Lima: Universidad Major San Marcos.

Jordan, R. & Fuentes, H. (1966*b*). Las poblacions de aves guaneras y su situacion acual. *Inf. Inst. Mar. Peru*, **10**, 1–31.

Kooyman, G.L., Davis, R.W., Croxall, J.P. & Costa, D.P. (1982). Diving depths and energy requirements of King Penguins. *Science*, **217**, 726–7.

Lasiewski, R.C. & Dawson, W.R. (1967). A re-examination of the relation between standard metabolic rate and body weight in birds. *Condor*, **69**, 13–23.

MacCall, A.D. (1983). Variability of pelagic fish stocks off California. Expert consultation to examine changes in abundance and composition by species of neritic fish resources. *FAO Fish. Rep.*, no. 291, part 2, 101–12.

May, R.M., Beddington, J.R., Clarke, C.W., Holt, S.J. & Laws, R.M. (1979). Management of multispecies fisheries. *Science*, **205**, 267–77.

Mitchell, B. & Sandbrook, R. (1980). *The management of the Southern Ocean*. London: International Institute for Environment and Development.

Murphy, R.C. (1925). *Bird islands of Peru*. New York: Putnams.

Murphy, R.C. (1936). *Oceanic birds of South America*. New York: American Museum of Natural History.

Myres, M.T. (1979). Long-term climatic and oceanographic cycles regulating seabird distribution and numbers. In *Conservation of marine birds of northern North America*, ed. J.C. Bartonek & D.N. Nettleship, pp. 123–39. Washington, D.C.: United States Fish and Wildlife Service.

Nagy, K.A., Siegfried, W.R. & Wilson, R.P. (1984). Energy utilization by free ranging Jackass Penguins *Spheniscus demersus*. *Ecology*, **65**, 1648–55.

Nelson, J.B. (1978). *The Sulidae: gannets and boobies*. Oxford: Oxford University Press.

Newman, G.G. & Crawford, R.J.M. (1980). Population biology and management of mixed species pelagic stocks off South Africa. *Rapp. P.-v. Reun. Cons. Perm. Int. Explor. Mer.*, **177**, 279–91.

Paulik, G.J. (1971). Anchovies, birds and fishermen in the Peru Current. In *Environment, resources, pollution, and society*, ed. W.W. Murdock, pp. 156–85. Stamford: Sinauer Press.

Rand, R. (1952). Guano enterprises in South West Africa. *Ostrich*, **23**, 169–85.

Rand, R.W. (1959). The biology of guano-producing seabirds. The distribution, abundance and feeding habits of the Cape Gannet, *Morus capensis*, off the southwestern coast of the Cape Province. *Invest. Rep. Div. Fish. Un. S. Afr.*, **39**, 36 pp.

Rand, R.W. (1960*a*). The biology of guano-producing seabirds. The distribution, abundance, and feeding habits of the Cape Penguin, *Spheniscus demersus*, off the southwestern coast of the Cape Province. *Invest. Rep. Div. Fish. Un. S. Afr.*, **41**, 28 pp.

Rand, R.W. (1960*b*). The biology of guano-producing seabirds. 3. The distribution, abundance and feeding habits of the cormorants Phalacrocoracidae off the southwestern coast of the Cape Province. *Invest. Rep. Div. Fish. Un. S. Afr.*, **42**, 32 pp.

Rand, R.W. (1963). The biology of guano-producing seabirds; composition of colonies on the Cape islands. *Invest. Rep. Div. Fish. Un. S. Afr.*, **43**, 32 pp.

Randall, R. & Ross, G.J.B. (1979). Increasing population of Gannets on Bird Island, Algoa Bay, and observations on breeding success. *Ostrich*, **50**, 168–75.

Ricker, W.E. (1975). Computation and interpretation of biological statistics of fish populations. *Bull. Fish. Res. Bd Can.*, **191**, 1–382.

Sanger, G.A. (1972). Preliminary standing stock and biomass estimates of seabirds in the Subarctic Pacific region. In *Biological oceanography of the northern North Pacific Ocean*, ed. A. Y. Takenouti, pp. 589–611. Tokyo: Idemitsu Shoten.

Schaefer, M. (1967). Dynamics of the fishery for the anchoveta *Engraulis ringens* off Peru. *Boln. Inst. Mar. Peru*, **1**, 189–304.

Schaefer, M. (1970). Men, birds and anchovy in the Peru Current – dynamic interactions. *Trans. Am. Fish. Soc.*, **99**, 461–7.

Schneider, D. & Hunt, G.L. (1982). Carbon flux to seabirds in waters with different mixing regimes in the southeastern Bering Sea. *Mar. Biol.*, **67**, 337–44.

Serra, J.S. (1983). Changes in the abundance of pelagic resources along the Chilean coast. Expert consultation to examine changes in abundance and compostion by species of neritic fish resources. *FAO Fish. Rep.*, no. 291, part 2, 255–84.

Shannon, L.V., Crawford, R.J.M. & Duffy, D.C. (1984). Pelagic fisheries and warm events, a comparative study. *S. Afr. J. Sci.*, **80**, 51–60.

Shelton, P.A., Crawford, R.J.M., Kriel, F. & Cooper, J. (1982). Methods used to census three species of South African seabirds, 1978–1981. *Fish. Bull. S. Afr.*, **16**, 115–20.

Siegel, S. (1956). *Non-parametric statistics for the behavioral sciences*. New York: McGraw-Hill.

Skud, B.E. (1982). Dominance in fishes: the relation between environment and abundance. *Science*, **216**, 144–9.

Smith, P.E. (1978). Biological effects of ocean variability: time and space scales of biological response. *Rapp. P.-v. Reun. Cons. Perm. Int. Explor. Mer*, **173**, 117–27.

Summerhayes, C.P., Hofmeyr, P.K. & Roux, R.H. (1974). Seabirds off the southwestern coast of Africa. *Ostrich*, **45**, 83–109.

Tovar, H. (1978). Las poblaciones de aves guaneras en los ciclos reproductivas de 1969/70 a 1973/74. *Inf. Inst. Mar. Peru*, **45**, 1–13.

Tovar, H. (1983). Fluctuaciones de poblaciones de aves guaneras en el litoral peruano, 1960–1981. Expert consultation to examine changes in abundance and composition by species of neritic fish resources. *FAO Fish. Rep.*, no. 291, part 3.

Valdivia, J. (1960). La cubicacion del guano de islas y la ecuacion de regresion. *Boln. Cia. Adm. Guano*, **36**, 8–10.

Vogt, W. (1940). Una depresion ecologica en la costa del Peru. *Boln. Cia. Adm. Guano*, **16**, 307–29.

Vogt, W. (1942). Aves guaneras. *Boln. Cia. Adm. Guano*, **18**, 1–132.

Walsh, J.J. (1978). The biological consequences of interactions of the climatic, El Niño, and event scales of variability in the eastern tropical Pacific. *Rapp. P.-v. Reun. Cons. Perm. Int. Explor. Mer*, **173**, 182–92.

Walsh, J.J. (1981). A carbon budget for overfishing off Peru. *Nature* (*Lond.*), **290**, 300–4.

Wiens, J.A. & Scott, J.M. (1975). Model estimation of energy flow in Oregon coastal seabird populations. *Condor*, **77**, 439–52.

Zuta, S., Tsukayama, I. & Villanueva, R. (1983). El ambiente marino y las fluctuaciones de las principales poblaciones pelagicas de la costa peruana. Expert consultation to examine changes in abundance and composition by species of neritic fish resources. *FAO Fish. Rep.*, no. 291, part 2, 179–254.

15

Seabirds as predators on marine resources, especially krill, at South Georgia

J. P. CROXALL and P.A. PRINCE
British Antarctic Survey, Natural Environment Research Council, High Cross, Madingley Road, Cambridge CB3 OET, UK

Introduction

Within the vast area of the Southern Ocean, sites where seabirds can breed are few, and nearly all of them have a rich and varied avifauna. The most diverse of these occur at the sub-Antarctic islands, with each main group having 20–30 breeding seabird species with total populations of many millions of birds. (For details see Croxall (1984) and appropriate chapters in Croxall, Evans & Schreiber (1984)).

South Georgia has 23 species of seabirds breeding (Prince & Croxall, 1983) (excluding the Sub-Antarctic Skua *Catharacta lonnbergi*, the Greater Sheathbill *Chionis alba* (which seldom take any marine prey) and the Rockhopper Penguin *Eudyptes chrysocome* which only breeds there occasionally) and a total population of over 70 million birds. The island is located further south than most other sub-Antarctic islands, being about 200 km south of the Antarctic Polar Front (Antarctic Convergence) and thus permanently within the domain of the Antarctic Surface Water mass. Oceanographically, the area is a complex one; the pattern in any particular season is dependent on the relative contribution of Bellingshausen and Weddell Sea currents and the position of the Weddell–Scotia Confluence and the Antarctic Polar Front (Priddle, Heywood & Theriot, 1986). The boundaries (fronts) between sub-Antarctic, Antarctic and Weddell Sea surface waters vary markedly (even by up to 150 km) and persistently, so that the system is in constant flux. These currents produce extensive local upwellings, particularly off the northwest end of the island, and several gyral systems exist which may retain water masses in position for periods of unknown duration.

The combination of major water transport systems and upwellings gives rise to large local concentrations of nutrients, which in turn produce major phytoplankton blooms supporting a rich population of zooplankton, especially Antarctic krill *Euphausia superba* but also copepods and amphipods. The area is usually regarded as primarily a feeding rather than a breeding ground for krill, transported thence by current systems but recent data suggest that this is probably an oversimplified view. Krill, whose very large local population is present at least throughout the summer, is the key organism in the food web, sustaining many fish, squid, seal, bird and whale species and populations. The substantial bird population is almost exclusively dependent on marine resources for its food and represents a very significant proportion of the local direct predatory impact on these resources, in conjunction with the large breeding populations of Southern Elephant Seal *Mirounga leonina*, Antarctic Fur Seal *Arctocephalus gazella* and the small (nowadays) whale populations and fish and squid (Croxall, Prince & Ricketts, 1985).

Estimates of the food consumed by the bird and seal populations have been dealt with in detail by Croxall, Ricketts & Prince (1984) and Croxall, Prince & Ricketts (1985), with other pertinent information and analyses for seabirds in Pennycuick, Croxall & Prince (1984). This chapter briefly reviews their conclusions and seeks to complement these with a more qualitative discussion of aspects of the dynamics of local seabird–prey interactions and a reassessment of the role of diet in potential ecological isolating mechanisms.

Prey consumption by South Georgia seabirds

The sources and methods by which the energy and food requirements of seabirds breeding at South Georgia were derived are given in full by Croxall, Ricketts & Prince (1984) and Croxall, Prince & Ricketts (1985). Briefly, the model uses extensive data on the timing and duration of breeding events, on the quantitative composition (by weight) of the diet, on breeding population size and on breeding success and frequency. Bioenergetic data are available on swimming costs for three penguin species, flight costs for one albatross species (Costa & Prince, 1986) and estimates of incubation costs (of penguins and albatrosses) and moult costs (penguins) for several species (Croxall, 1982). Activity budget data are available for most inshore-feeding species and for three albatrosses while at sea on foraging trips (Prince & Francis, 1984; see also Chapter 7). The values for some of the more important parameters are summarised in Tables 15.1 and 15.2. Most of the others are available in Croxall, Ricketts & Prince (1984), Croxall, Prince & Ricketts (1985) and Pennycuick *et al.* (1984).

Table 15.1. *Body weight, population size and timing and duration of breeding cycle events in South Georgia seabirds* (*R = resident*)

Species	Weight (g) ♂	Weight (g) ♀	Breeding population (pairs $\times 10^3$)	Pre-lay attendance (days)	Mean date of: Laying	Hatching	Fledging	Incubation period (days)	Chick-rearing period (days)
King Penguin *Aptenodytes patagonicus*	13450	13760	34	17	8/12:21/2	31/1:16/4	30/12:25/2	54	313
Gentoo Penguin *Pygoscelis papua*	5890	5890	100	R	25/10	30/11	19/2	36	81
Chinstrap Penguin *P. antarctica*	4000	3600	4	25	30/11	6/1	1/3	37	54
Macaroni Penguin *Eudyptes chrysolophus*	5000	4600	5000	20	23/11	27/12	24/2	34	59
Wandering Albatross *Diomedea exulans*	10580	9020	4.7	27	24/12	4/3	14/12	78	278
Black-browed Albatross *D. melanophrys*	3922	3694	60	16	27/10	3/1	29/4	68	116
Grey-headed Albatross *D. chrysostoma*	3751	3624	60	26	19/10	30/12	20/5	72	141
Light-mantled Sooty Albatross *Phoebetria palpebrata*	2840	2840	8	21	31/10	6/1	27/5	67	141
Southern Giant Petrel *Macronectes giganteus*	5035	3798	5	R	10/11	10/1	8/5	61	118
Northern Giant Petrel *M. halli*	4902	3724	3.5	60	30/9	1/2	22/3	62	112
Cape Pigeon *Daption capense*	442	407	20	40	8/12	22/1	11/3	45	48
Snow Petrel *Pagodroma nivea*	340	286	3	55	3/12	15/1	7/3	43	51
Antarctic Prion *Pachyptila desolata*	168	168	22000	52	16/12	30/1	22/3	45	51
Fairy Prion *P. turtur*	130	130	1	?	10/11	29/12	15/2	49	50
Blue Petrel *Halobaena caerulea*	193	193	70	52	26/10	11/12	29/1	46	49
White-chinned Petrel *Procellaria aequinoctialis*	1390	1280	2000	51	24/11	17/1	19/4	54	92
Wilson's Storm Petrel *Oceanites oceanicus*	34	34	600	45	14/12	24/1	20/3	41	55
Black-bellied Storm Petrel *Fregetta tropica*	53	53	10	45	25/12	3/2	11/4	40	67
Common Diving Petrel *Pelecanoides urinatrix exsul*	133	135	3800	21	25/10	18/12	10/2	54	54
South Georgia Diving Petrel *P. georgicus*	107	107	2000	21	13/12	28/1	14/3	46	45
Blue-eyed Shag *Phalacrocorax atriceps*	2867	2473	7.5	R	29/10	26/12	1/3	27	65
Antarctic Tern *Sterna vittata*	151	151	2.6	R	14/11	24/12	7/2	40	45

Table 15.2. *Dietary composition, provisioning rate and potential foraging range of South Georgia seabirds*

	Diet[a] (% by weight)								
Species	*Krill*	*Copepod*	*Amphipod*	*Other crustaceans*	*Fish*	*Squid*	*Meals (per chick per adult per day)*	*Potential foraging range (km)*	*Foraging type*
King Penguin	—	—	—	—	(30)	(70)	0·18	424	Offshore
Gentoo Penguin	68	—	—	—	32	—	0·50	35	Inshore
Chinstrap Penguin	100	—	—	—	—	—	0·50	68	Uniform
Macaroni Penguin	98	—	—	—	2	—	0·60	123	Offshore
Wandering Albatross	—	—	—	—	41	40	0·18	1478	Offshore
Black-browed Albatross	38	1	1	—	39	21	0·42	428	Offshore
Grey-headed Albatross	15	—	1	—	35	49	0·45	615	Offshore
Light-mantled Sooty Albatross	37	—	—	4	12	47	0·34	709	Offshore
Southern Giant Petrel	12	—	—	—	1	2	0·97	189	Inshore
Northern Giant Petrel	15	—	—	—	2	6	1·01	181	Inshore
Cape Pigeon	85	—	—	—	15	—	0·50	357	Offshore
Snow Petrel	80	—	—	—	10	10	0·50	357	Offshore
Antarctic Prion	58	31	8	—	2	1	0·66	244	Uniform
Fairy Prion	79	4	14	—	2	1	?	?	Uniform
Blue Petrel	82	4	5	—	8	1	0·40	670	Offshore
White-chinned Petrel	27	—	1	—	24	47	0·25	1218	Offshore
Wilson's Storm Petrel	40	10	40	—	10	—	0·75	189	Uniform
Black-bellied Storm Petrel	(40)	(10)	(40)	—	(10)	—	0·75	189	Uniform
Common Diving Petrel	15	68	17	—	—	—	0·80	243	Uniform
South Georgia Diving Petrel	76	20	4	—	—	—	0·90	216	Uniform
Blue-eyed Shag	—	—	—	10	70	20	10	12	Inshore
Antarctic Tern	(15)	(15)	(20)	—	(50)	—	40	2	Inshore

[a] Excluding carrion: parenthesis indicate estimated values.

The limitations of the data and approaches used were discussed by Croxall, Ricketts & Prince (1984). The main shortcomings are the potential inaccuracies of the population estimates, the need for bioenergetic and activity budget data for species other than albatrosses and penguins, the lack of information on diet outside the breeding season and the inability to quantify the actual feeding distribution of birds at sea. The model also ignores seasonal changes in predator body weight and in prey energy content. It does not consider post-fledging juveniles, immatures or non-breeding adults (and birds of at least the first two categories spend some time near the breeding colonies in summer). This is probably largely compensated for, however, by the assumption that all birds (except for well-known migrants like Black-browed Albatross *Diomedea melanophrys* and Wilson's Storm Petrel *Oceanites oceanicus*) stay in the general vicinity of South Georgia all year round. The over-estimate of residence time by breeding seabirds that this implies is probably compensated for by the presence of high-latitude bird populations around South Georgia in winter.

Taking all this into account, we estimate that the breeding seabirds, (with a total biomass of 32 846 tonnes), consume 7.8 million tonnes of prey each year, of which 73% is krill, 13% copepods, 6% squid, 5% fish and 3% amphipods (Table 15.3). This may over-estimate the importance of krill, as it is unlikely to be as important as fish and squid in species' winter diets, for which we have few data. Of the 22 main species of seabird (excluding the tiny population of Grey-backed Storm Petrels), only six take 1% or more of the total: namely Macaroni Penguin (51%), Antarctic Prion (30%), White-chinned Petrel (10%), Common Diving Petrel (4.5%), South Georgia Diving Petrel (2%), and Gentoo Penguin (1%). Thus penguins, forming 13% of the breeding numbers but 76% of the biomass, eat 53% of the food. The Macaroni Penguin takes 68% of the krill eaten by seabirds, followed by Antarctic Prion (24%), White-chinned Petrel (4%), South Georgia Diving Petrel (2%) and Gentoo Penguin and Common Diving Petrel (both 1%). Krill is also, however, of primary importance in the diet of Chinstrap Penguins and of all petrels except White-chinned Petrels, Giant Petrels (mainly scavengers of seal and penguin carcasses (Hunter, 1983)) and the Common Diving Petrel.

The copepods (chiefly *Rhincalanus gigas* and *Calanoides acutus*) are mainly taken by Antarctic Prions (72%), by forcing water through comb-like palatal lamellae, which act as filters along the side of the bill and by Common Diving Petrels (23%) which presumably feed in analogous ways to the small northern hemisphere alcids which they closely resemble. The main consumers of amphipods are Antarctic Prions (70%) and

Table 15.3. *Annual food consumption*[a] *(thousand tonnes) of seabirds breeding at South Georgia*

Species	Krill	Squid	Fish	Copepod	Amphipod	Other	Total
King Penguin	0	44·8	19·2	0	0	0	64·0
Chinstrap Penguin	2·5	0	0	0	0	0	2·5
Gentoo Penguin	61·2	0	28·8	0	0	0	90·0
Macaroni Penguin	3872·7	0	79·0	0	0	0	3951·7
Wandering Albatross	0	1·2	1·2	0	0	0	2·4
Black-browed Albatross	8·3	4·6	8·7	0·2	0·2	0	22·0
Grey-headed Albatross	6·3	20·6	14·7	0	0·4	0	42·0
Light-mantled Sooty Albatross	1·8	2·3	0·6	0	0	0·2[b]	4·9
Southern Giant Petrel	2·7	0·4	0·2	0	0	0	3·3
Northern Giant Petrel	1·5	0·6	0·2	0	0	0	2·3
Cape Pigeon	3·1	0	0·5	0	0	0	3·6
Snow Petrel	0·4	0·05	0·05	0	0	0	0·5
Antarctic Prion	1345·4	23·2	46·4	719·1	185·6	0	2319·7
Fairy Prion	0·08	0	0	0	0·01	0	0·1
Blue Petrel	6·6	0·08	0·6	0·3	0·4	0	8·0
White-chinned Petrel	210·1	365·8	186·8	0·0	7·8	0	770·5
Wilson's Storm Petrel	3·8	0	1·0	1·0	3·7	0	9·5
Black-bellied Storm Petrel	0·2	0	0·05	0·05	0·02	0	0·5
Common Diving Petrel	52·8	0	0	239·2	59·8	0	351·8
South Georgia Diving Petrel	125·6	0	0	33·0	6·6	0	165·2
Blue-eyed Shag	0	0·9	3·3	0	0	0·5[c]	4·7
Antarctic Tern	0·04	0	0·1	0·04	0·05	0	0·2
Total	5706	466	390	993	265	1	7820

[a] Does not include carrion and other non-marine-derived prey. [b] Decapods. [c] Isopods.

Common Diving Petrels (23%), although they are a major element in the diet of Wilson's Storm Petrels. *Themisto* (formerly *Parathemisto*) *gaudichaudii* is ubiquitous, occurring as a trace in the diet of almost every seabird, but a variety of other species are taken, particularly by Antarctic Prions and Blue Petrels (Prince, 1980*a*).

Squid and fish are much less important than krill to South Georgia seabirds. The main squid predators are the White-chinned Petrel (79%), King Penguin (10%), Antarctic Prion (5%) and Grey-headed Albatross (4%); squid also predominate in the diet of Light-mantled Sooty Albatrosses and are of considerable importance to Wandering Albatrosses. Fish are taken by many species but only Blue-eyed Shags and Antarctic Terns depend on them for their livelihood. The main consumers are the White-chinned Petrel (48%), Macaroni Penguin (20%), Antarctic Prion (12%), Gentoo Penguin (7%), King Penguin (5%), Grey-headed Albatross (4%), Black-browed Albatross (2%) and Blue-eyed Shag (1%). Otherwise, fish are only important in the diet (throughout the winter breeding season) of Wandering Albatrosses (J.P. Croxall & P.A. Prince, unpublished data).

Seasonal effects

Our dietary data indicate that there is usually little change in composition throughout the chick-rearing period. Information for the incubation period, however, is very sparse and for the non-breeding period it is non-existent. With these provisos, the broad pattern of changes in the food requirements of the breeding seabird community is shown in Fig. 15.1. This emphasises the role of krill and shows a marked increase in its consumption from spring to late summer with possibly some reduction thereafter. The seasonal pattern of avian krill consumption is dominated by the activities of Macaroni Penguins; the very marked drop in March is almost exclusively due to their highly synchronised onshore moult fast then. In fact this reduction is partly compensated for by their newly fledged juveniles (which equal nearly one-quarter of the adult population biomass) which are still in the vicinity of South Georgia. However, the true consumption in late February–early March is much greater than shown, because Macaroni Penguins double their body weight in a 14-day period at sea in preparation for moult and the model does not take into account these extra food requirements.

Consumption of fish and squid show no clear seasonal pattern, other than a reduction during spring and early summer when the main consumers are incubating, and a rise in January–February which is purely a consequence of Macaroni Penguins, even though they only take 2% by weight of fish in their diet.

Distributional effects

Observations of South Georgia seabirds at sea are unsuitable for quantitative estimation of the distribution of their foraging effort because coverage of such a large oceanic area is very patchy, because most species feed at night (observations are therefore chiefly of birds travelling to and from feeding grounds) and because the main consumers are penguins, which are difficult to observe at sea. To model possible distributional effects, Croxall Ricketts & Prince (1984) used data on the duration of foraging trips (derived from the frequency with which chicks are fed by their parents)

Fig. 15.1. Seasonal changes in food consumption by breeding populations of South Georgia seabirds: (a) all species; (b) Macaroni Penguin; (c) all species, fish only; (d) all species, squid only.

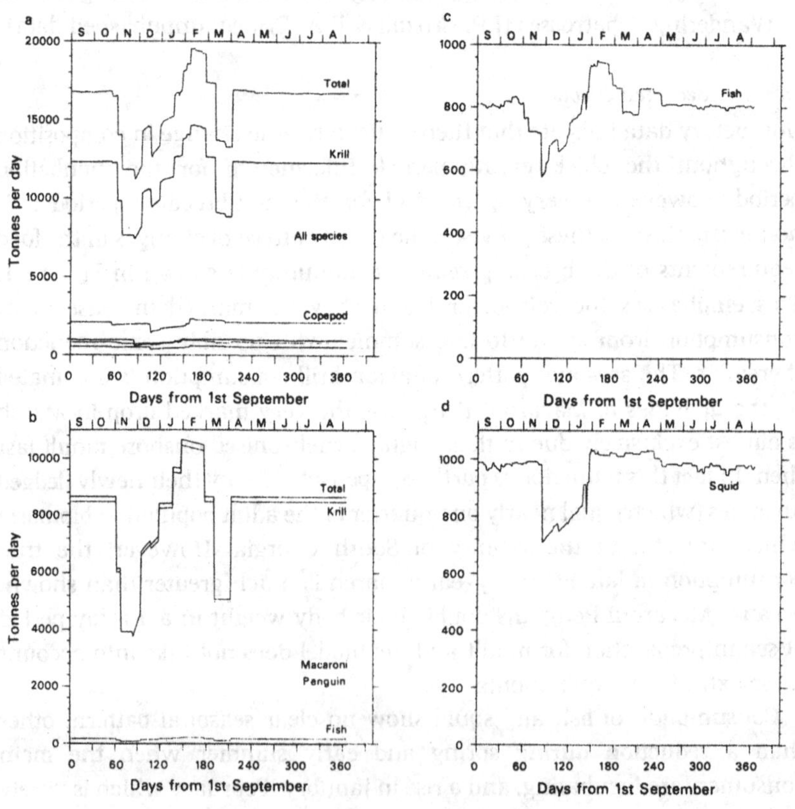

and extensive information on flight (and swimming) speeds and patterns (see particularly Pennycuick, 1982) to derive potential mean maximum foraging ranges. They then characterised species as inshore (with abundances declining exponentially with distance from the colony), offshore (with abundances declining exponentially from the mean maximum range towards the colony) and uniform (spread evenly to mean maximum range). On this basis, maximum food consumption by breeding seabirds is well within 80 km of South Georgia (due to Macaroni Penguins), with high levels out to 250 km (chiefly the large Antarctic Prion population), diminishing rather rapidly thereafter. Except for some inshore-feeding species, the favoured foraging areas themselves are, however, unknown at present.

Predator–prey interactions

Direct information on the nature of the actual interactions between a pelagic marine predator and its prey is very difficult to acquire. Our knowledge of aspects of this topic for the South Georgia area derives largely from the deployment of recording devices attached to birds and seals and partly from inferences based on the nature of the prey themselves. To assist interpretation, it is useful to consider data for two species (Adélie *Pygoscelis adeliae* and Chinstrap Penguins *P. antarctica*) from a site (South Orkney Islands) 600 km south of South Georgia and from Antarctic Fur Seals at Bird Island, South Georgia.

For adequate assessment of predator–prey relationships, we need to know how the level of predation relates to the total stock of prey, particularly that potentially available to the various predators, what is the status (e.g. sex, age) of each main prey taken, and how this relates to the prey species' own population dynamics.

Krill

Present estimates of krill stocks in the South Georgia area (0.5 million tonnes in December) are very difficult to reconcile with the seabird consumption (0.4 million tonnes in the same month) because the estimates are derived from what are effectively instantaneous surveys of small areas using echosounders that are unable to cover the top 10 m of the water column. The lack of knowledge of the rate at which krill are replenished in the area, via the major current systems is, however, probably the most serious problem.

From detailed studies of the diets of Adélie and Chinstrap Penguins, the sizes of krill that they catch through the summer were compared with similar data from scientific and commercial net hauls within, or near, the foraging ranges of penguins engaged in rearing chicks, when the trips to sea by adults are at their shortest (Croxall, Prince & Ricketts, 1985; Lishman, 1985*a*). The similarities are striking (Fig. 15.2), and suggest that

Fig. 15.2. Length-frequency distributions of krill eaten by various predators compared with those taken by net-hauls. (Data from Croxall & Prince, 1980*a*; BIOMASS, 1982; Croxall & Pilcher, 1984; Lishman, 1985*a*).

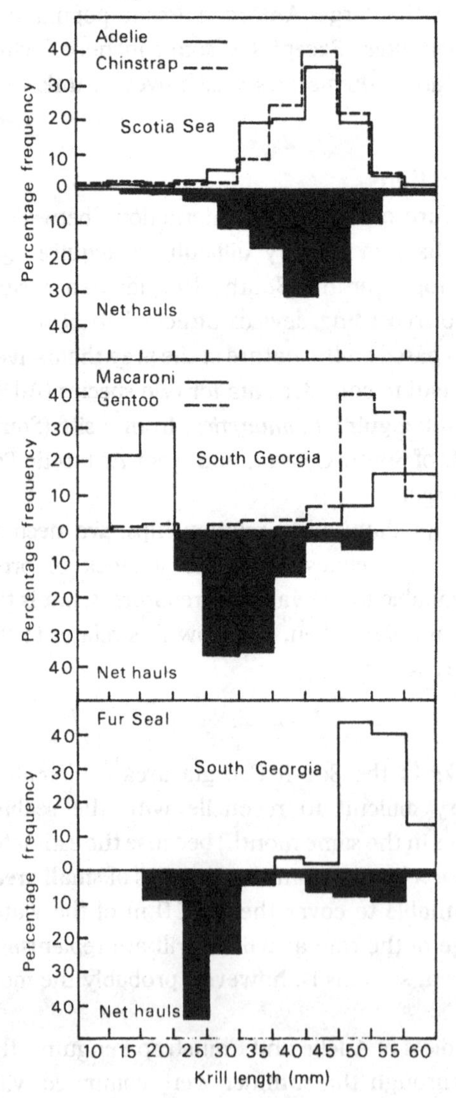

potential direct competition may exist between the natural and commercial predators in this area. At South Georgia, however, there is much less similarity (Fig. 15.2), with Fur Seals and Macaroni Penguins (and most other seabirds also) taking much mature krill, while net hauls, including those by nets identical to those used further south, mainly capture immatures. This is an important discrepancy, which might be explained by the predator and net samples coming from slightly different areas and needs to be resolved if we are to understand the dynamics of krill around South Georgia.

More detailed data on the dynamics of krill–predator interactions come from continuous records of the diving patterns of female Fur Seals engaged in 3–8-day foraging trips to sea (Fig. 15.3). Of over 4000 dives from 36 complete days at sea, 75% were made at night with peak activity around dawn and dusk. Night-time dives were much shallower than daytime ones and a comparison of diving patterns with diel changes in the vertical distribution of krill around South Georgia shows that the shallow dives coincide with krill being nearest the surface. There are two particularly important results. First, the seal's diving pattern is closely linked with the vertical movements of its prey. Second, although over 40% of the krill

Fig. 15.3. Antarctic Fur Seal diving activity and depth in relation to time of day and krill vertical distribution (after Croxall *et al.*, 1985a).

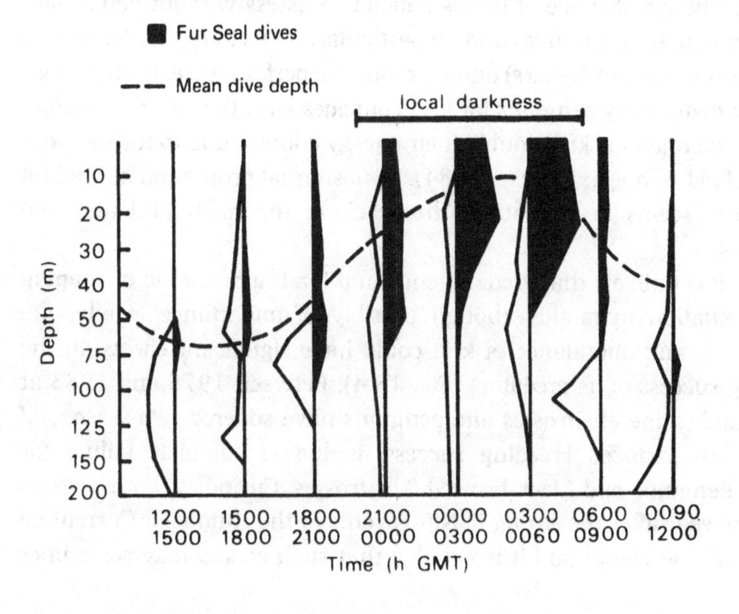

in the water column at any time of day was below 75 m, the seals made virtually no attempt to exploit this, only 3% of dives exceeding this depth, even though they are physiologically capable of diving deeper than 100 m. Thus, despite efficient exploitation of krill's vertical migration, a significant fraction of the population seems to be largely unexploited by the seals. This implies that, at least at one stage in the study season, and for a predator with considerable diving capacity, the local krill supply was more than sufficient to meet the energy demands of a predator engaged in rearing offspring.

Krill-eating penguins also mainly take prey from near the surface. The most detailed data are for Chinstrap Penguins (Lishman & Croxall, 1983); 40% of over 1000 dives were shallower than 10 m, 90% reached less than 40 m and only 1% exceeded 80 m. Such penguins need to catch 5–10 g (representing 5–20 individuals) of krill per dive to meet their energy requirements; this is equivalent to catching one krill every 5–10 s, if all dives are equally successful (see Chapter 6). For Gentoo Penguins, which take some bottom-dwelling fish in addition to krill, half the dives exceeded 95 m, a few reached 100–125 m, and there are strong indications that trips in which krill were caught chiefly comprised shallow dives whereas those in which fish were captured involved many deep dives (R.W. Davis and J.P. Croxall, unpublished data).

Most of the important predation on krill by seabirds (and seals) thus involves adult krill, including gravid females, in the top 50 m of the water column. The significance of this is difficult to assess without better data on krill population dynamics and, in particular, on its longevity (estimates range between 2 and 5 years) and reproductive performance. Even though the peak demand by penguins for krill coincides with the latter's breeding season, when female krill double their energy content due to the development of lipid-rich eggs (Clarke, 1984), a substantial proportion of the krill population seems to lie outside the reach of the main bird and seal predators.

While it is unlikely that predator consumption is a major factor shaping krill population dynamics, though possibly of importance locally, the distribution and abundance of krill could have significant effects on the breeding success of its predators (Fig. 15.4). Between 1975 and 1983 at Bird Island, some albatrosses and penguins have suffered two seasons of significantly reduced breeding success, including complete failure for Gentoo Penguins and Black-browed Albatrosses. Curiously, both seasons were one year after the strong El Niño events in the Humboldt Current off Peru (and elsewhere) and it is possible that such events may sometimes have repercussions in the Southern Ocean. In 1983–84, Fur Seals suffered the highest pup mortality yet recorded, survivors reached only 70% of their normal weaning weight, and the duration of the foraging trips to sea by females was doubled.

It would be premature to conclude from all this, however, that seabirds will automatically be good indicators of the state of local krill stocks and that they can be easily used to gauge the effect of commercial exploitation. Even in the two very anomalous years, different species reacted to different extents, and data from other seasons indicate considerable natural variation. It is certain that a variety of biological characteristics of each species would need to be measured over many seasons before any changes could be ascribed reliably to smaller-scale (e.g. harvest-induced) variations in the availability of prey.

Fish

The paucity of pelagic schooling fish in the Southern Ocean may account for the fact that their importance to predators is less than in the northern hemisphere. Fish found in predator stomachs is usually highly digested, and detailed studies using otoliths to identify the species present and to estimate their size and weight have only been carried out for a few seabirds. The two main fish-eating inshore-feeding seabirds, Gentoo Penguin and Blue-eyed Shag, both take juveniles of several species of Antarctic cod *Notothenia* which mature in coastal kelp beds. They are chiefly benthic-demersal species and therefore suitable prey for pursuit-

Fig. 15.4. Breeding success (chicks fledged from eggs laid) of various seabird species at Bird Island, South Georgia, 1975–83.

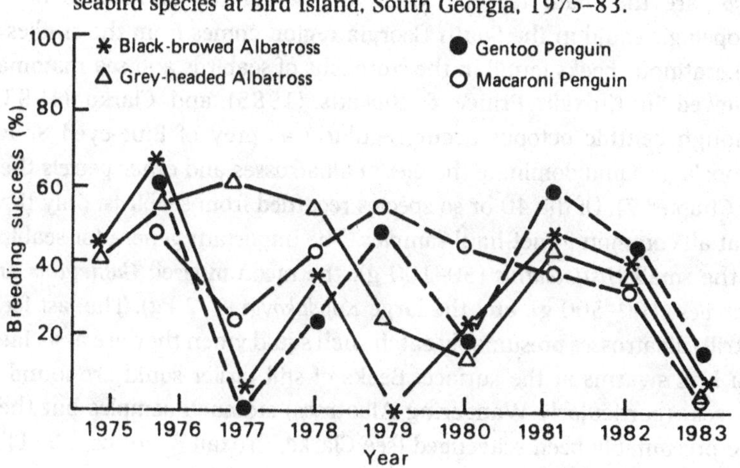

diving seabirds. Antarctic Terns, which feed by plunge-diving in the shallow waters of bays and inlets, principally take fingerlings.

The local commercial fishery exploited adult *N. rossii* for many years but its share of the catch dropped from 98% in 1969 to 0.4% in 1980, presumably due to overexploitation (Kock, 1985). The epipelagic (and krill-eating) *Champsocephalus gunnari* is now the main target (and its stock biomass has apparently declined significantly (Kock, 1985)); adults of this species are important prey for male Antarctic Fur Seals and are also taken by albatrosses, presumably when the fish are at the surface at night along with krill swarms. However, the size and identity of some of the fish in Wandering Albatross food samples suggest that many may be derived by scavenging behind fishing boats, particularly as fleets are operating in the South Georgia area in winter. Unfortunately, there are no data on diet composition before the local fishery developed but it is interesting that chick-rearing success in recent years has been higher than in the early 1960s (J.P. Croxall, unpublished data). White-chinned Petrels, which eat considerable quantities of fish, appear mainly to take lanternfish Myctophidae, which are also recorded in the diets of Blue Petrels and Wilson's Storm Petrels.

Squid

Squid are probably the most elusive of the major resources of the Southern Ocean, because all but the very largest nets catch only small species and juveniles of large ones. Consequently, there is at present no commercial squid fishery in the area, in contrast to the flourishing one on the Patagonian shelf off the Falkland Islands, where inshore species (*Illex*, *Loligo*) are the main targets. Much of the limited knowledge of the mesopelagic squid in the South Georgia region comes from the studies of the keratinous beaks found in the stomachs of seabirds and sea mammals (reviewed in Croxall, Prince & Ricketts (1985) and Clarke (1983)). Although benthic octopus occur regularly as prey of Blue-eyed Shags, mesopelagic squid dominate the diet of albatrosses and other petrels there (see Chapter 7). Of the 40 or so species recorded from seabirds, only three are at all common in net-haul samples. The important genera for seabirds are the small *Histioteuthis* (50–100 g), the medium-sized *Galiteuthis* and *Todarodes* (100–500 g), and the large *Kondakovia* (1–2 kg). The last feeds on krill; albatrosses presumably catch such squid when they are associated with krill swarms at the surface. Beaks of still larger squid are found in albatross (particularly Wandering Albatross) stomach samples but these have presumably been scavenged (see Clarke, Croxall & Prince, 1981).

The diving records of King Penguins, which feed extensively on medium-sized squid (about 400 g), show that over half their dives exceed 50 m and 1% reach 235 m (Kooyman *et al.*, 1982). This pattern is very different from that of the krill-eating penguins. Furthermore, King Penguins can meet their energy demands while rearing chicks by capturing prey on only 5–10% of dives.

The data from these predators show how widespread and abundant squid must be in sub-Antarctic waters and, as far as the principal species are concerned, we are likely to learn more about their biology and dynamics from predator samples than from net hauls.

Ecological isolating mechanisms

This topic was extensively discussed by Croxall & Prince (1980*b*) who concluded that temporal, dietary and geographical factors (i.e. foraging area or zone) are all important, in varying combinations and extents, in achieving some degree of ecological segregation between the members of the South Georgia seabird community. Fresh information over the last five years has not greatly changed this picture, which is summarised in Table 15.4. This mainly emphasises aspects of food and feeding ecology (and timing of breeding) rather than distinctions based on breeding-habitat preferences, because there is every indication that there are extensive areas of optimal breeding habitats still available for all South Georgia seabird species.

Although there is no direct evidence of interspecific competition for food (which could be regarded as potentially superabundant in summer), the ways in which the resources of the marine environment are partitioned suggest to us that access to **available** prey has been an important force shaping the nature of species' ecological adaptations. Considering that relatively few prey types are available (and that almost no seabird species takes any one of these exclusively), that there is no great diversity of feeding biotopes, and that the ability to adjust the breeding timetable is very restricted in the short Antarctic summer, the actual diversity of ecological patterns is considerable.

We might expect to find the best insights into these problems by considering the most closely related species (Table 15.5). South Georgia Diving Petrels and Fairy Prions occupy habitats on South Georgia not shared by any other species. Otherwise, apart from the distinction between surface and burrow-nesting species, there are no other significant habitat differences. Timing of breeding and diet, however, show more features of interest (Fig. 15.5). The Wandering Albatross is unique in rearing chicks

Table 15.4. *Principal potential mechanisms for ecological segregation in the breeding season for the main species of South Georgia seabirds*

	Diving species		Surface-feeding species
Winter			
	King Penguin		Wandering Albatross
Summer			
Inshore		Fish	Antarctic Tern[a]
Fish	Blue-eyed shag	Squid and fish	White-chinned Petrel
	Gentoo Penguin		Light-mantled Sooty Albatross
			Grey-headed Albatross
Krill	Gentoo Penguin		
Offshore		Copepods	Common Diving Petrel[b] (early)
Krill	Macaroni Penguin		Antarctic Prion (late)
Fish and squid	King Penguin		
		Amphipods	Wilson's Storm Petrel
		Krill	
		Inshore	Common Diving Petrel[b] (early)
			South Georgia Diving Petrel[b] (late)
			Fairy Prion (early)
			Antarctic Prion (late)
			Wilson's Storm Petrel (late)
		Offshore	Blue Petrel (early)
			Antarctic Prion (late)
			Black-browed Albatross
		Carrion	Northern Giant Petrel (early)
			Southern Giant Petrel (late)

[a] Plunge-diving is characteristic feeding method.
[b] Most feeding involves relatively shallow (< 10 m depth), wing-propelled dives.

Table 15.5. *Timing of breeding, diet and breeding habitat of closely related seabird taxa at South Georgia*

Breeding season (mean dates)			Diet (% by weight)										
						Crustacea					Penguin		
Egg laying	Hatching	Fledging	Fish	Squid	Lamprey	Krill	Amphipods	Copepods	Other	Small birds	& seal carrion	Habitat	References
Black-browed Albatross, *Diomedea melanophrys*													
27 Oct	3 Jan	28 Apr	38	21	1	39	1	—	—	—	—	Tussock slopes	1, 2
Grey-headed Albatross, *D. chrysostoma*													
19 Oct	30 Dec	19 May	24	49	10	16	1	—	—	—	—	Tussock slopes	1, 2
Wandering Albatross, *D. exulans*													
24 Dec	11 Mar	14 Dec	41	40	—	—	—	—	—	—	19	Tussock meadows	3
Southern Giant Petrel, *Macronectes giganteus*													
11 Nov	10 Jan	13 May (♂)	1	1	—	17	—	—	—	8	71	Tussock slopes	4
		7 May (♀)	0	1	—	26	—	—	—	10	63	& meadows	
Northern Giant Petrel, *M. halli*													
30 Sept	29 Nov	23 Mar (♂)	2	1	—	15	—	—	—	3	79	Tussock meadows	4
		19 Mar (♀)	5	3	—	33	—	—	—	11	47	(esp. coastline)	
Fairy Prion, *Pachyptila turtur*													
10 Nov	29 Dec	15 Feb	2	1	—	79	14	4	0	—	—	Boulder rock crevices	5
Antarctic Prion, *P. desolata*													
16 Dec	30 Jan	19 Mar	2	1	—	57	8	31	1	—	—	Tussock grassland	6
Blue Petrel, *Halobaena caerulea*													
24 Oct	11 Dec	27 Jan	8	1	—	78	4	1	7	—	—	Tussock grassland	6
Common Diving Petrel, *Pelecanoides urinatrix*													
25 Oct	18 Dec	9 Feb	—	—	—	15	17	68	—	—	—	Tussock slopes	7
South Georgia Diving Petrel, *P. georgicus*													
12 Dec	28 Jan	14 March	—	—	—	76	4	20	—	—	—	Fine scree	7

1, Tickell & Pinder, 1975. 2, Prince, 1980*b*. 3, Croxall & Prince, unpublished. 4, Hunter, 1983, 1984. 5, Prince & Copestake, unpublished. 6, Prince, 1980*a*, in press. 7, Payne & Prince, 1979.

through the winter. The differences from other albatrosses in its diet could simply be directly related to this – and the absence of krill almost certainly is – but its main squid prey is *Kondakovia* rather than *Todarodes* and we believe both to be available year-round in the area.

The two small albatrosses differ primarily in the composition of their diet (although they are very different demographically (Prince, 1985)) and this has now been confirmed over several years, including ones in which shortage of krill drastically reduced Black-browed Albatross breeding success, yet it did not switch to feeding its chicks fish and squid, which were being successfully exploited in the same year by Grey-headed Albatrosses (see Chapter 7).

Within each of the remaining three groups, species show quite marked differences in the timing of chick rearing and they also take different prey or a different balance of prey types. The giant petrels show very extensive dietary overlap; indeed there is more difference between sexes than

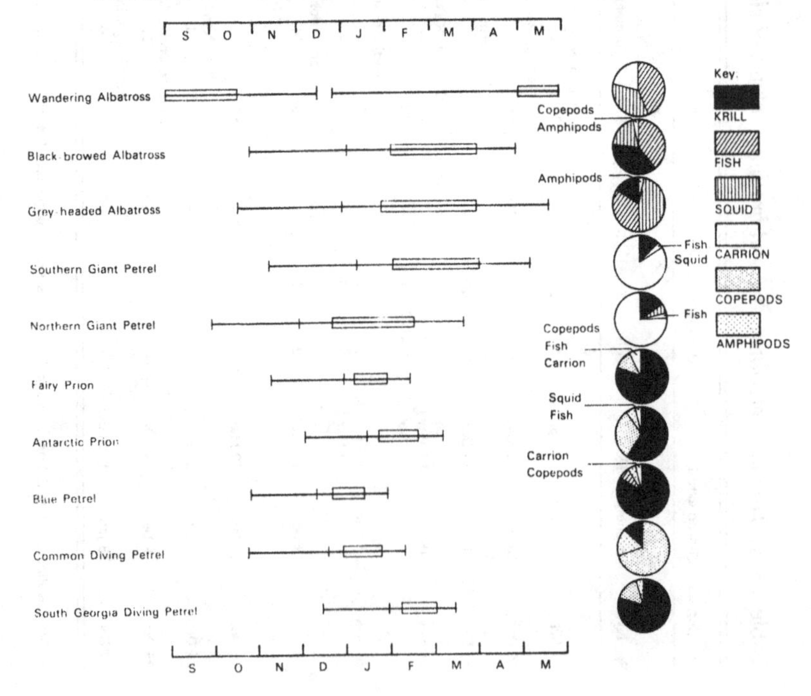

Fig. 15.5. Timing of breeding (horizontal line; vertical bars mark, in sequence, mean dates of laying, hatching and fledging) and diet (% by weight) in groups of related seabirds at South Georgia. Thickened horizontal line denotes period of maximum food requirements of chick.

between species and apparently considerable overlap in the chick-rearing periods. This is very similar to the results of another intensively studied species-pair, the Adélie and Chinstrap Penguins (Lishman, 1985*a,b*) where the only apparent distinctions are minor differences in prey and a few weeks' difference in the timing of breeding. Nevertheless, both sets of differences in breeding timetable are exactly sufficient to ensure that the peak energy (and hence food) demands by the chicks overlap only slightly or not at all. This appears to be a recurring theme because it applies also to the two diving petrels and to Blue Petrels and Antarctic Prions (although both these pairs have marked dietary differences as well). Fairy Prions, with a diet (see Chapter 7) and a breeding timetable intermediate between Blue Petrels and Antarctic Prions, might be regarded as the 'middle' species in a trio – which could partly account for their scarcity at South Georgia.

If we interpret differences in breeding timetable as adaptations to avoid coincidence of peak food demands in similar and/or closely related species, what then is the role of the dietary differences? It could be argued that they simply represent species taking a random selection of what is available at different times of year. However, of three early breeding species, two (Blue Petrel and Fairy Prion) apparently exploit mainly krill and one (Common Diving Petrel) mainly copepods, while of the later breeding species, South Georgia Diving Petrels take krill and Antarctic Prions take copepods. It is unlikely, therefore, that all the major prey categories are not available at least throughout this period of the summer, which suggests that the dietary differences might be additional ways further to reduce direct competition; it will, however, be difficult to obtain any direct evidence for this.

Conclusions

Although there is a considerable body of empirical data on the biology, feeding ecology and bioenergetics of South Georgia seabirds (which, despite the deficiencies summarised by Croxall, Ricketts & Prince (1984), are the best for any comparable marine system), the lack of equivalent information on the abundance, distribution and demography of the main prey species precludes any explicit quantitative conclusions on the role of seabirds in the food web.

In addition, of the other potentially important top predators, only for Antarctic Fur Seals (with a total population of about one million and a biomass of 26 000 tonnes) can similarly detailed models of food consumption be developed (Doidge & Croxall, 1985). We are almost entirely ignorant of the diet and energy requirements of the huge Southern

Elephant Seal populations (totalling some 350 000 individuals and a biomass of 184 000 tonnes (McCann, 1985)), or of the residual baleen whale population (for which we even lack reliable estimates of present numbers). As far as krill consumption is concerned, the role, and importance, of fish is imperfectly understood and that of squid unknown.

Our understanding of the dynamics of predator–prey interactions in the South Georgia marine system will depend as much on developments in these fields as on improvements of our knowledge of the seabirds that live there.

Acknowledgements We thank C. Ricketts for his original help in developing our models of seabird food requirements, the many colleagues whose field work contributed to the data on which these are based, S. Norris for typing the manuscript, P.C. Harper for many helpful comments on it and A. Sylvester for preparing the illustrations.

References

BIOMASS (1982). Post-Fibex data interpretation workshop. *Biomass Rep. Ser.*, **20**, 38 pp.

Clarke, A. (1984). Lipid content and composition of Antarctic krill *Euphausia superba* Dana. *J. Crust. Biol.*, **4**, Spec. No. 1, 283–92.

Clarke, M.R. (1983). Cephalopod biomass – estimation from predation. *Mem. Nat. Mus. Vict.*, **44**, 95–108.

Clarke, M.R., Croxall, J.P. & Prince, P.A. (1981). Cephalopod remains in regurgitations of the Wandering Albatross at South Georgia. *Bull. Br. Antarct. Surv.*, **54**, 9–22.

Costa, D.P. & Prince, P.A. In press. Foraging energetics of Grey-headed Albatrosses *Diomedea chrysostoma* at Bird Island, South Georgia. *Ibis*.

Croxall, J.P. (1982). Energy costs of incubation and moult in petrels and penguins. *J. Anim. Ecol.*, **51**, 177–94.

Croxall, J.P. (1984). Seabirds. In *Antarctic ecology*, vol. 2, ed. R.M. Laws, pp. 533–619. London: Academic Press.

Croxall, J.P., Evans, P.G.H. & Schreiber, R.W. (eds) (1984). *Status and conservation of the world's seabirds*. Cambridge: ICBP.

Croxall, J.P., Everson, I., Kooyman, G.L., Ricketts, C. & Davis, R.W. (1985). Fur seal diving behaviour in relation to vertical distribution of krill. *J. Anim. Ecol.*, **54**, 1–8.

Croxall, J.P. & Pilcher, M. (1984). Length-frequency distribution and sex ratio of krill eaten by Antarctic Fur Seals *Arctocephalus gazella* at South Georgia. *Bull. Br. Antarct. Surv.*, **63**, 117–25.

Croxall, J.P. & Prince, P.A. (1980*a*). The food of Gentoo Penguins *Pygoscelis papua* and Macaroni Penguins *Eudyptes chrysolophus* at South Georgia. *Ibis*, **122**, 245–53.

Croxall, J.P. & Prince, P.A. (1980*b*). Food, feeding ecology and ecological segregation of seabirds at South Georgia. *Biol. J. Linn. Soc.*, **14**, 103–31.

Croxall, J.P., Prince, P.A. & Ricketts, C. (1985). Relationships between prey life-cycles and the extent, nature and timing of seal and seabird predation in the Scotia Sea. In *Antarctic nutrient cycling and food webs*, ed. W.R. Siegfried, P.R. Condy & R.M. Laws, pp. 516–33. Berlin: Springer-Verlag.

Croxall, J.P., Ricketts, C. & Prince, P.A. (1984). Impact of seabirds on marine resources, especially krill, of South Georgia waters. In *Seabird energetics*, ed. G.C. Whittow & H. Rahn, pp. 285–318. New York: Plenum.

Doidge, D.W. & Croxall, J.P. (1985). Diet and energy budget of the Antarctic Fur Seal *Arctocephalus gazella*. In *Antarctic nutrient cycling and food webs*, ed. W.R. Siegfried, P.R. Condy & R.M. Laws. pp. 543–50. Berlin: Springer-Verlag.

Hunter, S. (1983). The food and feeding ecology of the giant petrels *Macronectes halli* and *M. giganteus* at South Georgia. *J. Zool., Lond.*, **200**, 521–38.

Hunter, S. (1984). Breeding biology and population dynamics of giant petrels *Macronectes* spp. at South Georgia. *J. Zool., Lond.*, **203**, 441–60.

Kock, K.-H. (1985). The state of exploited Antarctic fish stocks around South Georgia (Antarctic). *Arch. FischWiss.*, **36**, 155–83.

Kooyman, G.L., Davis, R.W., Croxall, J.P. & Costa, D.P. (1982). Diving depths and energy requirements of King Penguins. *Science*, **217**, 726–7.

Lishman, G.S. (1985*a*). The food and feeding ecology of Adelie Penguins *Pygoscelis adeliae* and Chinstrap Penguins *P. antarctica* at Signy Island, South Orkney Islands. *J. Zool., Lond.*, **205**, 245–63.

Lishman, G.S. (1985*b*). The comparative breeding biology of Adélie and Chinstrap Penguins at Signy Island, South Orkney Islands. *Ibis*, **127**, 84–99.

Lishman, G.S. & Croxall, J.P. (1983). Diving depths of the Chinstrap Penguin *Pygoscelis antarctica*. *Bull. Br. Antarct. Surv.*, **61**, 21–6.

McCann, T.S. (1985). Size, status and demography of Southern Elephant Seal *Mirounga leonina* populations. In *Sea mammals of southern latitudes*, ed. J.K. Ling & M.M. Bryden, pp. 1–17. Adelaide: South Australian Museum.

Payne, M.R. & Prince, P.A. (1979). Identification and breeding biology of the diving petrels *Pelecanoides georgicus* and *P. urinatrix exsul* at South Georgia. *N.Z. J. Zool.*, **6**, 299–318.

Pennycuick, C.J. (1982). The flight of petrels and albatrosses (Procellariiformes), observed in South Georgia and its vicinity. *Phil. Trans. R. Soc. Lond. B.*, **300**, 75–106.

Pennycuick, C.J., Croxall, J.P. & Prince, P.A. (1984). Scaling of foraging radius and growth rate in petrels and albatrosses. *Ornis Scand.*, **15**, 145–54.

Priddle, J., Heywood, R.B. & Theriot, E. (1986). Some environmental factors influencing phytoplankton in the Southern Ocean around South Georgia. *Polar Biol.*, **5**, 65–79.

Prince, P.A. (1980*a*). The food and feeding ecology of Blue Petrel *Halobaena caerulea* and Dove Prion *Pachyptila desolata*. *J. Zool. Lond.*, **190**, 59–76.

Prince, P.A. (1980*b*). The food and feeding ecology of Grey-headed Albatross *Diomedea chrysostoma* and Black-browed Albatross *D. melanophris*. *Ibis*, **122**, 476–88.

Prince, P.A. (1985). Population and energetic aspects of the relationships between Black-browed and Grey-headed Albatrosses and the Southern Ocean marine environment. In *Antarctic nutrient cycling and food webs*, ed. W.R. Siegfried, P.R. Condy & R.M. Laws, pp. 473–7. Berlin: Springer-Verlag.

Prince, P.A. (In press). Blue Petrel *Halobaena caerulea*. In *The handbook of Australian seabirds*, ed. S. Marchant.

Prince, P.A. & Croxall, J.P. (1983). Birds of South Georgia: new records and re-evaluations of status. *Bull. Br. Antarct. Surv.*, **59**, 15–27.

Prince, P.A. & Francis, M.D. (1984). Activity budgets of foraging Gray-headed Albatrosses. *Condor*, **86**, 297–300.
Tickell, W.L.N. & Pinder, R. (1975). Breeding biology of the Black-browed Albatross *Diomedea melanophris* and Grey-headed Albatross *D. chrysostoma* at Bird Island, South Georgia. *Ibis*, **117**, 433–51.

16

Conclusions

J. P. CROXALL
British Antarctic Survey, Natural Environment Research Council, High Cross, Madingley Road, Cambridge CB3 OET, UK

This volume was designed to provide an up-to-date conspectus of information on seabird adaptations for feeding in the marine environment and of the nature of recent and current studies that seek to integrate quantitative data on seabird trophodynamics into an ecosystem perspective. As such, it portrays the present state of the art and does not purport to provide new approaches or definitive conclusions.

Hunt & Schneider (Chapter 2) emphasise the diversity and dynamics of biological processes in oceans and seas and the variety of temporal and geographical scales at which they operate. The rest of the first part of this book testifies to the diversity of ways in which seabirds have adapted to cope with such an environment, involving an impressive spectrum of size (25–12 000 g) and physiological and behavioural adaptations (e.g. from penguins that dive to 250 m to frigatebirds that never even alight on the sea surface).

The second half of the book illustrates something of the diversity of approaches that have contributed to the quantitative study of the role of seabirds in marine ecosystems. The treatments by Harrison & Seki (Chapter 13) and Croxall & Prince (Chapter 15) are based on particularly detailed information, but essentially conventional models, of breeding season activities and ecology, but of the very contrasting seabird communities in tropical and polar oceanic environments. The former is dominated by terns, Pelecaniformes and albatrosses, and for them the activities of predatory fish that incidentally drive to the surface potential prey for seabirds is of considerable significance. The latter is dominated by penguins and small petrels, with the emphasis on the location and exploitation of swarming crustaceans.

In dealing with a coastal system Briggs & Chu (Chapter 12), use models of similar basic structure but combine data from breeding colonies and roosts with the results of very extensive ship and air surveys of seabird densities ever. Detailed at-sea surveys form the basis of Schneider, Hunt & Powers' (Chapter 11) comparison of energy flux in two boreal shelf systems. The variation within each system is greater than differences between them and, in terms of energy transfer, they are shown to be intermediate between open ocean systems (with fluxes an order of magnitude lower) and coastal upwellings (with fluxes two to three times greater).

A very different approach is made by Duffy & Siegfried (Chapter 14), who use guano production (estimated from its commercial exploitation) to examine interannual and longer-term (e.g. decadal) variations in food consumption by seabirds in the rich coastal upwelling systems of the Humboldt and Benguela Currents. Sanger (1972 and in press) has already described many of the quantitative links involving seabirds in the Gulf of Alaska. His review here (Chapter 10) investigates more precisely the trophic levels at which seabirds operate. Arctic seabirds function at between the third and fifth levels (inclusive), averaging 4.0 (i.e. second-order carnivores). This provides support for models (e.g. Evans, 1973) that have assumed that seabirds generally function at the fourth level, although it must not take the place of investigations of trophic relations of species at specific study sites, because considerable intersite and interspecies differences exist.

Several of the contributions here and a number of other similar studies have estimated the overall food consumption of certain seabird populations (Table 16.1). Despite the varied data sources and assumptions on which these estimates are based, they all indicate that seabirds consume far from regligible quantities of marine organisms as food. Two important questions derive from this. What influence do seabird predators have on their prey and what influence do changes in prey populations have on seabirds?

At a theoretical level, it has been argued that seabird populations may be regulated by a shortage of nest sites (Potts, Coulson & Deans, 1980; Duffy, 1983), social factors (Wynne-Edwards, 1963) or by density-dependent effects involving food shortage (Lack, 1954, 1966). Studies of seabird communities have produced some evidence of interspecific competition for nest sites (e.g. Lack, 1934; Williams, 1974). However, it has more often been concluded that the single most important mechanism in reducing interspecific competition is segregation of feeding ecology by some combination of temporal, spatial or dietary factors (Ashmole & Ashmole, 1967; Pearson, 1968; Diamond, 1978; Croxall & Prince, 1980).

In tropical regions, where prey availability is fairly uniform at most times of the year, it is logical that events during the breeding season, when adults have to find food within a restricted distance from their breeding colony, may be critical (Ashmole, 1963, 1971). In temperate regions, however, where many types of marine prey (e.g. zooplankton) are less abundant in winter than in summer, it is possible that winter food shortage could be the key factor regulating populations, despite the fact that seabirds are free (from parental duties) to forage where they choose in search of suitable food (and may make substantial transequatorial migrations, presumably in order to achieve this). Proponents of this idea usually regard the summer food supply as superabundant (e.g. Salomonsen, 1955; Beck, 1970) but this is not necessarily the same as readily **available** in surplus. Indeed, recent analyses suggest that the size and patterns of spatial distribution of colonies of British seabirds strongly support the hypothesis that breeding seabird populations can cause local prey depletion to the extent that their numbers may be limited by interspecific competition for food at this time (Furness & Birkhead, 1984).

Another adaptation usually regarded as having evolved in response to annual and seasonal variations in food availability, is brood-size reduction, whereby eggs are laid and chicks hatch asynchronously, providing the earliest chick or chicks to hatch with a clear growth, and therefore potential survival, advantage. This is widespread in seabirds (e.g. most Phalacrocoracidae and Pelecanidae, many Sulidae and Spheniscidae, some gulls and terns) where it is regarded principally as a mechanism for adjusting fledgling production to food supply (O'Connor, 1978). There have been few field investigations but studies of Herring Gulls *Larus argentatus* (Haymes & Morris, 1977) and Glaucous-winged Gulls *L. glaucescens* (Hunt & Hunt, 1976) showed that lowest levels of brood reduction were recorded in seasons when food was most abundant. However, as Shaw (1985) has pointed out, in most cormorants the bulk of the brood reduction takes place when the chicks are still very small and when the food requirements of the whole brood are 5–10 times less than the demands of the surviving one or two chicks will be in a few weeks time. This suggests that relations between food availability and brood reduction are not necessarily simple and direct and may even involve selective feeding by parents, implying their ability to assess the relative favourability of local conditions in any given season.

There are a number of studies, however, that have produced more direct, albeit circumstantial, evidence of the influence of food availability on reproductive success or population trends. Thus Harris & Hislop

Table 16.1. *Food consumption by certain seabird populations*

Area	No. of species	Population size (millions)	Population status	Food consumption ('000 tonnes)	Main species	References
Bering Sea	14	53	All	580–1150	Brunnich's Guillemot *Uria lomvia* Common Guillemot *U. aalge* Short-tailed Shearwater *Puffinus tenuirostris*	Hunt, Burgeson & Sanger, 1981
Oregon	4	4·4	All	62·5	Short-tailed Shearwater Common Guillemot Brandt's Cormorant *Phalacrocorax penicillatus*	Wiens & Scott, 1975
California	21	0·85	Breeding	193	Sooty Shearwater *Puffinus griseus* Brown Pelican *Pelecanus occidentalis* Common Guillemot	See Chapter 12
	*c.*75	1·8–7·0	Non-breeding			
Hawaiian Islands	18	0·5	Breeding	410	Laysan Albatross *Diomedea immutabilis* Black-footed Albatross *D. nigripes* Bonin Petrel *Pterodroma hypoleuca* Wedge-tailed Shearwater *Puffinus pacificus* Sooty Tern *Sterna fuscata*	See Chapter 13

North Sea	22	2·6	Breeding	250+	Common Guillemot	Furness, 1982, 1984;
		?2·6	Non-breeding		Black-legged Kittiwake	Bourne, 1983
					Rissa tridactyla	
					Northern Fulmar	
					Fulmarus glacialis	
					Herring Gull	
					Larus argentatus	
Peru	10–15	17	1960–65	2600	Guanay Cormorant	Schaefer, 1970
					Phalacrocorax bougainvillii	
		4·5	1965–68	700	Peruvian Booby	
					Sula variegata	
S.W. Africa	3	0·15	Breeding	16·5	Jackass Penguin	Furness & Cooper, 1982
					Spheniscus demersus	
		1·0	Non-breeding		Cape Gannet	
					Sula capensis	
					Cape Cormorant	
					Phalacrocorax capensis	
South Georgia	22	70	Breeding	7800	Macaroni Penguin	Croxall, Ricketts & Prince, 1984
					Eudyptes chrysolophus	See Chapter 15
					Antarctic Prion	
					Pachyptila desolata	
					White-chinned Petrel	
					Procellaria aequinoctialis	

(1978) showed that Atlantic Puffin *Fratercula arctica* chicks had fastest growth and highest fledging weights in years when they were fed Sprats *Sprattus sprattus* (their preferred food), than in years when Whiting *Merlangus merlangus*, of lower energy content, was the main diet. Various other studies, particularly of alcids (e.g. Vermeer & Cullen, 1979; Wehle, 1983) have recorded significant variations between years in chick growth rates and fledging weight, which correlated well with similar variations in provisioning rate (as estimated by frequency (and/or size) of meals delivered by parents). This suggests that the annual variation in growth rate relates directly to food supply. Similarly, krill availability has had a marked influence on reproductive performance in a number of seabirds for which it is the main food resource at South Georgia (see Chapter 15), although the magnitude of the effect differs between species and seasons.

As regards more long-term effects, Lid (1981) attributed the drastically reduced breeding success of Atlantic Puffins, Common Guillemots *Uria aalge*, and Razorbills *Alca torda* in Norway since 1969 to reduction in the local Herring *Clupea harengus* stocks from 10 million tonnes in 1958 to 0.5 million tonnes in 1969 and an order of magnitude increase in catches of Sandeel *Ammodytes marinus* during the 1960s. These two fish are the main prey of all three alcids over large areas of Norway and only in areas where they rely on other prey has breeding success been maintained (Barrett, Anker-Nilssen & Folkestad, 1985).

The changes in fishery practice and seabird populations in the Benguela Current system have been extensively reviewed (Frost, Siegfried & Cooper, 1976; Crawford & Shelton, 1981; Furness, 1982; Burger & Cooper, 1984; Duffy & Siegfried: Chapter 14) and need only be summarised here. Before the mid-1960s, the South West African purse-seine fishery took chiefly South African Pilchard *Sardinops ocellata*. At this time, the main prey of the three dominant seabirds of the region, Jackass Penguin *Spheniscus demersus*, Cape Cormorant *Phalacrocorax capensis* and Cape Gannet *Sula capensis* were pilchards and Cape Anchovy *Engraulis capensis*. The estimated pilchard stock biomass was reduced by 50% in the late 1960s; catches fell rapidly, dropped again in 1974 and stayed low thereafter; anchovy (and Horse Mackerel *Trachurus trachurus*) catches increased over this period. Broadly similar events took place in the southern African purse-seine fishery. The reductions in fish stocks and a decrease in the average size of fish available showed some correlation with decreases in populations of all three seabirds (Crawford & Shelton, 1981), especially Jackass Penguin which, being flightless, has the most restricted foraging range. The diet of Jackass Penguins also changed markedly over the same period, except along the southeast coast of South Africa, where no purse-seine fishery has operated (Randall & Randall, 1986), reflecting the changes in the nature of the commercial fish catches.

Obviously the overfishing for pilchard, to which is attributed the collapse of its fishery, has had an effect on seabirds, especially Jackass Penguins. However, the situation is not as simple as it is sometimes portrayed as being. While Cape Gannet populations off Namibia have undoubtedly decreased, South African populations have remained stable. Cape Cormorants in Namibia have increased, though this may relate to greater availability of guano-platform nest sites. Even with Jackass Penguins, Crawford & Shelton's (1981) data show that, **prior** to the 1969 pilchard population collapse, four west-coast colonies were already decreasing, two were stable and one increasing. After 1969, the increasing one declined dramatically, the two stable ones increased very greatly and the rest decreased! As Burger & Cooper (1984) noted, the factors affecting seabirds are clearly not identical to those affecting fishery harvests and the data are largely insufficient to conclude unequivocally that there has been an overall decrease in the quantities of fish available to seabirds in southern Africa.

Another well-studied interaction between a seabird and its prey that is commercially exploited involves the Brown Pelican *Pelecanus occidentalis* and the Northern Anchovy *Engraulis mordax* in the Southern California Bight (SCB) (Anderson *et al.*, 1980; Anderson, Gress & Mais, 1982; Anderson & Gress, 1984). Up to 1979, there were moderate to high anchovy levels in the SCB. Pelican reproductive rates, but not numbers of breeding pairs at the colonies, were highly correlated with anchovy abundance and with the commercial anchovy catches. After 1979, however, despite large increases in the anchovy harvest quotas, catches in the USA were well below these quota levels. However, during this latter period, the Mexican commercial fishery, which harvests the same anchovy sub-population, has increased greatly and there are signs of reduced pelican reproductive success in part of the SCB since 1979.

Overall then, there is a wealth of information to indicate that the nature and availability of prey can influence seabird reproductive success. There have been few attempts to quantify such interactions more critically but, in an ingenious model of Common Guillemot feeding rates and chick growth, Ford *et al.* (1982) concluded that the breeding success of a guillemot population would fall substantially with a reduction in food density of only 10–30% from some long-term average. They suggested that a reduction of over 40% would lead to total reproductive failure.

To what extent, then, do seabirds influence their prey populations? A number of studies quantifying prey consumption by seabird populations have concluded that they may consume 20–30% of the annual fish production of certain areas (e.g. Oregon, 22% (Wiens & Scott, 1975); Foula, Shetland, 29% (Furness, 1978); Benguela system, 20% (Furness & Cooper, 1982); Peruvian system, 28% (Schaefer, 1970; McCall, 1984; but see Chapter 14)). These seabird–fishery interactions have been described in detail in various publications (principally Furness, 1982; Nettleship, Sanger & Springer, 1984) and are usually interpreted as indicating that seabirds are major predators of pelagic fish. They may also be significant competitors with commercial fisheries and predatory fish, at least in some areas (Furness, 1984). Naturally, this has very important implications for the potentially conflicting requirements of fisheries' management and seabird conservation. However, alternative interpretations have suggested that at least some of these estimates of seabird consumption may be too high by nearly an order of magnitude (Bourne, 1983; see also Chapter 14).

There are many reasons why a variety of interpretations of the results and data of existing models are possible and we are unlikely to resolve these discrepancies until better information is available. Because estimates of natural mortality (e.g. consumption by predators such as birds, seals and fish themselves) are an integral part of many methods of assessing fish stock sizes, it is particularly important not only that the seabird and fishery data should be as accurate as possible but also that they should be based on similar spatial and temporal scales. Failure to do this probably accounts for the different interpretations of the impact of seabirds on fisheries, especially as it is very difficult in practice to partition total mortality into natural mortality and that due to the fishery.

In estimating food consumption by seabirds, the main limitations are the accuracy of the data on population size and on various aspects of bioenergetics, especially adult energy requirements for existence and for foraging, an understanding of which requires good knowledge of activity budgets. Even with well-studied populations of largely inshore-feeding species, with realistic activity budgets, the 95% confidence interval of estimates of population energy requirements is no better than 30% of the mean (Furness, 1982). In comparison, estimates of breeding population size are certainly no better than 20% and those of non-breeding populations are considerably less still. However, there is no reason to believe that estimates of fish population sizes (even of commercially exploited species) are usually significantly more accurate than those of the better-studied seabird species. Therefore, assessments of both predator requirements and prey stock might be inaccurate by half an order of magnitude.

Most of the remaining problems concern our ability directly and realistically to compare seabird activities with the appropriate fishery parameters. Because quantitative studies of seabird diets are a relatively recent development, they often refer only to a single site within a species' or even a population's foraging range, at a particular time of year (usually during chick rearing) and are thus not necessarily typical of the year-round diet. Furthermore, only in a very few cases do we have detailed information on the age structure (or length-frequency) of the prey population exploited by the birds, which is then available for critical comparison with similar data for any commercial fishery and for the natural prey population.

It is obviously of crucial importance to know whether seabirds and commercial fisheries are potentially competing for fish of the same sizes or ages or whether the fishery is catching adults (the normal situation) and the birds are taking juveniles. In this case, the 'impact' of the seabirds is greatly diluted because of the many other factors causing mortality of young fish. Thus, Little Penguins *Eudyptula minor* take pilchards *Sardinops neopilchardus* and anchovies *Engraulis australis* that are significantly smaller than those caught commercially, at least in data from the 1940s (see Chapter 6). For Jackass Penguins in the 1950s, just under half the pilchards *S. ocellata* and anchovies *E. capensis* taken were of commercial size but nowadays there would seem to be a greater overlap between the anchovies taken by penguins and the commercial fishery, judging from the dietary data in Wilson (1985). In contrast, Croxall & Lishman (Chapter 6) show that Adélie *Pygoscelis adeliae* and Chinstrap Penguins *P. antarctica* and commercial fishery nets take krill of essentially the same size, around the South Orkney and South Shetland Island. However, at South Georgia a similar comparison shows significant differences between krill caught by nets and by Macaroni *Eudyptes chrysolophus* and Gentoo Penguins *Pygoscelis papua*.

A related problem concerns increases in food availability as a direct result of a fishery. The large quantities of offal and rejected by-catch, the inevitable products of even single-species target fishing, provide a potentially rich resource for seabirds scavenging behind fishing vessels. The extent to which such food is used by seabirds essentially reduces the level of direct competition between birds and the fishery, although it may, of course, permit increases in bird populations.

Even if there were cases for which we knew both the overlap between predator diet and fishery catches in terms of the age-classes of prey taken and also the population structure and dynamics of the prey, we would still find it difficult to quantify either the exact nature of the competition between the natural and commercial predators or their impact on their common prey. This is because of important uncertainties concerning the area and time over which we should integrate predator–prey interactions.

Both predators and prey are mobile, although to different extents, and this mobility may change substantially according to the season of the year and the age and reproductive status of the animal. Individual predators and prey are themselves parts of a population and, ideally, we should only be comparing discrete, or self-contained populations, usually known as stocks, which will rarely, if ever, be geographically congruent. In practice it is also exceedingly difficult to define these stocks, but it may well be misleading to extrapolate to larger areas seabird consumption estimates derived from a restricted area. In particular, most studies of seabird food consumption have concentrated on communities rich in species and individuals. Inevitably these are located in areas of permanent or seasonally high production and, especially where major current systems are involved, it may be difficult not only to determine the prey biomass available at any one time but also to assess what relationship this bears to the whole stock, only a fraction of which may come within the foraging range, either vertical or horizontal, of the seabirds.

One of the keys to realistic assessment of seabird food consumption in relation to commercial fishery operations and to the fish stocks themselves is, therefore, comparison at appropriate spatial scales. The influence of temporal effects is probably no less important but even more difficult to define. Obviously there is substantial variation in fish stocks on a variety of temporal scales, ranging from long-term trends and fluctuations (e.g. in the Peruvian Anchoveta fishery) to medium-term effects (e.g. those mediated by environmentally induced changes in juvenile survival and recruitment rate) and to seasonal changes. Most comparisons with seabird consumption are based on single estimates of bird energy requirements and on some 'average' value for fish stocks and fishery catches. This must be a gross over-simplification of the real situation.

It will be clear from the foregoing that we are only in the very early stages of critical study of the dynamics of seabird–fishery–prey interactions in marine environments. It is not surprising, therefore, that, depending on the assumptions made, rather different views of the relative significance of the important links should emerge.

There is considerable scope for improving the data on which the present models are based. Substantial progress towards a better understanding of the main interactions can only sensibly proceed from fully integrated (and probably long-term) research by fisheries, seabird and marine biologists, placed in appropriate oceanographic contexts and scales. If this volume contributes to this progress it will have served its purpose.

References

Anderson, D.W. & Gress, F. (1984). Brown Pelicans and the anchovy fishery off southern California. In *Marine birds: their feeding ecology & commercial fisheries relationships*, ed. D.N. Nettleship & P.F. Springer, pp. 128–35. Ottawa: Can. Wildl. Serv. Spec. Publ.

Anderson, D.W., Gress, F. & Mais, K.F. (1982). Brown Pelicans: influence of food supply on reproduction. *Oikos*, **39**, 23–31.

Anderson, D.W., Gress, F., Mais, K.F. & Kelly, P.R. (1980). Brown Pelicans as anchovy stock indicators and their relationships to commercial fishing. *Calif. Coop. Oceanic Fish. Invest. Rep.*, **21**, 54–61.

Ashmole, N.P. (1963). The regulation of numbers of tropical oceanic birds. *Ibis*, **103b**, 458–73.

Ashmole, N.P. (1971). Seabird ecology and the marine environment. In *Avian biology*, vol. 1, ed. D.S. Farner & J.R. King, pp. 112–286. London & New York: Academic Press.

Ashmole, N.P. & Ashmole, M.J. (1967). Comparative feeding ecology of seabirds of a tropical oceanic island. *Bull. Peabody Mus. Nat. Hist.*, **24**, 1–131.

Barrett, R.T., Anker-Nilssen, T. & Folkestad, A.O. (1985). Monitoring of breeding auks and kittiwakes in Norway. In *Population and monitoring studies of seabirds. Proceedings of the Second International Conference of the Seabird Group*, ed. M.L. Tasker, pp. 13–15. Aberdeen: N.C.C.

Beck, J.R. (1970). Breeding seasons and moult in some smaller Antarctic petrels. In *Antarctic ecology*, vol. 1, ed. M.W. Holdgate, pp. 542–50. London: Academic Press.

Bourne, W.R.P. (1983). Birds, fish and offal in the North Sea. *Mar. Poll. Bull.*, **14**, 294–6.

Burger, A.E. & Cooper, J. (1984). The effects of fisheries on seabirds in South Africa and Namibia. In *Marine birds: their feeding ecology and commercial fisheries relationships*, ed. D.N. Nettleship & P.F. Springer, pp. 150–60. Ottawa: Can. Wild. Serv. Spec. Publ.

Crawford, R.J.M. & Shelton, P.A. (1981). Population trends for some southern African seabirds related to fish availability. In *Proceedings of the symposium on birds of the sea and shore*, ed. J. Cooper, pp. 15–41. Cape Town: African Seabird Group.

Croxall, J.P. & Prince, P.A. (1980). Food, feeding ecology and ecological segregation of seabirds at South Georgia. *Biol. J. Linn. Soc.*, 814, 103–31.

Croxall, J.P., Ricketts, C. & Prince, P.A. (1984). The impact of seabirds on marine resources, especially krill, at South Georgia. In *Seabird energetics*, ed. G.C. Whittow & H. Rahn, pp. 285–317. New York: Plenum Press.

Diamond, A.W. (1978). Feeding strategies and population size in tropical seabirds. *Am. Nat.*, **12**, 215–23.

Duffy, D.C. (1983). Competition for nesting space among Peruvian guano birds. *Auk*, **100**, 680–8.

Evans, P.R. (1973). Avian resources of the North Sea. In *North Sea science*, ed. E.D. Goldberg, pp. 400–12. Cambridge, Mass.: Princeton University Press.

Ford, R.G., Wiens, J.A., Heinemann, D. & Hunt, G.L. (1982). Modelling the sensitivity of colonially breeding marine birds to oil spills: guillemot and kittiwake populations on the Pribilof Islands, Bering Sea. *J. Appl. Ecol.*, **19**, 1–32.

Frost, P.G.H., Siegfried, W.R. & Cooper, J. (1976). Conservation of the Jackass Penguin *Spheniscus demersus* (L.). *Biol. Conserv.*, **9**, 79–99.

Furness, R.W. (1978). Energy requirements of seabird communities: a bio-energetics model. *J. Anim. Ecol.*, **47**, 39–53.

Furness, R.W. (1982). Estimating the food requirements of seabird and seal populations and their interactions with commercial fisheries and fish stocks. In *Proceedings of the symposium on sea and shore birds*, ed. J. Cooper, pp. 1–14. Cape Town: African Seabird Group.

Furness, R.W. (1984). Seabird–fisheries relationships in the northeast Atlantic and the North Sea. In *Marine birds: their feeding ecology and commercial fisheries relationships*, ed. D.N. Nettleship, G.A. Sanger & P.F. Springer, pp. 162–9. Ottawa: Can. Wildl. Serv. Spec. Publ.

Furness, R.W. & Birkhead, T.R. (1984). Seabird colony distributions suggest competition for food supplies during the breeding season. *Nature*, **311**, 655–6.

Furness, R.W. & Cooper, J. (1982). Interactions between seabird populations and fish stocks of the Saldanha region, South Africa. *Mar. Ecol. Prog. Ser.*, **8**, 243–50.

Harris, M.P. & Hislop, J.R.G. (1978). The food of young Puffins, *Fratercula arctica*. *J. Zool., Lond.*, **185**, 213–36.

Haymes, G.T. & Morris, R.D. (1977). Brood size manipulations in Herring Gulls. *Can. J. Zool.*, **55**, 1762–6.

Hunt, G.L., Burgeson, B. & Sanger, G.A. (1981). Feeding ecology of the eastern Bering Sea. In *The eastern Bering Sea shelf: oceanography and resources*, vol. 2, ed. D.W. Hood & J.A. Calder, pp. 629–47. Seattle: University of Washington Press.

Hunt, G.L. & Hunt, M.W. (1976). Gull chick survival: the significance of growth rates, timing of breeding and territory size. *Ecology*, **57**, 62–75.

Lack, D. (1934). Habitat distribution in certain Icelandic birds. *J. Anim. Ecol.*, **3**, 81–90.

Lack, D. (1954). *The natural regulation of animal numbers*. Oxford: Oxford University Press.

Lack, D. (1966). *Population studies of birds*. Oxford: Oxford University Press.

Lid, G. (1981). Reproduction of the Puffin on Rost in the Lofoten Islands in 1964–1980. *Cinclus*, **4**, 30–9.

McCall, A.D. (1984). Seabird–fishery trophic interactions in eastern Pacific boundary current: California and Peru. In *Marine birds: their feeding ecology and commercial fisheries relationships*, ed. D.N. Nettleship, G.A. Sanger & P.F. Springer, pp. 136–48. Ottawa: Can. Wildl. Serv. Spec. Publ.

Nettleship, D.N., Sanger, G.A. & Springer, P.F. (1984). *Marine birds: their feeding ecology and commercial fisheries relationships*. Ottawa: Can. Wildl. Serv. Spec. Publ.

O'Connor, R.J. (1978). Brood reduction in birds: selection for fratricide, infanticide and suicide? *Anim. Behav.*, **26**, 79–96.

Pearson, T.H. (1968). The feeding biology of seabird species breeding on the Farne Islands, Northumberland. *J. Anim. Ecol.*, **37**, 521–52.

Potts, G.R., Coulson, J.C. & Deans, I.R. (1980). Population dynamics and breeding success of the Shag, *Phalacrocorax aristotelis*, on the Farne Islands, Northumberland. *J. Anim. Ecol.*, **49**, 465–84.

Randall, R.M. & Randall, B.M. (1986). The diet of Jackass Penguins *Spheniscus demersus* in Algoa Bay, South Africa and its bearing on population declines elsewhere. *Biol. Conserv.*, **37**, 119–34.

Salomonsen, F. (1955). The food production of the sea and the annual cycle of Faeroese marine birds. *Oikos*, **6**, 92–100.

Sanger, G.A. (1972). Preliminary standing stock and biomass estimates of seabirds in the subarctic Pacific region. In *Biological oceanography of the northern North Pacific Ocean*, ed. A.Y. Takenouti *et al.*, pp. 589–611. Tokyo: Idemitsu Shoten.

Sanger, G.A. (In press). Diets and food web relationships of seabirds in the Gulf of Alaska and adjacent marine regions. In *Environmental assessment of the Alaskan continental shelf. Final reports of the principal investigators.* Juneau, Alaska: NOAA, Office of Marine Pollution Assessment.

Schaefer, M.B. (1970). Men, birds and anchovies in the Peru current – dynamic interactions. *Trans. Am. Fish. Soc.*, **9**, 461–7.

Shaw, P. (1985). Brood reduction in the Blue-eyed Shag *Phalacrocorax atriceps*. *Ibis*, **127**, 476–94.

Vermeer, K. & Cullen, L. (1979). Growth of Rhinoceros Auklets and Tufted Puffins, Triangle Island, British Columbia. *Ardea*, **67**, 22–7.

Wehle, D.H.S. (1983). The food, feeding ecology and development of young Tufted and Horned Puffins in Alaska. *Condor*, **85**, 427–42.

Wiens, J.A. & Scott, J.M. (1975). Model estimation of energy flow in Oregon coastal seabird populations. *Condor*, **77**, 439–52.

Williams, A.J. (1974). Site preferences and interspecific competition among guillemots *Uria aalge* (L.) and *Uria lomvia* (L.) on Bear Island. *Ornis Scand.*, **5**, 113–21.

Wilson, R.P. (1985). Seasonality in diet and breeding success of the Jackass Penguin *Spheniscus demersus*. *J. Ornithol.*, **126**, 53–62.

Wynne-Edwards, V.C. (1963). *Animal dispersion in relation to social behaviour*. Edinburgh: Oliver & Boyd.

INDEX